THE GROWTH OF SINGLE CRYSTALS

SOLID STATE PHYSICAL ELECTRONICS SERIES

Nick Holonyak, Jr., *editor*

ANKRUM, *Semiconductor Electronics*
BURGER AND DONOVAN, eds., *Fundamentals of Silicon Integrated Device Technology:*
Vol. I: *Oxidation, Diffusion and Epitaxy*
Vol. II: *Bipolar and Unipolar Transistors*
GENTRY, GUTZWILLER, HOLONYAK, AND VON ZASTROW, *Semiconductor Controlled Rectifiers: Principles and Applications of p-n-p-n Devices*
LAUDISE, *The Growth of Single Crystals*
NUSSBAUM, *Applied Group Theory for Chemists, Physicists, and Engineers*
NUSSBAUM, *Electromagnetic and Quantum Properties of Materials*
NUSSBAUM, *Semiconductor Device Physics*
PANKOVE, *Optical Processes in Semiconductors*
ROBERTS AND VANDERSLICE, *Ultrahigh Vacuum and Its Applications*
VAN DER ZIEL, *Solid State Physical Electronics*, 2nd ed.
WALLMARK AND JOHNSON, eds., *Field-Effect Transistors: Physics, Technology and Applications*
WESTINGHOUSE ELECTRIC CORPORATION, *Integrated Electronic Systems*

PRENTICE-HALL INTERNATIONAL, INC., *London*
PRENTICE-HALL OF AUSTRALIA, PTY. LTD., *Sydney*
PRENTICE-HALL OF CANADA, LTD., *Toronto*
PRENTICE-HALL OF INDIA PRIVATE LIMITED, *New Delhi*
PRENTICE-HALL OF JAPAN, INC., *Tokyo*

THE GROWTH OF SINGLE CRYSTALS

R. A. Laudise

Bell Telephone Laboratories
Murray Hill, New Jersey

Prentice-Hall, Inc.

Englewood Cliffs, New Jersey

Current printing (last digit):
10 9 8 7 6 5 4 3 2 1

13–365320–X

Library of Congress Catalog Card No [illegible]04173

Printed in the United States of America

To Joyce

PREFACE

This book is intended for all those who want to grow crystals. It is intended both for those who consider crystal growth their principal professional activity and for those who want crystals in order to study their properties and find that they must grow them themselves. Background through a first course in physical chemistry is assumed but is not essential particularly if the reader devotes special attention to Chapters 2 and 3 and the references contained therein.

The book is intended to be used in several ways:

1. As a text or collateral reading in a senior level or graduate level course in a materials science curriculum where the contents, if presented with an appropriate laboratory or research problem, can be used to train that rare and greatly needed individual, the professional crystal grower.

2. As a professional crystal grower's *vade mecum*. Chapters 1, 2, and 3 summarize material on means of determining crystallinity, thermodynamics and kinetics not readily available in a form useful for the grower elsewhere, and the remaining chapters discuss each of the major methods and are arranged in a manner to assist the grower in a logical choice of growth method and to view each method in as logical a theoretical framework as is presently possible.

3. As a source book about particular growth methods and the growth of particular materials. Each of the major methods is related to theory, equipment is described (suppliers of unusual equipment are mentioned),

procedures are discussed and accounts of the growth of representative crystals are given.

If the reader does not elect to read the book cover to cover, it is suggested that he read or at least scan the introductory material in the chapter where the method or crystal he is especially interested in is described.

This book is not encyclopedic. Brevity, critical evaluation, the inclusion of background material and an attempt to present crystal growth techniques as a logical, coherent body of knowledge, have caused the author to abandon attempts to list every crystal ever grown, although most materials of present-day development and research importance have been included. Nevertheless, a search of the periodical literature before beginning experimental work is imperative.

Chapters 1, 2, and 3 are written from the viewpoint of the crystal *grower's* needs. If your interests are primarily perfection studies, thermodynamics or theory of crystallization, then entire works devoted to these subjects should be consulted.

Any work is largely a product of the environment in which it is produced. This book is especially the product of more than a decade's association with the preparation of single crystals for research and development at Bell Telephone Laboratories. Representative materials and illustrative procedures have in the main been chosen from this experience. In making this choice, I regret that a lack of familiarity with much good work in other laboratories may have, in some cases, caused its omission. No value judgment is intended. Similarly, the author is not a historian of science. Lengthy historical sections have been omitted unless they served a valid tutorial purpose. References are meant mainly to lead the reader to good recent expositions of material which could not be adequately covered in the text and not as unmitigated priority judgments.

Parts of the book have been critically reviewed by several of my colleagues whose suggestions were invaluable. I would particularly like to thank R. L. Barns, J. R. Carruthers, W. C. Ellis, K. A. Jackson, K. Nassau, J. W. Nielsen, W. G. Pfann, C. D. Thurmond, and J. H. Wernick for their suggestions on the manuscript. Some of the material was developed for courses delivered at the Hebrew University, Jerusalem, The University of California at Los Angeles, and the Massachusetts Institute of Technology. Professor Michael Schieber of Hebrew University and Professor Harry Gatos of M.I.T. are especially thanked for making these opportunities available to the author. I have learned much from day-by-day association with colleagues not mentioned above and would especially like to mention A. A. Ballman, J. G. Bergman, G. T. Kohman, E. D. Kolb, J. P. Remeika and L. G. Van Uitert, in this respect. Miss Kathleen Donnelly ably carried out the prodigious amount of secretarial work involved; my wife, Joyce Laudise, lent constant encouragement and many hours of time in literature work. A. G.

Chynoweth, J. H. Scaff, and N. B. Hannay have often been sources of advice and encouragement. What is good in this work is due in large measure to these friends; the faults are the author's.

The following publishers and publications are to be thanked for permission to reproduce the illustrations used: American Chemical Society; American Institute of Physics; The American Mineralogist; Electrochemical Society, Inc.; Interscience Publishers, Inc.; John Wiley & Sons, Inc.; Journal of the American Ceramic Society; Journal of Physics and Chemistry of Solids; Macmillan (Journals) Ltd.; McGraw-Hill Book Company; The Royal Society; Van Nostrand Reinhold Company; W. H. Freeman and Company and *Zeitschrift für Naturforschung*.

R. A. Laudise

Murray Hill, N.J.

CONTENTS

LIST OF TABLES

1

SINGLE CRYSTALS

Crystals have interested man because of their beauty and rarity since prehistoric times, but their large-scale use has been brought about mainly by the demands of solid-state physics for materials for research and devices.

This book is intended to tell how to grow crystals and presents the necessary background concerning the perfection of crystalline materials, thermodynamics, kinetics of crystallization processes, and theory of the various methods to make one effective as a crystal grower.

Chapter 1 is devoted to the question of recognizing crystallinity in a material and determining the perfection of crystals. Chapters 2 and 3 discuss thermodynamics and kinetics of crystallization and the remaining chapters discuss in detail the various growth methods and the growth of specific materials.

1.1 What Is a Crystal?

Matter may exist in three states of aggregation—*solid*, *liquid*, or *gas*. In the gaseous state, the molecules are separated by comparatively large distances (about 30 Å at 1 atm). This large separation results in comparatively negligible interactions between the molecules and the molecules are thus free to move in any direction. Therefore, a gas has a very low viscosity and expands to fill completely a containing vessel of any size or shape. The arrangement of molecules in a gas is essentially completely disordered. In the liquid state the molecules (or atoms) are separated by about 1 Å, and their interactions are consequently much stronger than in a gas. Thus a liquid exhibits higher viscosity and does not expand to fill completely its container. There is short-range order in a liquid, but it does not persist more than a few atomic diameters from a given atom. In the solid state, the atomic separation is about the same as in a liquid, but the interactions between atoms are stronger. Thus the atoms are able to move only in vibrations of extremely low amplitude about fixed positions relative to one another. As a result, solids have rigidity, fixed shape, and mechanical strength. In addition, a *crystalline* solid is characterized by long-range order extending over many atom diameters. Upon increasing the internal energy in a crystalline solid by heating it, melting occurs at a fixed temperature for a given pressure or in a few cases sublimation to the gaseous state occurs. A further increase in internal energy will volatilize the material. At every temperature, gas or vapor of a material will exist in equilibrium with the material at a definite pressure. Thus in terms of internal energy for a particular material, the internal energy of the gaseous state $>$ the internal energy of the liquid state $>$ the internal energy of the solid state.

There is another class of materials often called *amorphous solids*, including glasses, waxes, and pitches, that possess such a high viscosity as to behave essentially as solids. Such materials do not have fixed melting points, and they exhibit the short-range order characteristic of liquids. It is often convenient to think of these substances as supercooled liquids. Figure 1.1 shows schematic representations of a gas, a liquid, and a solid. The representation of an amorphous solid would be identical to that of the liquid. A useful means of illustrating the difference between a crystalline solid and a liquid is shown in Fig. 1.2. Figure 1.2 shows the radial-distribution function, that is, the number of atoms encountered as a function of distance from a given atom for solid (crystalline) and for liquid potassium. A high degree of order even at

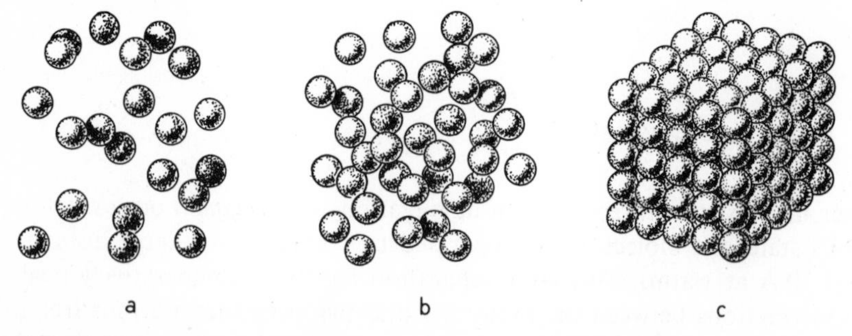

Fig. 1.1 Schematic representations of a gas (a), a liquid (b), and a solid (c).

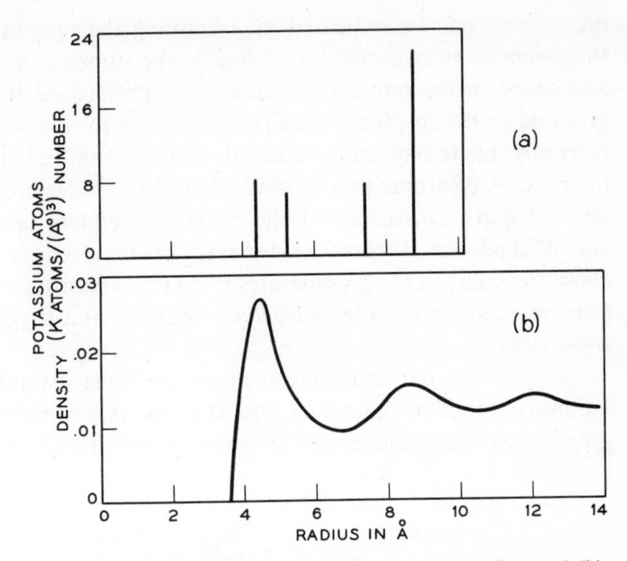

Fig. 1.2 Radial distribution function for (a) crystalline and (b) liquid potassium (after Thomas and Gingrich, 1938).

distances longer than those shown would occur in the crystal, while, as can be seen, the order in the liquid (or the amorphous solid) is short-range.

It is interesting to point out that there is a class of materials called liquid crystals (Brown and Shaw, 1957) whose properties are intermediate between those of liquids and crystals. These materials exhibit the flow behavior of liquids but are not isotropic in all of their properties, as is the case with true liquids. One way of viewing liquid crystals is to consider that they have one- or at most two-dimensional order, while true crystals exhibit three-dimensional order.

Let us examine the nature of the ordered array of atoms in a crystal. We will give here only a brief outline. Texts such as Buerger (1942) and Lipson and Cochran (1966) should be consulted for details. We may describe a crystal in terms of the pattern or arrangement of its constituent atoms. This pattern is often described in terms of the *unit cell*, which is an imaginary parallelepiped containing atoms of the crystal that, if moved or translated and repeated over and over again, will reconstruct the pattern of all of the atoms in the crystal. Repetition by translation is a kind of movement that can be represented by a vector. The origin of the vector describing the translation may be taken as any point within the crystal convenient for the problem at hand. If it is taken as some arbitrary reference point in the pattern to be repeated, the repetitive action of the translation obviously reproduces this reference point (as well as all the others in the system) as a three-dimensional pattern of points in space. This pattern is a *point-space lattice*. The grid, or line lattice, can be indicated by specifying the magnitude and direction of three representative grid lines, that is, the dimensions or *lattice parameters* and shape of the unit cell. It should be emphasized that the point lattice is given us by nature; we choose the line lattice for convenience. It is geometrically possible to have only a limited number of spatial arrangements of points in space or of atoms in a crystal. These can be described in terms of the 14 *Bravais space lattices* or of the 32 *crystal classes* or *point groups*. The 32 crystal classes are further divided into 230 *space groups*. One of the simplest classifications of crystals separates them into seven systems: *cubic, tetragonal, hexagonal, orthorhombic, monoclinic, triclinic*, and *trigonal*. Figure 1.3 shows these systems.

It has been found convenient to describe crystals by the use of the methods of analytic geometry and to adopt axes of reference called *crystallographic axes*. These axes are shown as heavy lines in Fig. 1.3. The axes are

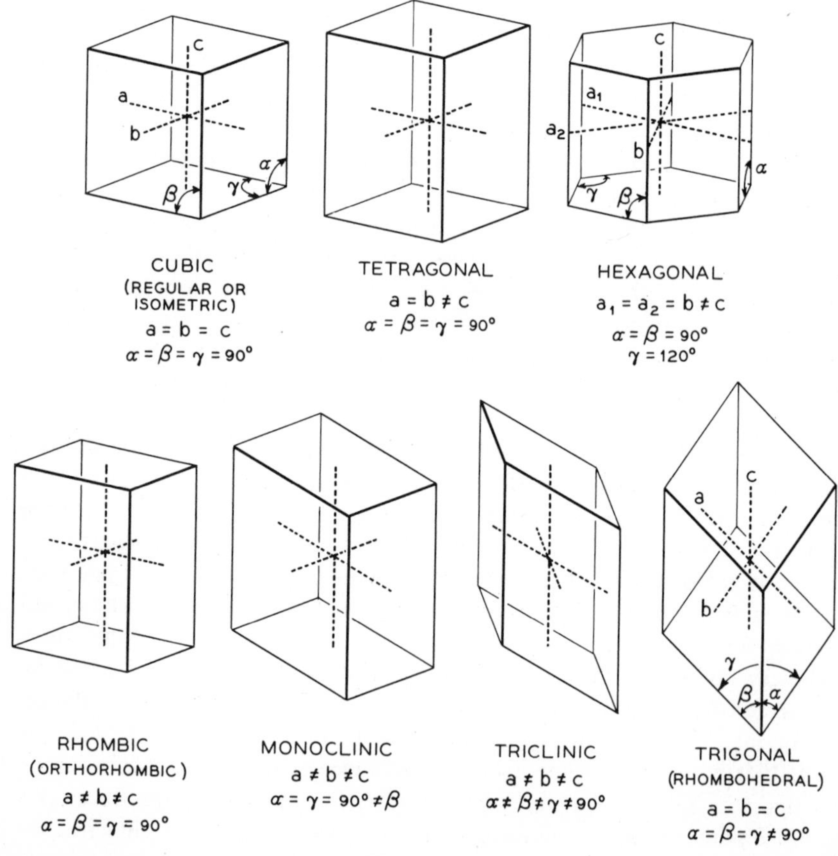

Fig. 1.3 Crystal systems.

generally chosen to correspond with the edges of the unit cell. Various methods have been devised to express the intercepts of crystal planes on the crystal axes. The most universally employed system is that of Miller. The *Miller indices* of a plane are the reciprocals of the intercepts of the plane with the crystallographic axes.† The Miller indices are always expressed as whole numbers and if the reciprocals of the intercepts come out as fractional values the fractions are cleared. The indices of a plane are always placed in brackets. The shape of the brackets indicates the following:

(100)	A particular 100 plane, that is, a plane whose intercepts with the axes are 1, ∞, ∞ (it is parallel to two axes)
{100}	All the planes in the crystal equivalent to (100) (all of the {100} planes will describe a crystal of a given shape or habit; {100} is sometimes called a form symbol)
[100]	A direction parallel to the line from the origin to the point where the reciprocals of the coordinates expressed as small integers are 1, 0, 0‡
$\langle 100 \rangle$	All the directions in the crystal equivalent to [100]

Certain planes in a crystal are sometimes described as belonging to the same *zone*. All those planes that are parallel to a given direction are called a zone and the direction to which they are parallel is the zone axis.

In addition to describing the symmetry of a crystal by means of translation operations, rotation and reflection operations are often useful in discussing crystal symmetry. Thus crystals may have *centers of inversion*, *axes of rotation*, and *reflection planes*. In a crystal with a center of inversion the properties of the structure are the same at vector distance $\vec{r}$ from a point as they are at vector distance $-\vec{r}$. In a crystal with an axis of rotation the structure is reproduced by rotation through an angle of $360°/n$, where $n = 2, 3, 4$, or 6. A structure has a reflection plane if half the structure is related to the other half as an object is related to its mirror image. Generally, to be called a *single crystal*, it will not have any macroscopic region misoriented (crystallographically) with respect to any other region by more than a few degrees, although even a definition as loose as this may be disputed by some.

In most of the solid materials of nature and of commerce, the individual crystals are rather small and the materials contain many of these *crystallites*. Each of these crystallites is misoriented with respect to its neighbors to a greater or lesser degree. Often these crystallites are called *grains*, and the regions between crystallites are called *grain boundaries*. Figure 1.4 shows a

---

†Thus in all the systems except the hexagonal, three indices are used to describe a plane. In the hexagonal system four indices can be used to describe a plane, but three are sufficient to identify uniquely a given plane. The redundant index is often indicated by a dot. Thus $(11\bar{2}0)$ is equivalent to $(11\cdot0)$.

‡In the cubic system, this will be a direction normal to (100).

(a)

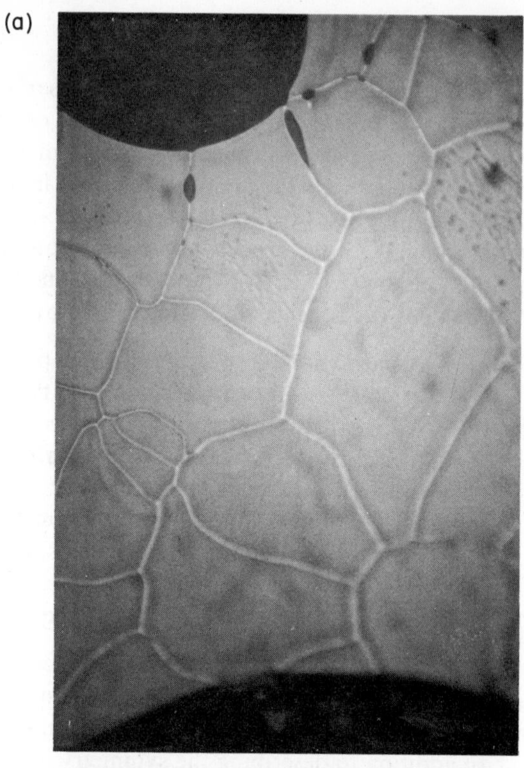

(b)

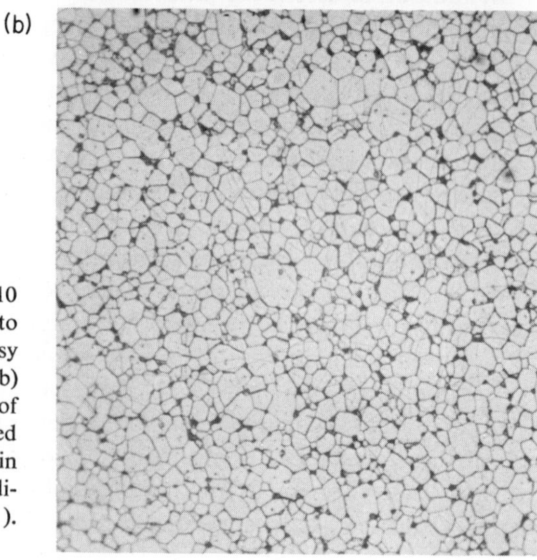

Fig. 1.4 (a) Pt metal etched 10 minutes in molten Na_2CO_3 to reveal grains (500×). (Courtesy of H. J. Levinstein.) (b) Polycrystalline sample of $Mg_{0.675}Mn_{0.525}Fe_{1.8}O_4$ etched 3 minutes at room temperature in HF–HNO_3–H_2O showing individual crystallite grains (500×). (Courtesy of F. R. Monforte.)

metallic and an inorganic oxide sample in which etching has been used to delineate the grains. In the case of inorganic oxides and salts, such *polycrystalline* materials are usually called *ceramics*.

The principal purpose of this book is to discuss and describe the techniques used in the preparation of single crystals. The term *single crystal* is difficult to define but is usually thought of as a crystallite that has either been found in nature, separated from a polycrystalline mass, or deliberately prepared. The crystallite should be of size sufficient for the esthetic, technological, or scientific purpose for which it is intended. Our main concern will be with single crystals deliberately prepared and of a size greater than about 1 mm^3 because this size is a practical lower limit for convenient manipulation and for most measurements.

The rest of this chapter is devoted to a brief survey of the methods used to establish that a material is crystalline. Such studies involve the determination of the nature and concentration of the *imperfections* present in crystals. This field (which also includes the study of noncrystalline and polycrystalline materials) is emerging as a discipline in its own right and is called *characterization*. A material is completely characterized when the identity and position of all of its constituent atoms are known.† A whole spectrum of techniques is employed to do this, and the remainder of this chapter can do no more than outline some of them. However, the importance of characterization for both the grower and the user of crystals cannot be overemphasized, and the neglect of the field by the user tends more and more to make it an area where the grower should take responsibility. For instance, characterization is not the measurement of conductivity and mobility in a semiconductor, no matter how important the user may feel these parameters may be to either the device uses or the basic physics of the material. Such measurements may be essential in combination with other measurements in characterizing the material, that is, in learning the identity and location of its constituent atoms. In any case, the identity and location of the constituent atoms determine the semiconductor and, of course, all the other properties of the material.

1.2 Experimental Evidence for Crystallinity

The principal identifying characteristic that distinguishes all crystalline solids from amorphous solids is long-range atomic order. Most crystals exhibit a definite melting point and often (depending principally on the method of preparation) are bounded by planar, regular external faces. The angles between these faces (interfacial angles) are rather accurately constant (Steno's law) and can even be used in identification. Some crystals *cleave* in a regular manner. That is, they may be broken along sets of smooth-plane surfaces everywhere parallel to each other throughout the body of the crystal. Crystals that possess *cleavage planes* may usually be cleaved by first scribing a groove parallel to the cleavage plane and then placing a sharp knife or razor blade

†By this definition, no material has ever been completely characterized, but many have been characterized to a point where at least some of their properties are understood in a fundamental way.

in the groove and tapping the back of the blade briskly. Mica, sodium nitrate, gallium arsenide, and sodium chloride are examples of crystals that cleave readily. Another property that crystals sometimes exhibit is transparency at visible wavelengths. Polycrystalline masses are usually translucent or opaque because of light scattering at voids at the grain boundaries and elsewhere and because of scattering due to birefringence.† All of these properties are often used as tests of crystallinity. However, heating can cause decomposition, polymorphic-phase changes, and devitrification, and none of the above properties, except long-range order, are always exhibited by all crystals and some of the properties are exhibited by amorphous materials. Thus it is usually safer to identify materials as crystalline because of some property more directly a measure of their long-range atomic order.

All crystalline materials (except those in the cubic crystal system), including liquid crystals, are optically anisotropic, that is, the effect of the crystal on light depends on the direction with which the light passes through the crystal. This fact is the basis for a convenient and powerful method of proving crystallinity, studying perfection and otherwise characterizing transparent materials.

A most useful tool for this type of analysis is the petrographic (polarizing) microscope, which has proved to be invaluable to the crystal grower. Many texts (Wahlstrom, 1969; Hartshorne and Stuart, 1960, 1964; Blass, 1961; Winchell and Winchell, 1964; Ford, 1958) describe its use, especially for the identification of crystals. Petrographic methods can be applied to powders as well as large single crystals.

Frequently, petrography is an indispensable adjunct to X-ray methods in crystal studies. Some questions, such as symmetry and crystal orientation, may sometimes be more conclusively answered by petrography than by X-ray examination. However, the most powerful technique for studying order in materials is X-ray diffraction.

No real crystal is perfect. That is, no real crystal is perfectly ordered and free from impurities. However, it will be useful to begin by discussing an idealized perfect crystal. It is not appropriate to give a detailed discussion here of the techniques used in X-ray diffraction; nevertheless, several remarks should be made. Because the spacing of the atoms in a crystal is comparable to the wavelength of X-rays, a crystal will act like a three-dimensional diffraction grating for X-rays. The fundamental equations involved are from von Laue and Bragg. Bragg considered X-ray diffraction in terms of reflection from atomic planes and one form of his equation is

$$n\lambda = 2d \sin \theta \tag{1.1}$$

where $n = 1, 2, 3, 4, \ldots$, the order of the reflection; λ is the wavelength of

---

†Polycrystalline ceramics such as Al_2O_3 can be transparent when made dense enough to be free of voids.

the incident radiation; d is the spacing between the atomic planes from which the reflection takes place; and θ is the angle between the incident beam and the atomic plane causing the reflection. Because Snell's law† holds and the refractive index for X-rays for any substance is very nearly equal to one, θ is also the angle between the reflected beam and the atomic plane.

In the study of crystals, X-ray diffraction is used for five purposes: (1) to prove crystallinity, (2) to determine structure, (3) to orient crystals, (4) to determine perfection, and (5) to determine lattice parameters, and hence composition, because lattice parameters depend on composition. If two crystals A and B are mutually soluble in each other and the lattice parameter is a linear function of composition, Vegard's law‡ is obeyed and the determination of composition from lattice-parameter measurements is straightforward. We will consider purposes (1) and (4) and will touch only briefly on (5) as it relates to characterization, because the determination of structure, orientation, and lattice parameters is amply treated elsewhere (Buerger, 1942; Lipson and Cochran, 1966; Wood, 1963; Bond, 1960; Barns, 1967; Barrett and Massalski, 1966). The following sections describe the four classes of X-ray techniques most applicable in proving crystallinity. Section 1.4 describes methods of studying perfection.

1.2.1 ROTATING-CRYSTAL METHODS

When X-rays are used with a strong monochromatic component, Eq. (1.1) will be satisfied for discrete angles that occur when a crystal is rotated in the beam. In some variations of rotating-crystal methods, the film is also moved in such a manner that the reflections are recorded in a readily interpreted way. These methods are used principally for determining crystal structures but are also useful for orienting crystals. Rotating crystal methods are described in detail in Barrett and Massalski (1966).

1.2.2 POWDER METHODS

If one uses X-rays having strong monochromatic components and a sample consisting of many tiny, randomly oriented crystals, Eq. (1.1) will be

---

†Snell's law states that $\sin i/\sin r = n$, where i is the angle of incidence, r is the angle of refraction, and n is the index of refraction.

‡Vegard's law may be expressed as

$$a = a_A + n_B(a_B - a_A)$$

where a is a lattice parameter of a crystal composed of a solid solution of A and B, a_A is the lattice parameter of A, a_B is the lattice parameter of B, and n_B is the mole fraction of B in the crystal. Gschneidner and Vineyard (1962) give a good treatment of the departures that occur from Vegard's law and their implications.

fulfilled because some of the crystals will be oriented to give reflections without need for rotation. The most common powder technique is the Debye–Scherrer–Hull method (Azaroff and Buerger, 1958; D'Eye and Wait, 1960). The powdered crystals are formed into a cylindrical rod less than $\frac{1}{2}$ mm in diameter. The film for recording the diffracted rays is mounted in a camera designed to hold it in a cylindrical form, so that when a collimated beam of X-rays is incident on the specimen, the diffracted rays from the lattice plane intercept the film in a set of curves and form (as shown in Fig. 1.5) the so-called Debye–Scherrer rings. An estimate of the crystallite grain size (in the polycrystalline sample) may be made from an examination of these rings. When the particle size of the crystallites exceeds about 10^{-4} cm, the rings are spotty, as shown in Fig. 1.5. With dimensions between about 10^{-4} and 10^{-5} cm the lines are sharp and when dimensions are less than about 10^{-5} cm the lines become broader and more diffuse. The diffraction pattern from a liquid or an amorphous solid consists of a few diffuse bands, while that from a crystalline solid exhibits a larger number of sharp bands. Figure 1.5 shows the typical X-ray diffraction patterns obtained from α-quartz (crystalline SiO_2) and vitreous silica. It is unusual to find more than two or three diffuse bands in an amorphous solid. X-ray powder diffraction is a common way of distinguishing crystals from glasses. Fourier analysis of the intensity distribution in the Debye–Scherrer patterns from liquids and amorphous solids enables one to calculate radial distribution functions like that of Fig. 1.2.

Quantitative measurement of line broadening of Debye–Scherrer patterns has been applied to studies of grain size. In addition, the line width is affected by stacking faults and strains and these imperfections have been studied at least semiquantitatively by means of careful measurements of line broadening (Bennil and Geibe, 1959). Anomalies in the intensities of the reflections can also occur if the crystallites in a polycrystalline sample are not randomly oriented.

Modern powder methods often employ diffractometers, which use quantum counters instead of film as the detector (Barrett and Massalski, 1966). Guinier cameras, which differ from Debye–Scherrer–Hull cameras in that monochromators are incorporated, are being increasingly applied principally

Fig. 1.5 Top: X-ray powder picture of crystalline α-quartz. Bottom: X-ray powder picture of vitreous silica. (Courtesy of J. R. Carruthers and M. Grasso.)

because they provide higher resolution and greater interplanar-spacing accuracy (Barrett and Massalski, 1966; Kikuta et al., 1966).

1.2.3 LAUE METHODS

If a single crystal larger than the beam (typically about 1 mm dia.) is held stationary in a beam having a wide range of X-ray wavelengths, each set of planes will, in effect, choose its own wavelength to satisfy Eq. (1.1). The diffracted rays are recorded on a flat film normal to the incident beam. They appear as a geometric arrangement of spots. If the material has crystallites smaller than the beam size, the geometric arrangement does not appear. If macroscopic regions of the material are misoriented with respect to one another the spots break up into several closely spaced spots. This effect is used in studying perfection by the Schulz technique discussed below.

1.2.4 LOW-ANGLE SCATTERING

X-rays that are diffracted through only a few degrees from the center beam in, for example, a geometry similar to that used for Debye–Scherrer pictures, are caused by short-range order in the crystal. Such small-angle scattering has been used to study small inclusions (about 10^{-6} cm) in crystals, small slowly varying strains, and order in liquids and glasses. The use of low-angle scattering has been used (Brady and Petz, 1961; Debye, 1959; Zimm, 1950; and Frisch and Brady, 1962) to study clustering in liquids near the consolute temperature and to study the structure of solutions. Light-scattering techniques have been used to study fluctuations greater than 10^{-6} cm, while X-rays are useful for dimensions below this. Low-angle scattering is particularly useful in the study of order in liquids, glasses, and solutions. While low-angle scattering is not generally used to prove crystallinity, it is appropriate for the study of the onset of crystallization (devitrification) in an amorphous medium.

1.3 Imperfections in Crystals

Imperfections in crystals may be divided into two categories—*chemical imperfections* and *physical imperfections*. A chemical imperfection is the presence of a foreign atom or a *vacancy*† in the crystal. Physical imperfections include *strain*, *dislocations*, *grain boundaries*, *twin planes*, and *stacking faults*.

---

†Some authors consider a vacancy a physical imperfection; the distinction will not be important for our purposes.

If the impurity atom occupies a site ordinarily occupied by an atom in a crystal it is said to be a *substitutional impurity*. The impurity atom may fit itself between the atoms comprising the crystal; it is then said to be an *interstitial impurity*. Whenever the atoms in a particular region of a crystal are displaced from their ideal positions, the crystal is said to be *strained.* Strain can be caused by chemical imperfections when a "foreign" atom does not fit the lattice properly and dislocations are partly relieved by strain associated with them. Stress and thermal treatment often introduce strain into a crystal.

1.3.1 VACANCIES

Vacancies are holes or vacant lattice sites in the crystal lattice caused by the absence of atoms. Two major types of vacancy defects can occur; *Frenkel defects*, in which certain of the atoms or ions have migrated to interstitial positions leaving behind the holes that they vacated, and *Schottky defects*, where the atoms that might occupy the vacancies are not present in the crystal. Figure 1.6 shows schematic representations of the various defects so far discussed. Because an increase in vacancies results in an increase in disorder with a consequent increase in entropy, it can be shown (see, for instance, Swalin, 1962) that the fraction of total sites vacant, X_v, in a crystalline lattice is given by

$$X_v = \exp\left(-\frac{\Delta G_v}{kT}\right) \tag{1.2}$$

where ΔG_v is the change of the Gibbs free energy in the formation of a single vacancy, k is the Boltzmann constant, and T is absolute temperature. Thus vacancies are stable in any crystal at all temperatures greater than absolute zero. ΔG_v can be obtained from a measurement of X_v (obtained by a comparison of X-ray and pycnometric densities) as a function of temperature and

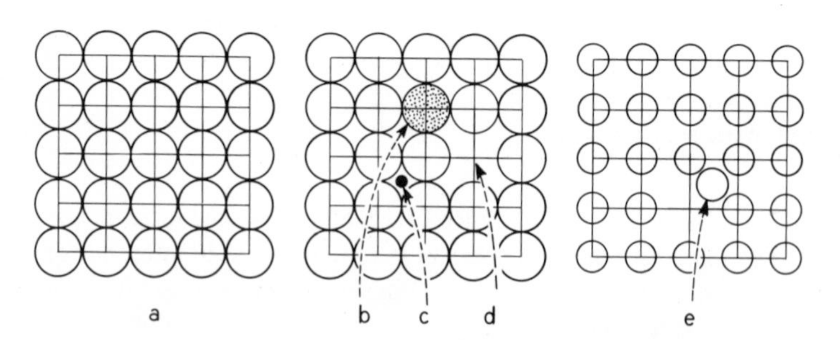

Fig. 1.6 Two-dimensional schematic representations: (a) perfect crystal; (b) substitutional impurity; (c) interstitial impurity; (d) Schottky defect; (e) Frenkel defect.

may be estimated in some cases from bond energies and other considerations. Vacancies occur even in elements and stoichiometric or daltonide compounds but their concentration is usually higher in nonstoichiometric or berthollide compounds, where their negative free energy of formation, $-\Delta G_v$, is greater because they are easier to charge-compensate. See Sec. 2.5 for a discussion of charge compensation. In nonstoichiometric compounds an "excess" of a particular atom can be accommodated either substitutionally or interstitially.

1.3.2 DISLOCATIONS

There are two principal kinds of dislocations—*edge dislocations* and *spiral* or *screw dislocations*. Figure 1.7 is a two-dimensional representation of an edge dislocation. For simplicity the spheres representing atoms are omitted. They would be located at all the intersection points of the lines of Fig. 1.7. The deformation may be thought of as caused by the adding of an extra plane of atoms, *AB*, in the upper half of the crystal. Planes in the upper half of the crystal are closer together than their normal equilibrium positions (they are strained compressively), while those in the lower half are farther apart than normal (they are strained in tension). Thus we see that near the dislocation the crystal is highly strained. Figure 1.8 shows the edge dislocation of Fig. 1.7 in three dimensions. A configuration such as this could be achieved, conceptually at least, by inserting the extra plane *ABB'A'* or by shearing the crystal by pushing on *CDEF* while pulling on *GCFH*. *AA'* is called the *dislocation line* and is a boundary between that part of the crystal where *glide* has occurred and the part that is unaltered. Glide is translation of a part of

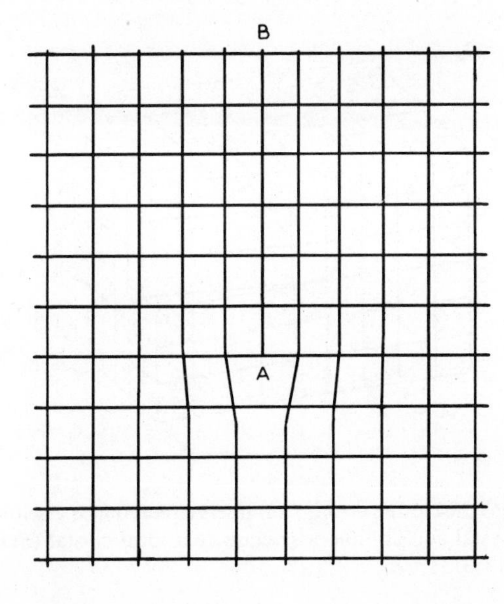

Fig. 1.7 Edge dislocation—shown schematically in two dimensions.

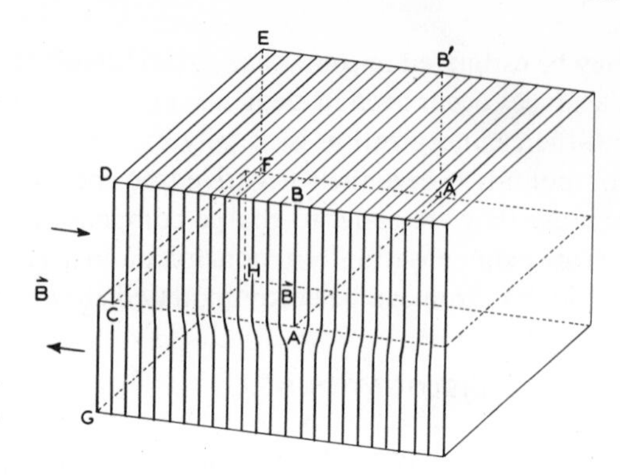

Fig. 1.8 Edge dislocation—shown schematically in three dimensions.

the crystal parallel to itself along a plane in the crystal. The plane $ACFA'$ along which motion has occurred is called the *slip plane* or *glide plane*, and the magnitude and direction of motion are indicated by a vector called the Burgers vector, $\vec{B}$, shown in Fig. 1.8 (Burgers, 1939; Frank, 1951; Van Bueren, 1960). In an edge dislocation $\vec{B}$ is at right angles to the dislocation line.†

If the direction of shear is applied to the crystal so that the direction of motion and hence the Burgers vector is parallel to the dislocation line, the imperfection formed is a screw dislocation. Figure 1.9 shows a screw dislocation. AA' is the dislocation line. The dislocation can be thought of as having been created by pushing on $ABCD$ while pulling on $AEFG$. Every plane of atoms passing through AA' can be thought of as a glide plane.

In addition to the above *pure dislocations*, mixed dislocations are possible in which $\vec{B}$ is oriented in a position intermediate between perpendicular and parallel to the dislocation line. Dislocations may be formed by straining

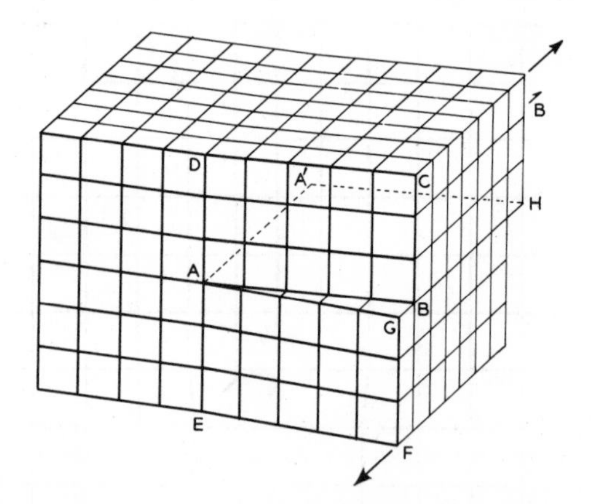

Fig. 1.9 Screw dislocation in three dimensions. (From *Dislocations in Crystals* by W. T. Read, Jr., 1953. Used with permission of Mc-Graw Hill Book Company.)

† $\vec{B}$ is more rigorously defined as the closure error in a Burgers circuit that is essentially a comparison between the real crystal and an otherwise equivalent ideal crystal (Frank, 1951; Van Bueren, 1960; Read, 1953).

crystals by physical manipulation, by thermal treatment, and by other means; and as we shall see, dislocations, except under unusual circumstances, are always formed during crystal growth.

Some dislocations are formed by the congealing of vacancies. Indeed vacancies often move in a manner to form dislocations, but not all dislocations are formed by the congealing of vacancies. The equilibrium-dislocation concentration in crystals of reasonable size at room temperature is less than one per crystal. This can be shown in the following way (Swalin, 1962).

The change in the Gibbs free energy during a process in a crystal that tends to form an imperfection is

$$\Delta G = n\Delta H_v - T(\Delta S_c + n\Delta S_v) \tag{1.3}$$

where n is the number of imperfections, ΔH_v is the enthalpy of formation of an imperfection, ΔS_c is the configurational or mixing entropy associated with the formation of n imperfections, ΔS_v is the vibrational entropy resulting from the disturbance of the nearest neighbors by an imperfection, and T is absolute temperature. For the imperfection to be stable, ΔG must be negative. Thus $T(\Delta S_c + n\Delta S_v) > n\Delta H_v$ because ΔS_c predominates over ΔS_v and ΔH_v. Therefore, for small values of n, point defects are thermodynamically stable. However, the enthalpy associated with dislocation formation is large and the entropy is small so that dislocations are ordinarily not thermodynamically stable. Thus it is possible to prepare dislocation-free but not vacancy-free crystals. Vacancies result in charge imbalance in the crystal. For instance, a cation vacancy leaves the crystal with net-negative charge. Substitutional atoms of charge different from the atom ordinarily present at that lattice site and interstitial atoms also cause charge imbalance. Thus the concentration of vacancies, interstitials, and appropriate substitutional atoms is related by appropriate equilibria that follow the law of mass action and one type of imperfection may sometimes be caused to decrease at the expense of an increase in another.

1.3.3 GRAIN BOUNDARIES

When the angle between adjacent crystallites is small, the resultant low-angle grain boundary consists of an array of dislocations (Burgers, 1940). Figure 1.10 is a two-dimensional representation of a pure *tilt boundary*, *AB*. As can be seen, the misorientation is achieved by means of a number of edge dislocations shown as inverted T's. In an analogous manner, screw dislocations are involved in twist-low-angle grain boundaries (Read, 1959).

When one bends a crystal plastically, edge dislocations may be formed. These dislocations usually arrange themselves as shown in Fig. 1.11(a). Figure 1.11 shows the lattice planes of a crystal immediately after bending (a) and after dislocation movement (b). Annealing is usually required to acceler-

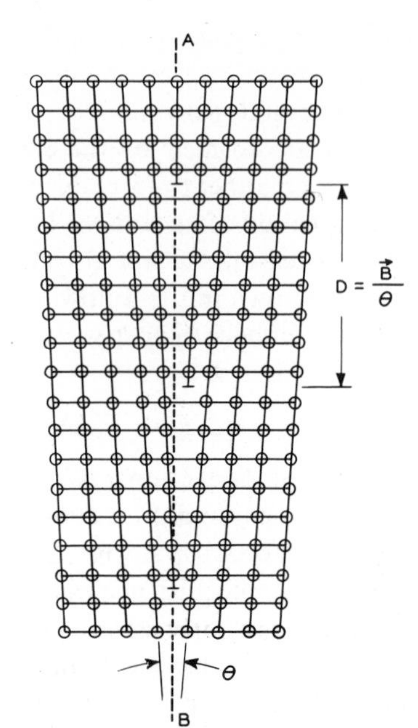

Fig. 1.10 Low-angle grain boundary (after Burgers, 1940). (Courtesy of the Institute of Physics and the Physical Society.)

ate dislocation movement and some dislocations annihilate one another as they come together during movement. The atoms resume their normal planar configuration because the dislocations rearrange themselves by climbing in a plane perpendicular to the plane of the paper in Fig. 1.11. This rearrangement takes place by vacancy movement in the glide plane. The result resembles polygons, so the crystal is said to be *polygonized* and the creation of low-angle grain boundaries by any process is often termed *polygonization*. In some crystals, it has been noted that the surfaces consist of a mosaic of regions each of which is quite perfect but misoriented slightly with respect to its neighbors. If the misorientation between the individual regions is small and if their size is small, the regions are referred to as *mosaic blocks*. Mosaic blocks are

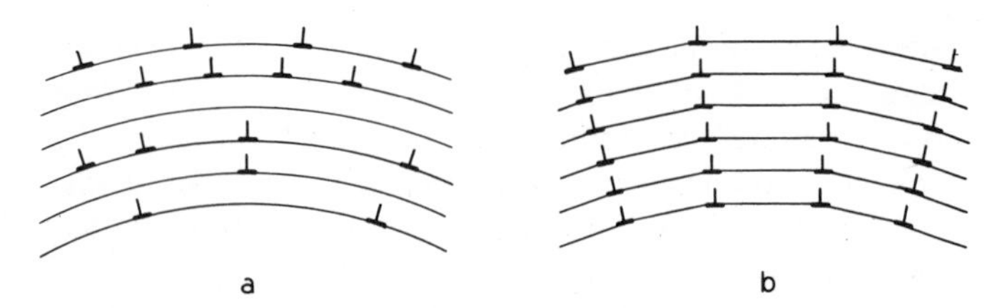

Fig. 1.11 Movement of dislocations in polygonization (a) after bending, (b) after movement. (From *Mineralogy: Concepts, Descriptions, Determinations*, by L. G. Berry and Biran Mason, W. H. Freeman and Company, Copyright 1959.)

usually considered to be from a few hundred to a few thousand lattice distances in extent and to be misoriented by a few minutes of arc. If the size and misorientation are larger, they are usually considered to be conventional grains bounded by conventional low-angle grain boundaries. Thus there is a continuous series of states of order between mosaic structures and structures showing conventional low-angle grain boundaries.

Under certain conditions, vacancies may move within a crystal and collect together with the creation of a partial dislocation. As we have seen, a given boundary can be caused by a collection of dislocations, and if the angles between grain boundaries are large enough we have a polycrystal. There is thus a continuous series of states of decreasing order from an ideal perfect crystal to polycrystalline material. The genesis of dislocations and their role in crystal growth are discussed in Chapter 3.

In a similar manner, there can be a series of states of order from a perfect crystal to a glass. For example, if we consider the devitrification of a glass, the long-range order will increase gradually as devitrification proceeds until the resultant product formed is clearly crystalline. Indeed, partly-ordered glasses, such as Pyroceram, in which the degree of order is controlled by means of added crystal-nucleating agents and controlled-heat treatment, are important articles of commerce.

1.3.4 TWINS

Two other types of imperfections that are important in crystals are *twins* and *stacking faults*. Composite crystals in which the individual parts are related to one another in a definite crystallographic manner are *twinned crystals*. Twinned crystals are called *simple twins* if composed of two parts in related orientation and *multiple twins* if composed of more than two parts. In *contact twins*, the two individuals meet along a definite plane called the *composition plane*. In some cases the orientations of the two individuals of the crystal are related by reflection across a lattice plane common to both individuals. This plane is called the *twin plane* and it may or may not coincide with a composition plane.

In a *polysynthetic twin* the twinning pattern is repeated and more than one composition plane is present. If an ordinary twin is represented by $A \mid B$, where A and B are regions of different orientation bearing a twin relationship and $\mid$ is the composition plane, then $A \mid B \mid A \mid B$, etc., is a polysynthetic twin.

In some twinned crystals the twin parts are related by rotation about some crystallographic direction. This direction is called the *twin axis*.

Twinned crystals may be classified according to their method of formation. Twins formed during the growth process are *growth twins*. Twins may also be produced by *mechanical deformation*. By *gliding* a crystal of calcite or sodium nitrate a deformation twin can be formed. Glide occurs when the atoms are caused to move in a *glide plane* by, for instance, pressing a knife blade against the crystal so that the blade is parallel to the glide plane.

One should not confuse gliding or a gliding plane with a glide plane of symmetry. The term *glide plane of symmetry* is used in describing the symmetry of a crystal lattice. An array of atoms has a glide plane of symmetry if the arrangement on one side of the symmetry plane can be made by (1) translating the arrangement on the other parallel to the symmetry plane by a distance of half the distance necessary to repeat the pattern and then (2) reflecting the arrangement of (1) normal to the glide plane of symmetry.

When a material of a given chemical composition may exist in several structures, these structural modifications are called *polymorphs*. When the polymorphic forms are the mirror images of one another they are called *enantiomorphs*. Each polymorphic form has its range of conditions for stability. A pair of polymorphs of a given substance will generally have a pressure–temperature line along which they coexist and across which one will transform to the other. For instance,

$$\alpha\text{-quartz} \underset{1\ \text{atm}}{\overset{573°\text{C}}{\rightleftharpoons}} \beta\text{-quartz}$$

In the α-quartz/β-quartz inversion the transformation usually begins at many centers. Figure 1.12 (Berry and Mason, 1959) shows a projection of the Si atoms on the (00.1) plane in (a) low-temperature α-quartz, (b) high-temperature β-quartz, and (c) a possible result of cooling β-quartz below 573°C. β-quartz has higher symmetry than α-quartz and when the transformation began in (c) at two points, one on either side of the plane *AB*, the result was twinning with a composition plane *AB*, which is the plane where the growth regions met. Such a twin whose orientational possibilities are influenced by the structure of the material from which it transformed is called a *Dauphiné twin*. In quartz, Dauphiné twinning is called *electrical twinning* because the

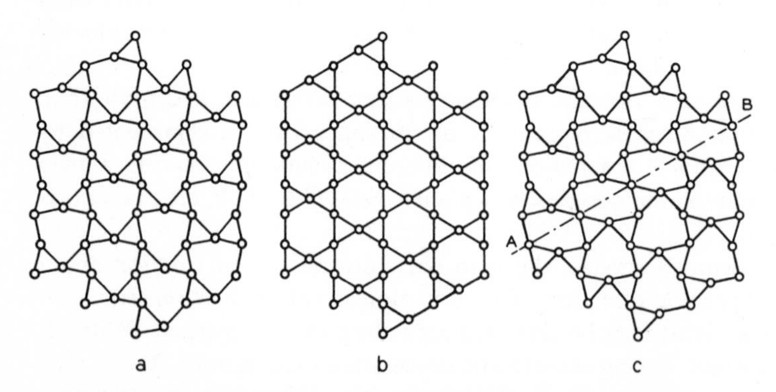

Fig. 1.12 (a) α-quartz, (b) β-quartz, (c) Dauphiné twin (after Berry and Mason, 1959).

effect is to alter the piezoelectric properties. The twins are related to one another by 180° rotation about ⟨00.1⟩. The crystal axes are parallel but the polarity of the electric axis is reversed. Dauphiné twinning has no effect on the optical properties of quartz.

Another type of twinning in quartz and other materials is called *optical twinning* because it may be detected in polarized light (optical twinning also affects the electrical properties of quartz). In optical twinning (sometimes called *Brazil twinning*) in quartz, the twinned parts are related by reflection over $\{11\bar{2}0\}$. Again the crystal axes are parallel but the electrical polarity is reversed and the sign of the rotation of polarized light is opposite in each region. Twinning that can be detected by electric-axis reversal is called electrical twinning, while that detected by examination between crossed polarizers is optical twinning. Crystals that lack a center of symmetry are piezoelectric. Most piezoelectric crystals exhibit electrical twinning and many crystals that lack centers of symmetry exhibit optical twinning.

1.3.5 STACKING FAULTS

Crystalline structures can be formed, at least mentally, by stacking layers and there are a number of imperfections that can result in crystals of certain symmetries because of the improper order of these layers.

The polymorphs of silicon carbide can be pictured as different stacking sequences of the same silicon–carbon layer. This silicon–carbon layer consists of a hexagonal close-packed layer of silicon that has, immediately above the silicon atoms, a layer of carbon atoms. These layers may be stacked in three different orientations with respect to one another. As shown in Fig. 1.13, if we call layers of the given orientation *A*, *B*, and *C* when they are stacked in the sequence *ABC–ABC–ABC* repeated throughout the crystal (which for brevity we will indicate as ···–*ABC*–···), the polymorph is called β-SiC (cubic). There are a number of other ways in which the layers can be stacked. For instance, when the stacking sequence is ···–*ABCA*–···, the polymorph is called SiC-III or SiC-4H; when the sequence is ···–*ABCACB*–···, the compound is SiC-II or SiC-6H. The 4 and 6 refer to the number of layers in the repeating pattern and H indicates hexagonal structure (Ramsdell, 1944, 1945, 1947). Polymorphs that differ only in the order in which their layers are stacked are called *polytypes*. The regularity in the stacking order in some polytypes is remarkable. In SiC-I or SiC-15-R the sequence is ···–*ABCBACABACBCACB*–···. The repeat distance is thus 15 layers and the structure is rhombohedral.

Strictly speaking the term *polytype* should be reserved for crystals of the same structure-lattice type that differ in the stacking sequence. Thus 4-H and 6-H are polytypes, while 4-H and 15-R are polymorphs of SiC. A polytype designated as 87-R is known and 270-R has been reported (Zhdanov and Minervna, 1947).

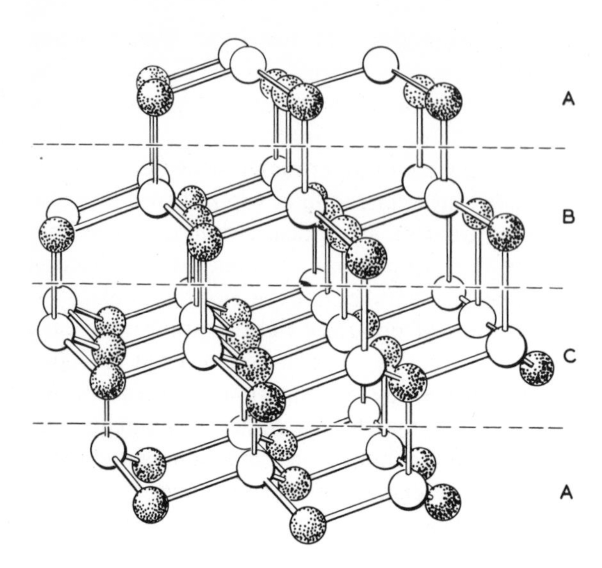

Fig. 1.13 Structure of β-S.C.

Often in polytypes an error in the stacking sequence can occur, resulting in a thin layer of material of one structure "sandwiched" between layers of material of another structure. For instance, in a cubic crystal such as cubic ZnS (sphalerite) the stacking sequence is $\cdots$–*ABC*–$\cdots$, while in hexagonal ZnS (wurtzite) the sequence is $\cdots$–*AB*–$\cdots$. If in a ZnS crystal the structure is $\cdots$–*ABC*–$\cdots$–$\underline{AB}$–$\cdots$–*ABC*–$\cdots$ the underlined region is a stacking fault or a local layer of macroscopic dimensions of hexagonal ZnS in a cubic ZnS crystal. Often stacking faults occur as a series of parallel layers of material of one polytype in another.

If the layers are of macroscopic thickness the coalescence of the two types may be evident from an examination of the shape or crystal *morphology*, for each region will exhibit the morphology of the polytype present in that region. In some cases the crystal will contain interfacial angles greater than 180° (*reentrant angles*) between the faces of adjacent polytypic regions. The term *syntaxis* is sometimes used to describe the coalescence of polytypic substances.

We have discussed twin planes above and it would be instructive to describe a twin plane in terms of the stacking sequence involved. In sphalerite this would be $\cdots$–*ABC* | *CBA*–$\cdots$, where the vertical line is the twin plane.

1.3.6 VICINAL FACES

When the orientation of a crystal face is determined with an optical goniometer, it will sometimes be found to be misoriented slightly from its "true" position. Such faces are called *vicinal* faces. Whether they are slightly-misoriented faces of low Miller index or faces of high index lying close in orientation to the "true" low-index faces is a question that so far has not been answered unequivocally. X-ray studies seem to show that the faces are not low-index faces, but the evidence is not overwhelming (Terpstra and Codd, 1961; Bond and Andrus, 1952; Schubnikov and Brunowski, 1931). The explanation may be found eventually in terms of low-angle grain boundaries.

1.3.7 OTHER IMPERFECTIONS

When atoms dissolve in a crystalline lattice and form a solid solution, they may choose random substitutional sites or random interstitial sites to occupy or may enter the lattice in an ordered manner. If the atoms in substitutional or interstitial positions are highly ordered, we may speak of a *superlattice* of these atoms and the degree of order of this superlattice may be studied by X-ray and electron microscope techniques. There will obviously be possibilities for many of the imperfections that occur in an ordinary lattice to occur in this superlattice.

A term used in a mineralogical context to describe crystals that have been disordered because of radiation damage is *metamict*. Metamict is usually used to describe minerals that have been damaged by natural radiation, which can be so severe that an essentially-amorphous solid results. Damage from man-made radiation sources can cause similar imperfections in crystalline materials.

1.4 Methods of Studying Perfection

1.4.1 X-RAY METHODS

The theory of reflection from large perfect crystals predicts that the angular range of reflection will be of the order of a few seconds of arc and that the intensity will be proportional to the structure factor. In most crystals, the range of reflection is of the order of minutes of arc and the intensity is proportional to the square of the structure factor. The reason for these discrepancies is that most crystals are not perfect; tiny regions are misoriented with respect to one another because of dislocations, grain boundaries, and mosaic blocks.

Thus the diffraction of X-rays provides a means of determining the perfection of crystalline materials and even of mapping individual dislocations. We can give here only a brief outline of the techniques presently used. For a more complete discussion the recent literature should be consulted

(Newkirk and Wernick, 1962; Weissmann, 1963; Warren, 1959; Barrett and Massalski, 1966). The techniques most often used are as follows:

1. *Double-crystal spectrometer*. The integrated intensity diffracted by a crystal oriented near the Bragg angle and the width of the Bragg peak both depend strongly on crystal perfection. A number of techniques in which a crystal is displaced through a small angle while the reflections are observed have been used to study perfection. The most useful results have been obtained when two crystals are used. Characteristic X-rays are reflected at the Bragg angle from the reference crystal to the crystal to be studied, as shown in Fig. 1.14. The reflecting planes of the crystal to be studied are usually parallel to those of the reference crystal.

If the intensity of the reflection is plotted as a function of θ, the "rocking angle," a curve similar to that shown in Fig. 1.14, is obtained. The width of the peak is a measure of the perfection. This technique has been used to study a variety of materials, including α-quartz, Si, W, and others and is well described in the literature (Bond and Andrus, 1952; Batterman, 1959; Webb, 1962).

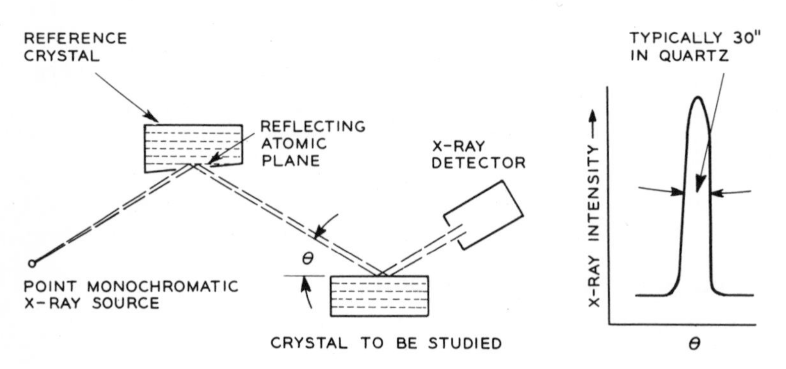

Fig. 1.14 Double crystal spectrometer for studying perfection (after Bond and Andrus, 1952). (Courtesy of *American Mineralogist*.)

2. *Schulz technique*. The geometry of this method is shown in Fig. 1.15. As employed by Schulz (1954), the single-crystal sample is placed at an angle of about 25° to a nearly parallel X-ray beam. White radiation is used. The several spots produced on the film are simply Laue spots, about the same size and shape as the specimen. Each of these spots will be made up of subunits with slightly different orientation, each corresponding to a region in the crystal oriented slightly differently.

The Schulz technique is probably the easiest X-ray method of studying perfection in practice because very simple apparatus can be used, no special crystal orientation is necessary, and exposure time can be as short as 3 min; but spatial and angular resolution are only fair.

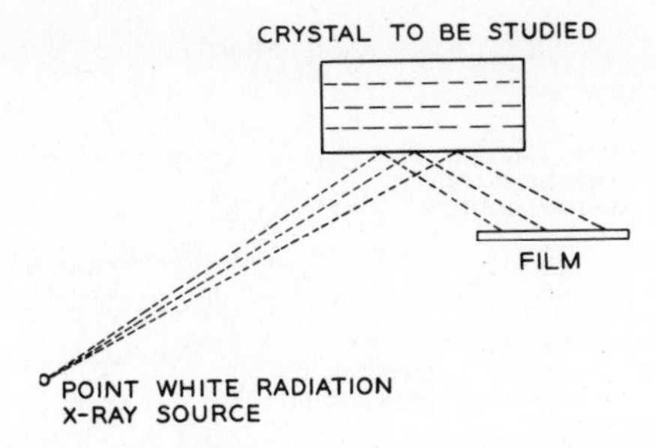

Fig. 1.15 Schulz technique for studying perfection (after Webb, 1962). (Courtesy of John Wiley & Sons, Inc.)

3. *Back-reflection Berg–Barrett technique.* A schematic diagram illustrating this technique is shown in Fig. 1.16. Because a region around a dislocation is strained and a strained region reflects more effectively, dislocations will appear on the film as dark regions of increased reflected intensity.

4. *Transmission (Diffraction) Berg–Barrett technique.* This technique is illustrated in Fig. 1.17. Here the diffracted beam traverses the crystal, and because regions containing dislocations diffract more strongly, they appear as dark regions on the film.

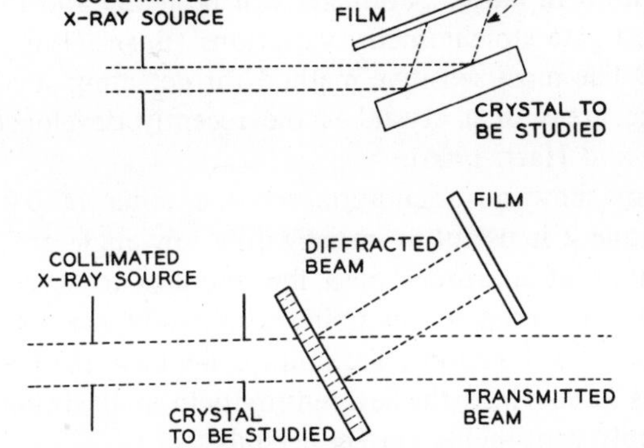

Fig. 1.16 Back-reflection Berg–Barrett technique for studying perfection (after Webb, 1962). (Courtesy of John Wiley & Sons, Inc.)

Fig. 1.17 Transmission Berg–Barrett technique for studying perfection (after Webb, 1962). (Courtesy of John Wiley & Sons, Inc.)

5. *Borrmann anomalous-transmission technique.* Figure 1.18 illustrates this technique wherein both the transmitted and the diffracted beams are recorded. In this method both the transmitted intensity and the diffracted intensity are reduced at dislocations.

6. *Lang technique.* The experimental arrangement for the Lang technique is generally similar to that for the transmission Berg–Barrett method. A highly collimated narrow beam of X-rays is diffracted through the crystal and the Bragg reflection from the parallel planes approximately normal to

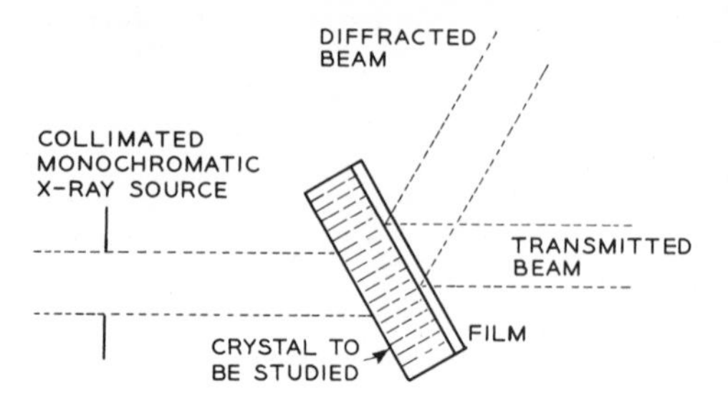

Fig. 1.18 Borrmann anomalous-transmission technique for studying perfection (after Webb, 1962). (Courtesy of John Wiley & Sons, Inc.)

the face of the specimen is recorded on the film. The film and specimen are moved simultaneously and a complete "map" of the dislocation network can be obtained. The dislocations are observed as regions of intensified reflecting power through their action in altering the distribution of energy between multiply reflected primary and diffracted beams. The result is a dark image on the film corresponding to dislocations. Suitable changes in angular orientation of the crystal allow the dislocation network to be determined in three dimensions.

7. *Lattice-parameter measurements.* Many sorts of imperfections in crystals affect the lattice parameter of macroscopic regions. These lattice-parameter variations can be readily determined by Bond's method (1960), which is a technique for measuring interplanar spacings with an accuracy of a few parts per million. Because the diffracting volume is quite small ($1 \times 0.025 \times 0.005$ mm), maps of variation in lattice parameter can be made and the method can be used to investigate stoichiometry variations (Barns, 1967).

8. *X-ray interferometer.* The most sensitive method for detecting small distortions over macroscopic areas of a crystal is the recently developed X-ray interferometer (Bonse and Hart, 1966).

Techniques 1 and 8 can show misorientations of the order of 0.01 seconds of arc, while technique 2 is useful in investigating low-angle grain boundaries and misorientations of the order of a few minutes to several degrees. Techniques 3–6 and 8 are used for mapping dislocation networks and even investigating individual dislocations. Techniques 3–6 have the best spatial resolution. Techniques 5 and 8 have the best sensitivity to small strains, but technique 8 is effective only with highly perfect crystals.

Perfection may also be studied by comparison of the density of a material as determined by X-ray methods, which gives the "ideal" density of the lattice and the density determined pycnometrically, that is, by measurement of the volume and weight of the sample. The difference in these values gives a measure of vacancy and substitutional and interstitial impurity concentrations. Uncertainties in atomic weights and Avogadro number make the technique somewhat more reliable for relative than for absolute measurements (Smakula et al., 1955).

(a)

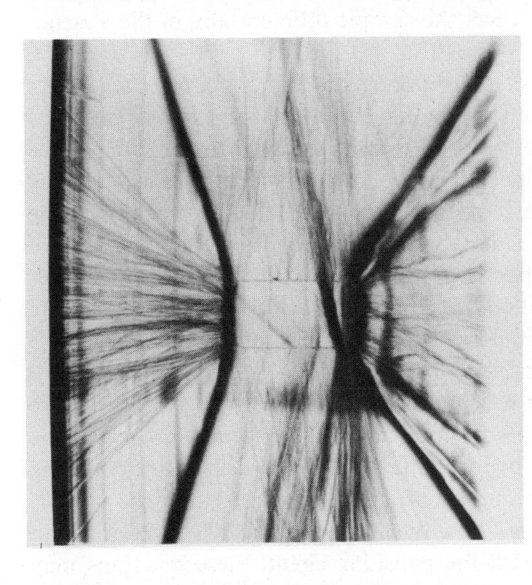

(b)

Fig. 1.19 Pictures taken in studies of perfection of materials: (a) Schulz picture of Verneuil grown sapphire—spot size and shape correspond to sample size and shape. Approximately rectangular region in spot is misoriented region in sample (courtesy of R. L. Barns); (b) Lang picture of hydrothermally grown quartz (courtesy of W. J. Spencer).

Figure 1.19 shows several X-ray pictures taken in recent studies of the perfection of materials.

1.4.2 ETCHING

Another powerful method for investigating perfection in crystalline materials is etching followed by microscopic examination. The individual grains in ceramics are often revealed by suitable solvents or etchants that differentially attack the material. Similarly, low-angle grain boundaries and even individual dislocations can be made visible. In addition, etching often reveals particulate impurities and second phases.

A grain boundary is a region where impurities, dislocations, and strains tend to concentrate. The solubility and more importantly the rate of solution will be affected by the presence of impurities and strains, and thus a grain boundary will be preferentially attacked by an etchant. Similarly, because a dislocation has a region of strain associated with it, etching proceeds more vigorously at the dislocation. Some etchants have the characteristic that their attack on a material is insensitive to the presence of impurities or strains. These are the so-called *polishing etches*. They are useful for uniformly removing surface damage before examination of the properties of a material. It is often necessary to lap or polish a surface carefully with a series of increasingly finer abrasive agents before polish etching. In some cases, polishing is achieved by the simultaneous use of an etchant and mechanical rubbing. This technique is sometimes called *chemical lapping*. If a polishing etch is not available, a smooth surface sometimes can be achieved by chemical lapping, using an etch that would ordinarily attack the sample differentially in the absence of rubbing.

The choice of a sequence of polishing agents is more complicated than it would at first seem. Although it is true that surface damage can be removed by a polishing etch (provided one can be found) or by means of some of the techniques mentioned below, mechanically lapping and polishing in a way to minimize damage will greatly reduce experimental complications. One problem that is often overlooked in lapping is the possibility of formation of a Beilby layer (Bowden and Tabor, 1964). When a solid is polished, no matter how lightly loaded or how well cooled, there will exist protuberances where the loading is high. Here frictional heat will often generate temperatures high enough to melt several atomic layers of the material. Surface tension will tend to cause this melted layer to smooth over valleys in the surface before it freezes. The result will be an amorphous surface layer or a microcrystalline surface layer. If the material is brittle it may break and tear under polishing before melting, but if the polishing agent is lower-melting than the crystal, the polishing agent may melt and flow over the surface. If the material can oxidize, decompose, or react with the polishing agent, these reactions may take place at local hot spots and the surface will be a "fudge" of reaction products. Thus the role of frictional heat should not be neglected in the choice of polishing agents and techniques. These considerations show that cutting, which is characteristic of metallographic polishing methods as opposed to optical polishing methods, which tend to smear, is to be preferred. The techniques useful in the preparation of highly polished glass surfaces should not be applied blindly to crystals. In glasses the formation of a Beilby layer is often advantageous, but in crystals its formation is usually disastrous.

Recent work (Rabinowicz, 1965, 1968) shows that, at least in the case of high-melting-point materials, polishing does not necessarily involve a Beilby layer. For instance, the Beilby mechanism would predict that polishing by melting of the material being polished would be effective only when the melting point of the polishing powder is higher than that of the material being polished. This correlation is observed to be true only for relatively low-melting-point materials. Samuels (1967) has shown that some polished surfaces that look smooth under ordinary light exhibit small scratches when observed by phase-contrast microscopy. Such scratches would not be likely if polishing were by the Beilby mechanism. Rabinowicz (1965, 1968) has shown by a study of materials with a profile meter before and after polishing that polishing by abrasion is not a usual mechanism. Rabinowicz explains polishing as caused by the dislodging of atoms at high points on the surface. These atoms are subject to greater downward displacement during polishing and break free from the material when the load is removed because of elastic rebound. It is likely that the Rabinowicz mechanism is responsible for polishing of high melters, but the possibility of Beilby effects, at least in lower-melting materials, is still considerable.

Electrolytic polishing and spark cutting (Hall, 1956) have been used to prepare surfaces of conductors without gross damage. Chemical and wetstring sawing can also be used to prepare surfaces with minimum damage.

Once the surface is suitably prepared, the crystal is subjected to an etchant that in some manner differentially attacks strained regions, regions high in impurities, or regions containing dislocations. The choice of an etchant for a given crystal is almost entirely a matter of empiricism. Often the etchant contains an oxidizing agent and a solvent for the oxide that is formed. Thus it may be more proper to think of the etchant as a reagent that reacts with the crystal rather than as a solvent.

The exact mechanism by which an etchant may either remove material uniformly or preferentially attack a dislocation is obscure in the case of most etches. However, several etches whose compositions are chemically simple have been studied in some detail and their action is fairly well understood. We can probably safely infer from these studies that the mechanisms of at least some of the chemically more complicated etches are analogous. For instance, the mechanism of etching of LiF by aqueous FeF_3 has been studied by Gilman and coworkers (1958) and Sears (1960). They conclude that the first removal of an atom from the surface occurs where a dislocation intersects the surface. This is caused by the fact that strain energy and "core" energy cause atoms to be less tightly bound at the dislocation. The role of ferric fluoride is to inhibit the motion of steps across the surface relative to the initiation of new steps at the dislocation. This has the effect of increasing the rate of dissolution in the pit (normal to the crystal surface) relative to the rate parallel to the crystal surface. The FeF_3 atoms are chemisorbed at kinks† in steps where they make strong bonds to F^-. In the absence of Fe^{3+} (or Al^{3+}, which acts similarly), the pits do not form, and when Fe^{3+} is less than about 10^{-7} molar, the pits are shallow. Thus in many cases the etchant contains an impurity whose role is to be preferentially adsorbed or chemisorbed at a kink in a step. Table 1.1 contains a partial list of etchants that have been used to study crystalline perfection.

The differential volatility of strained and unstrained regions in a technique called *thermal etching* may in some cases be used to reveal dislocations. The crystal is heated and either differential sublimation or differential reaction with a reactive-gas phase leads to the delineation of surface features.

Etch pits are often bounded by planar surfaces that correspond to low-index faces of the crystal. Such faces may be of aid in determining the symmetry of a crystal, in orienting a crystal, in determining whether a crystal is twinned, and in identifying the face of a crystal.

The problem of establishing that each etch pit is associated with a single dislocation has not been completely resolved. In only a few materials for specific crystallographic faces etched with particular etchants has a one-to-one correspondence been fully demonstrated. The techniques that have been used to interrelate etch pits and dislocations are as follows:

1. *Plastic deformation.* Gilman and Johnston (1957, 1958) succeeded in observing the individual dislocation lines in LiF. After a given deformation, etch pits were observed along slip lines. Following a second deformation, the crystal was etched again. New, smaller etch pits were revealed along the slip band.‡ The new pits were displaced from the old pits, while the old pits had etched further but had begun to develop flat bottoms. It was concluded that the pits produced after the first etching delineated individual dislocations that had migrated farther under the influence of stress to points where etching took place during the second etching. The fact that the original pits began to develop flat bottoms during the second etch was interpreted as an indication that they no longer delineated dislocations. Using this technique, Gilman et al. (1958) followed the motion of individual dislocations and determined their velocity as a function of applied stress.

Similar studies were performed by Dorendorf (1957) on germanium sur-

---

†For a discussion of the role of steps and kinks in growth, see Chapter 3. Figure 3.2 shows a diagram of a typical step and kink.

‡Many crystals deform partially by means of *translational slip* or movement of one part of a crystal as a unit across a neighboring part. The line of intersection of the slip surface with the outer surface of the crystal is the *slip band.* Detailed studies of slip have usually shown that a slip band is composed of a lamella of atomic planes, each of which has slipped slightly from the next; the summation of this slip is the macroscopically observed slip. This macroscopic slip of the individual planes is identical with *glide,* described previously.

Table 1.1

REPRESENTATIVE CHEMICAL ETCHANTS USED TO REVEAL PERFECTION

Material	*Etch*	*Results*	*Reference*
$BaTiO_3$	0.5% HF + HNO_3; 1.5 min	Reveals domain, grain, and twin boundaries on polished surface	Tennery and Anderson, 1958
LiF	1 part HF + 1 part HAc (glacial) + 20 ppm FeF_3; 0.5–1.5 min	Reveals dislocations as pits on {100}, {111}, and other faces	Gilman and Johnston, 1956, 1957
Ge	HF–HAc–HNO_3–liquid Br_2 (50: 50: 80: 1 by volume) called CP-4; 1–2 min	Reveals dislocations as pits on {100}, {111}, and other faces	Vogel et al., 1953; see also Vogel, 1955
Si	50% HF–70% HNO_3–HAc (glacial) (1: 3: 10 by volume); 1–2 hr	Reveals dislocations as pits on {100}, {111}, and other faces	Dash, 1958
NaCl	Methanol	Reveals screw dislocations and edge dislocations	Amelinckx, 1954
GaAs	30% HNO_3; 15 min	Triangular etch pits at dislocations	See Faust, 1962, for review of etchants for III–V compounds
$CaWO_4$	Saturated aqueous CrO_3–concentrated HF (2: 1 by volume); 2–25 min	Reveals dislocations on {100} and {001}	Levinstein et al., private communication
Al	HNO_3–HCl–HF (47%–50%–3% by volume)	Reveals dislocations associated with Fe impurity	Hendrickson and Machlin, 1955
SiC	Fused borax at about 800°C	Reveals dislocations on several faces	Amelinckx and Strumane, 1960
SiC	Fused Na_2CO_3 at 1000°C	Reveals dislocations	Horn, 1952
AgCl	3–5 *N* $Na_2S_2O_3$	Reveals dislocations on {100} and {111}	Jones and Mitchell, 1957
$CaF_2 \cdot 3\,Ca_3P_2O_8$ (apatite)	Concentrated tartaric or citric acid at 60–70°C	Reveals dislocations on basal plane	Lovell, 1958
$CaCO_3$ (calcite)	90% formic acid; 15 sec	Reveals dislocations emerging on cleavage faces	Keith and Gilman, 1960

faces. In his experiments the dislocations were caused to move by means of annealing following an initial deformation.

2. *Etch-pit count along low-angle grain boundary.* Figure 1.10 shows a low-angle tilt boundary. As stated above, the boundary may be considered an array of edge dislocations having the same sign. The spacing between the dislocations, D, is related to the rotational angle, θ, of one part of the crystal with respect to the other and to the Burgers vector by the equation

$$D = \frac{\vec{B}}{\theta} \tag{1.4}$$

Equation (1.4) is valied for $\theta < 15°$. Thus if etch-pit separation obeys this relationship, the pits are probably associated with dislocations. The average distance between pits is measured under the microscope, the rotational angle is determined by X-ray studies, and the magnitude of the Burgers vector is determined from crystallographic considerations. This technique was used by Vogel (1955) to show that the etch pits made by the etch CP-4 (see Table 1.1 for composition) on the {100} faces of germanium were associated with dislocations. The Burgers vector was estimated in the following way: Because the low-angle grain boundary (of the order of 1′) was shown to be a pure tilt boundary, the edge dislocations had to lie parallel to the axis of tilt. Thus the line of dislocation was a [100]. The interaction energy of an array of dislocations of like sign with parallel Burgers vectors was shown by Read (1953) to be minimized when they align themselves in a plane normal to the Burgers vectors. Because the boundaries occupied $(0\bar{1}1)$ planes, the Burgers vectors had to be the shortest lattice translations normal to $(0\bar{1}1)$. Thus $\vec{B} = \frac{1}{2}$ of the distance between {100} planes or $\sqrt{2/2}a_0 = 4$ Å. This was consistent with experimentally measured Burgers vectors in slip.

For cases where the contact plane forms an angle θ with the symmetry plane, the result is an asymmetrical tilt boundary and Eq. (1.4) is modified. The application of Eq. (1.4) is also modified, but the principle would be similar, for a twist boundary.

3. *Dislocation density in intersecting lineage boundaries.* Tilt boundaries often intersect in the form of T's or L's. The sum of the tilt angles, θ, encountered in a circuit about such an intersection must be zero provided the boundaries are pure tilt boundaries. Thus the dislocation densities in the intersecting boundary branches must be simply related.

For instance, in the germanium lattice on the (001) plane, Pfann and Lovell (1955) showed that

$$\rho_A + \rho_B = 0 \tag{1.5}$$

for an L intersection, and

$$\rho_A + \rho_B + \rho_C = 0 \tag{1.6}$$

for a T intersection, where ρ is the dislocation density and A, B, and C indicate the branch of the intersection. CP-4 was used as the etch and the signs of the dislocations were picked to fulfill Eqs. (1.5) and (1.6). Excellent agreement was obtained, furnishing strong support that the etch pits were indeed associated with dislocations.

4. *Correlation with other measurements.* Identification of etch pits with dislocations has often been made by means of studies of the dislocations by other means. X-ray studies, as we have seen, can be used to map dislocations. When the location of the dislocation is corroborated by etch-pit studies, there is justification in assuming that the pits are associated with the dislocations. Similarly, the direct examination of crystal surfaces, optical methods, and electron-optical methods, all of which are discussed below, may be used to study perfection, and the results in a given material may be compared with etching studies.

A detailed method for evaluating the perfection of single-crystal silicon by etching has been published (ASTM-F47-64T, 1967). Typical etch pits and etching behavior associated with dislocations, lineage (low-angle grain boundaries), slip, large-angle grain boundaries, twins, and grossly polycrysttalline material are illustrated, and detailed procedures for preparation of etches, etching, and evaluation are given.

1.4.3 DECORATION

It should be remembered that the etch-pit method, X-ray methods, and many of the other techniques for studying perfection usually give only a two-dimensional picture of the surface of the crystal under study. It is true that by lapping followed by etching and by taking X-ray pictures in several orientations, information about the three-dimensional dislocation network can be obtained, but with some experimental complications. Decoration, on the other hand, provides direct information about dislocations in the volume of the crystal. The technique consists of treating the crystal in such a manner as to cause some material to precipitate along the dislocations as observable particles. Indeed, it is the tendency for impurities to segregate along dislocations that aids in the differential attack of etches and that may be responsible for the differential scattering power of X-rays in some systems. Thus because impurities often tend to diffuse toward dislocations, annealing, which accelerates this diffusion, often is performed before etching. Actually, some etches are only effective on crystals containing certain impurities. In the case of the direct observation of dislocations by decoration, the impurity is grown into the crystal or is diffused in after growth. A difficulty associated with annealing to accelerate diffusion and exsolution is that it may alter the perfection of the crystal. Table 1.2 lists a number of decoration techniques for various materials. For materials not transparent to visible light, infrared and electron microscopy can be used to examine the decorated crystal.

Table 1.2

REPRESENTATIVE DECORATION TECHNIQUES USED TO REVEAL PERFECTION

Material	*Technique*	*Results*	*Reference*
Si	Evaporate aqueous $Cu(NO_3)_2$ on surface; anneal in H_2 at 950° for 30 min; quench; view with infrared image converter	Reveals dislocations because of preferential precipitation of Cu	Dash, 1958
AgBr and AgCl	Anneal and expose to light—crystals must be very pure; observe in in transmission	Reveals dislocations because of precipitation of colloidal silver	Hedges and Mitchell, 1953
NaCl or KCl	Heat crystal containing ~1% AgCl in H_2 at 650°C for ~2–3 hr; observe in dark-field microscope	Reveals dislocations by precipitation of colloidal silver halide	Amelinckx et al., 1955; Amelinckx and Strumane, 1960
CaF_2	Evaporate Ag layer, anneal in H_2 (e.g., 3 hr at 1200°C); observe in ultramicroscope	Reveals dislocations by Ag precipitation	Bontinck and Dekeyser, 1956; Bontinck and Amelinckx, 1957
NaCl and KBr	Heat crystal in evacuated tube containing gold halide at 500° for 2–3 hr	Reveals dislocations by Au precipitation	Barber et al., 1957

1.4.4 DIRECT OBSERVATION OF SURFACE

Low-power magnification and even naked-eye observation of crystal surfaces often reveal structure associated with imperfections. For instance, reentrant angles on a crystal surface suggest twinning. As we shall see in Chapter 3, spirally terraced hills often develop during crystal growth. The geometry of such surface structures reveals information concerning the perfection. For instance, simple spirals are often associated with a single-screw dislocation, and measurement of certain characteristics of the spiral can be used to deduce the Burgers vector. The details of the surface structures to be expected are discussed in Chapter 3.

Multiple-beam interferometry has been used with much success in investigating surface topology by Tolansky (1955, 1960).

1.4.5 OBSERVATION WITH THE ELECTRON MICROSCOPE

In a few substances with large lattice spacings, it is possible to resolve the individual lattice planes and even in materials with small lattice spacings it is possible to see dislocations directly in the image formed in the electron microscope. Dislocations have been observed indirectly in the moiré patterns when two thin sheets of material with similar lattice spacings are superimposed and examined in transmission (Oster and Nishijima, 1963).

The usual way in which dislocations are observed in electron microscopy is by means of the differential diffraction of the electron beam in strained and unstrained regions. Perfection can, in principle, be studied in any material that can be produced undamaged in sheets transparent to an electron beam of about 100 kV and not subject to electron damage. For most materials, 1000 Å is a useful thickness (Mitchell, 1962). If a crystal is nearly perfect (angular distortion $\sim 10^{-3}$ rad) and sufficiently thick ($\sim$1000 Å for germanium), Kikuchi lines (Kikuchi, 1928; Heidenreich, 1964) whose width is proportional to the crystal-plane curvature will be produced in electron diffraction. Kikuchi patterns are thus useful in studying perfection in crystals of high perfection.

1.4.6 OBSERVATION WITH FIELD-ION MICROSCOPE

The field-ion microscope is capable of resolving the individual atoms at the sharp point of a metal wire. In essence, the wire is made the cathode with a high-electron field at the tip. The emitted ions form an image on a fluorescent screen. The magnification is proportional to the ratio of screen distance to tip radius and can be as large as 10^6, so that resolutions of 20 Å can be obtained. Individual atoms are easily visible, but so far only a few high-melting materials can be studied. Vacancies, interstitials, impurity atoms, and dislocations have been observed (Müller, 1962), but the method is so far limited to refractory metals.

1.5 Why Are We Interested in Crystals?

Single crystals find important uses in research. As we have seen, according to one classification all true solids are crystals. Thus for an understanding of the physics and chemistry of the solid state, crystals are a prerequisite. One may use a polycrystalline sample rather than a single crystal for many studies, but often a reasonably large single crystal will be required. Polycrystals contain grain boundaries. If one desires a knowledge of some bulk property of a material and he measures that property on a polycrystalline specimen he will, in many cases, measure the property of the grain boundaries and not the bulk material. A notable example of a property where single crystals are essential is the electrical conductivity of semiconductors, which is particularly impurity-sensitive. Impurities tend to segregate at grain boundaries and thus single crystals are almost always required for a determination of any conductivity-dependent property in a semiconductor. Another common effect of grain boundaries and associated voids is light scattering, and thus single crystals are often required in optical studies.

Many properties of crystals depend on the crystallographic direction in which the measurement is made, because the spatial arrangement of the constituent atoms is not in general the same in all directions. Consequently, if one determines a directionally dependent property in a polycrystalline specimen where the crystallites are randomly oriented, he will measure an average value of the property in which the directional dependence is masked.

Single-crystal solids have important practical applications in technology. For example, much better frequency stability and lower acoustic losses can be achieved in single crystals than in polycrystalline aggregates. Thus single-crystal piezoelectric crystals, such as quartz, are used for frequency-control elements. Conductivity and mobility requirements dictate single-crystal semiconductors for transistors. The existence of lasers and masers has created severe new demands for single crystals for research and applications. The applications that single crystals find in technology and in device research include:

1. Transistors—Si, Ge, GaAs
2. Tunnel diodes, parametric diodes, signal diodes—GaAs
3. Strain gauges—Si, Ga(AsP)
4. Microwave limiters and tunable filters—yttrium–iron garnet
5. Lasers—$CaWO_4$, CaF_2, yttrium–aluminum garnet, ruby, GaAs, InP, InSb, InAs, Ga(AsP)
6. Electromechanical transducers—quartz, Rochelle salt, ammonium dihydrogen phosphate, CdS, GaAs
7. Filters and oscillators–quartz
8. Optical uses—CaF_2, quartz, LiF, calcite
9. Radiation detectors—anthracene, KCl, Si, GaAs, NaI doped with Tl, Ge doped with Li, triglycine sulfate, barium strontium niobate
10. Ultrasonic amplifiers—CdS
11. Industrial bearings—sapphire
12. Cutting and abrasives—sapphire, diamond, SiC
13. Rectifiers—Si and Ge
14. Laser modulators, harmonic generators, and parametric devices—KDP, $LiNbO_3$, $LiTaO_3$ barium strontium niobate, barium sodium niobate
15. Electroluminescent devices—GaP, GaAs, Ga(AsP)

There is a final reason we are interested in crystals. Since prehistoric times, man has treasured single-crystal gem stones for their beauty. Photomicrographs of snow crystals have pleased people since the late nineteenth century. Anyone who has grown or observed a single crystal knows the joy to be found in the symmetry and colors of inanimate nature.

REFERENCES

Amelinckx, S., *Acta Met.* **2**, 849 (1954).

Amelinckx, S., and G. Strumane, *Silicon Carbide—A High Temperature Semiconductor* Ed. by J. R. O'Connor and J. Smiltens, Pergamon, New York, 1960, pp. 162ff.

Amelinckx, S., W. Van der Vorst, R. Gevers, and W. Dekeyser, *Phil. Mag.* **46**, 450 (1955); Amelinckx, S., in *Dislocations and Mechanical Properties of Crystals*, Ed. by R. M. Fisher, Wiley, New York, 1957, pp. 3ff; S. Amelinckx and R. Strumane, *Acta Met.* **8**, 312 (1960).

ASTM-F47-64T, *1967 Book of ASTM Standards*, Part 8, pp. 527ff, American Society for Testing and Materials, Philadelphia, 1967.

Azaroff, L. V., and M. J. Buerger, *Powder Method in X-Ray Crystallography*, McGraw-Hill, New York, 1958.

Barber, D. J., K. B. Harvey, and J. W. Mitchell, *Phil. Mag.* **2**, 704 (1957).

Barns, R. L., *Mater. Res. Bull.* **2**, 276 (1967).

Barrett, C. S., and T. B. Massalski, *Structure of Metals*, McGraw-Hill, New York, 1966.

Batterman, B. W., *J. Appl. Phys.* **30**, 508 (1959).

Bennil, W. P., and I. G. Geibe, in *Methods of Experimental Physics*, Vol. 6. Ed. by K. Lark-Harowitz and Vivian A. Johnson, Academic Press, New York, 1959, pp. 203ff.

Berry, L. G., and Brian Mason, *Mineralogy*, Freeman, San Francisco, 1959, p. 137.

Blass, F. D., *An Introduction to the Methods of Optical Crystallography*, Holt, Rinehart and Winston, New York, 1961.

Bond, W. L., *Acta Cryst.* **13**, 814 (1960).

Bond, W. L., and J. Andrus, *Am. Mineralogist* **37**, 622 (1952).

Bonse, U., and M. Hart, *Z. Physik* **190**, 455 (1966).

Bontinck, W., and W. Dekeyser, *Physica* **22**, 595 (1956); W. Bontinck and S. Amelinckx, *Phil. Mag.* **2**, 94 (1957).

Bowden, F. P., and D. Tabor, *The Friction and Lubrication of Solids*, Oxford Univ. Press, New York, 3rd ed., 1964.

Brady, G. W., and John I. Petz, *J. Chem. Phys.* **34**, 332 (1961).

Brown, Glenn H., and Wilfred G. Shaw, *Chem. Rev.* **57**, 1049 (1957).

Buerger, M. J., *X-Ray Crystallography*, Wiley, New York, 1942.

Burgers, J. M., *Konikl. Ned. Akad. Wetenschap.* **42**, 293, 378 (1939).

Burgers, J. M., *Proc. Phys. Soc.* (*London*) **52**, 23 (1940).

Dash, W. C., in *Growth and Perfection of Crystals*, Ed. by R. H. Doremus, B. W. Roberts, and David Turnbull, Wiley, New York, 1958, pp. 361ff.

Debye, P., *J. Chem. Phys.* **31**, 680 (1959).

D'Eye, R. W. M., and E. Wait, *X-Ray Powder Photography in Inorganic Chemistry*, Academic Press, New York, 1960.

Dorendorf, Von Heinz, *Z. Angew. Physik* **9**, 513 (1957).

Faust, J. W., *Compound Semiconductors*, Vol. I, Ed. by Robert K. Willardson and Harvey L. Goering, Reinhold, New York, 1962, pp. 445ff.

Ford, W. E., *Dana's Textbook of Mineralogy*, 4th ed., Wiley, New York, 1958.

Frank, F. C., *Phil. Mag.* **42**, 809 (1951).

Frisch, H. L., and G. W. Brady, *J. Chem. Phys.* **37**, 1514 (1962) and references therein.

Gilman, J. J., and W. G. Johnston, *J. Appl. Phys.* **27**, 1018 (1956); also *Metal Progr.* **71**, 76 (1957).

Gilman, J. J., and W. G. Johnston, in *Dislocations and Mechanical Properties of Crystals*, Ed. by R. M. Fisher, Wiley, New York, 1957, p. 116.

Gilman, J. J., W. G. Johnston, and G. W. Sears, *J. Appl. Phys.* **29**, 747 (1958).

Gschneidner, K. A., and G. H. Vineyard, *J. Appl. Phys.* **33**, 3444 (1962).

Hall, R. C., *Metal Progr.* **70**, 78 (1956).

Hartshorne, N. H., and A. Stuart, *Crystals and the Polarizing Microscope*, Arnold, London, 1960.

Hartshorne, N. H., and A. Stuart, *Practical Optical Crystallography*, Arnold, London, 1964.

Hedges, J. M., and J. W. Mitchell, *Phil. Mag.* **44**, 223 (1953).

Heidenreich, R. D., *Fundamentals of Transmission Electron Microscopy*, Wiley, New York, 1964, pp. 189ff.

Hendrickson, A. A., and E. S. Machlin, *Acta Met.* **3**, 64 (1955).

Horn, F. H., *Phil. Mag.* **43**, 1210 (1952).

Johnston, W. G., and J. J. Gilman, *J. Appl. Phys.* **30**, 129 (1958).

Jones, D. A., and J. W. Mitchell, *Phil. Mag.* **2**, 1047 (1957).

Keith, R. E., and J. J. Gilman, *Acta Met.* **8**, 1 (1960).

Kikuchi, S., *Proc. Imp. Acad. Japan* **4**, 271 (1928).

Kikuta, S., K. Kohra, and Y. Sugita, *Japan. J. Appl. Phys.* **5**, 1047 (1966).

Levinstein, H. J., G. M. Loiacono, and K. Nassau, private communication.

Lipson, H., and W. Cochran, *The Determination of Crystal Structures*, Cornell Univ. Press, 3rd ed., Ithaca, N.Y., 1966.

Lovell, L. C., *Acta Met.* **6**, 775 (1958).

Mitchell, J. W., in *Direct Observation of Imperfections in Crystals*, Ed. by J. B. Newkirk and J. H. Wernick, Wiley-Interscience, New York, 1962, pp. 3ff.

Müller, Erwin W., in *Direct Observation of Imperfections in Crystals*, Ed. by J. B. Newkirk and J. H. Wernick, Wiley-Interscience, New York, 1962, p. 77.

Newkirk, J. B., and J. H. Wernick, Eds., *Direct Observations of Imperfections in Crystals*, Wiley-Interscience, New York, 1962.

Oster, Gerald, and Yasunori Nishijima, *Sci. Am.* **208** (5), 54 (1963).

Pfann, W. G., and L. C. Lovell, *Acta Met.* **3**, 512 (1955).

Pfann, W. G., and F. L. Vogel, Jr., *Acta Met.* **5**, 377 (1957).

Rabinowicz, E., *Friction and Wear of Materials*, Wiley, New York, 1965.

Rabinowicz, E., *Sci. Am.* **218** (6), 91 (1968).

Ramsdell, L. S., *Am. Mineralogist* **29**, 431 (1944); **30**, 519 (1945); **32**, 64 (1947).

Read, W. T., Jr., *Dislocations in Crystals*, McGraw-Hill, New York, 1953.

Read, W. T., Jr., in *Methods of Experimental Physics*. Vol. 6A, Ed. by K. Lark-Harowitz and Vivian A. Johnson, Academic Press, New York, 1959, pp. 322ff.

Samuels, L. E., *Metallographic Polishing by Mechanical Methods*, Pitman, London, 1967.

Schubnikov, A., and B. Brunowski, *Z. Krist.* **77**, 337–345 (1931).

Schulz, L. G., *Trans. AIME* **200**, 1082 (1954).

Sears, G. W., *J. Chem. Phys.* **32**, 1317 (1960).

Smakula, A., J. Kalnajs, and V. Sils, *Phys. Rev.* **99**, 1747 (1955).

Swalin, Richard A., *Thermodynamics of Solids*, Wiley, New York, 1962, pp. 218ff.

Tennery, V. J., and F. R. Anderson, *J. Appl. Phys.* **29**, 755 (1958).

Terpstra, P., and L. W. Codd, *Crystallometry*, Academic Press, New York, 1961, pp. 75, 360.

Thomas, D. E., and N. S. Gingrich, *J. Chem. Phys.* **6**, 411 (1938).

Tolansky, S., *Introduction to Multiple Beam Interferometry*, Longmans, London, 1955.

Tolansky, S., *Surface Microtopography*, Wiley-Interscience, New York, 1960.

Van Bueren, H. G., *Imperfections in Crystals*, North-Holland, Amsterdam, 1960, pp. 45ff.

Vogel, F. L., *Acta Met.* **3**, 245 (1955).

Vogel, F. L., Jr., W. G. Pfann, H. E. Corey, and E. E. Thomas, *Phys. Rev.* **90**, 489 (1953).

Warren, B. E., *Progr. Metal Phys.* **8**, 147 (1959).

Webb, W. W., in *Direct Observations of Imperfections in Crystals*, Ed. by J. B. Newkirk and J. H. Wernick, Wiley-Interscience, New York, 1962, pp. 31–32, 37.

Weissmann, Sigmund, *Am. Scientist*, **51**, 193 (1963).

Winchell, A. N., and H. Winchell, *The Microscopical Characters of Artificial Inorganic Solid Substances*, Academic Press, New York, 1964.

Wohlstrom, E. E., *Optical Crystallography*, 4th ed., Wiley, New York, 1969.

Wood, Elizabeth A., *Crystal Orientation Manual*, Columbia Univ. Press, New York, 1963.

Zhdanov, G. S., and Z. V. Minervna, *Zh. Esperim. i Teor. Fiz.* **17**, 3 (1947).

Zimm, B. H., *J. Phys. Chem.* **54**, 1306 (1950).

2

CRYSTAL-GROWTH EQUILIBRIA

2.1 Classification of Growth Processes

Crystal-growth equilibria is, in a sense, a contradiction in terms. Crystals do not grow at equilibrium. Nevertheless, before studying the kinetics of any process, it is essential to have some understanding of the equilibria involved. Equilibrium thermodynamics can shed considerable light on crystal-growth processes, and a consideration of the equilibria that can be shifted to make growth take place enables us to make a very useful classification of growth processes.

Crystal growth is a heterogeneous chemical reaction of the type (a) solid → crystal, (b) liquid → crystal, or (c) gas → crystal. It may take place in a system where, except for traces of impurities or low concentrations of deliberately added dopants,† the only component‡ present is the material being crystallized. We shall call growth under these conditions *monocomponent crystallization*. Growth may take place in a system where the impurity concentration or the level of added dopant is high, in which case the material to be crystallized is dissolved in a solvent§ or is formed by means of a chem-

†We shall use *dopant* to mean an impurity whose presence is desired in the grown crystal usually because it imparts some desirable mechanical, electrical, magnetic, or optical property.

‡Component is used in the sense that it is used in the Gibbs phase rule (Findley et al., 1951). See Sec. 2.6.

§That is, the dopant concentration is high enough that it may be considered the solvent.

ical reaction. Such growth is thus in a system where a component or components other than the component that forms the crystal is present. We shall call this *polycomponent crystallization.* Although there is no clear-cut boundary between mono- and polycomponent growth that can be set at some arbitrary concentration of second component, the classification is of use in a consideration of growth theories and in classifying growth techniques.

We shall restrict monocomponent growth to mean growth under conditions of low-doping and low-impurity concentration without the deliberate addition of other components to act as solvents or to react with the crystal-forming component. In polycomponent growth, the additional component may be an undesired impurity present in high concentration or a dopant, which must be present in high concentration in order to obtain a desired concentration in the grown crystal. More often, however, the additional component is added for some reason connected with the growth process. Usually the component is deliberately added to lower the melting point or to raise the volatility of the material being grown.

Some of the reasons for growing a crystal below its melting point or at the lowest possible temperature are

1. To avoid undesired polymorphic phase changes
2. To avoid high vapor pressure at the melting point
3. To avoid incongruent melting
4. To avoid decomposition
5. To avoid high-impurity solubility associated with high temperature
6. To increase experimental feasibility or convenience
7. To lower vacancy concentration and hence dislocation density and to reduce thermal strains and hence low-angle grain boundaries
8. To achieve an impurity distribution not possible at a higher temperature (e.g., volatility of impurity would cause loss)

Low-temperature growth is usually achieved by adding a component that lowers the melting point (growth from solution) or raises the volatility (growth by gas-phase reaction). Polycomponent growth poses the problems of second-component solubility in the crystal, of diffusion of the component forming the crystal to the growing interface, and of the other components away from the interface.

Growth processes may thus be classified according to the scheme of Table 2.1. The details of the various growth techniques mentioned in Table 2.1 will be discussed later.

2.2 Equilibria in Crystal Growth

Before considering the kinetics of any process one should first examine the equilibria. For crystal growth it is important to know

1. What is the stable solid phase?

Table 2.1

- I. Monocomponent
  - A. Solid–solid
    1. Strain annealing
    2. Devitrification
    3. Polymorphic-phase change
  - B. Liquid–solid
    1. Conservative
       - (a) Directional solidification—(Bridgman–Stockbarger)
       - (b) Cooled seed—(Kyropoulos)
       - (c) Pulling (Czochralski)
    2. Nonconservative
       - (a) Zoning (horizontal, vertical, float zone, growth on a pedestal)
       - (b) Verneuil (flame fusion, plasma, arc image)
  - C. Gas–solid
    1. Sublimation–condensation
    2. Sputtering

- II. Polycomponent
  - A. Solid–solid
    1. Precipitation from solid solution
  - B. Liquid–solid
    1. Growth from solution (evaporation, slow cooling, and temperature differential)
       - (a) Aqueous solvents
       - (b) Organic solvents
       - (c) Molten-salt solvents
       - (d) Solvents under hydrothermal conditions
       - (e) Other inorganic solvents
    2. Growth by reaction (media as above—temperature change, concentration change)
       - (a) Chemical reaction
       - (b) Electrochemical reaction
  - C. Gas–solid
    1. Growth by reversible reaction (temperature change, concentration change)
       - (a) Van Arkel (hot-wire) processes
    2. Growth by irreversible reaction
       - (a) Epitaxial processes

2. What is its solubility or what is the value for the equilibrium constant involved in its formation?

3. What is the solubility of other components in the stable solid phase?

For monocomponent crystal growth, only questions 1 and 3 (if doped crystals are being prepared) are pertinent. Ordinarily for any crystallization process to proceed, the crystal grown must be the thermodynamically stable solid phase at the pressure and temperature of crystallization. There are cases where a phase may be formed or a crystal grown under metastable conditions.

Under such conditions, equilibrium thermodynamics are not applicable and we shall consider such behavior beyond the scope of our present discussion.

2.3 Monocomponent Solid–Solid Equilibria

In crystal growth by strain annealing, a material is first strained by plastically deforming it (working it) or by subjecting it to a thermal treatment that strains it because of expansion–contraction brought about by thermal gradients in the material and/or rapid temperature changes. The strain is then annealed out of the material by carefully heating it, often under isothermal conditions, without rapid temperature changes. The unstrained single-crystal regions in the material grow at the expense of strained regions during the anneal with the result that the grain size increases. For any process at equilibrium $\Delta G = 0$, where ΔG is the change in the Gibbs free energy. For a process to take place spontaneously, $\Delta G < 0$. For any process

$$\Delta G = \Delta H - T\Delta S \tag{2.1}$$

and hence at equilibrium for any process

$$\Delta H = T\Delta S \tag{2.2}$$

where ΔH is the enthalpy change, ΔS is the entropy change, and T is the absolute temperature. In almost all crystal-growth processes, the product is more highly ordered than the starting materials†; thus ΔS is usually negative so that ΔH is negative and crystallization would ordinarily be expected to be an exothermic process.

Now for any process

$$\Delta H = \Delta E + \Delta(pv) \tag{2.3}$$

where ΔE is the change in internal energy in a system, $\Delta(pv)$ is the change in the pressure–volume product, and

$$\Delta E = q - w \tag{2.4}$$

where q is the heat added to the system and w is the work done by the system. Thus when a crystalline material is strained isothermally we may formulate the following expressions:

$$\underset{\text{Material (unstrained)}}{\text{state 1}} \longrightarrow \underset{\text{Material (strained)}}{\text{state 2}} \tag{2.5}$$

$$\Delta E_{1\rightarrow 2} = w - q \tag{2.6}$$

where w is the work added to the material by strain, q is the heat released, and $w > q$.

$$\Delta H_{1\rightarrow 2} = \Delta E_{1\rightarrow 2} \tag{2.7}$$

†A notable exception is growth by polymorphic transition, where the product may be of higher or lower symmetry than the starting material.

because Δpv is negligible and

$$\Delta G_{1\to 2} = w - q - T\Delta S \tag{2.8}$$

At low temperatures $T\Delta S$ is negligible, so that†

$$\Delta G_{1-2} \simeq w - q \tag{2.9}$$

Therefore, straining the crystal is not a spontaneous process, while annealing it is. The temperature is raised during annealing only to increase the rate. Quantitative equilibrium calculations of strain-annealing processes are difficult to make because w is not easily estimated. The strain energy about a single dislocation has been estimated, and the interfacial energy of grain boundaries has been determined (Swalin, 1962) but the energetics of a complex array such as occurs in a polycrystalline mass that has been worked are not easy to calculate.

In crystal growth by devitrification of glasses, the driving force arises because the glass is metastable with respect to the crystal.

In an ideal system the enthalpy of devitrification ΔH_D at a particular temperature, T, is given by

$$\Delta H_D = \Delta H_C + \Delta C_p(T_M - T) \tag{2.10}$$

where ΔH_C is the enthalpy change of crystallization at the melting point, ΔC_p is the change in the constant-pressure heat capacity, and T_M is the melting point.‡ Equation (2.10) is valid only when ΔH_C does not depend strongly on temperature or when $T_M - T$ is small.

Other thermodynamic quantities can be calculated by substitution of ΔH_D for ΔH in the appropriate equation; for instance, for devitrification,

$$\Delta G = \Delta H_D - T\Delta S \tag{2.11}$$

Devitrification does not take place at equilibrium, because the glassy state is unstable with respect to the crystalline state, and thus for devitrification,

$$\Delta H_D \neq T\Delta S \tag{2.12}$$

In growth by *polymorphic-* or *allotropic-*§phase change, two types of polymorphism may take place, reversible and irreversible. In *reversible polymorphism* the two forms are in reversible equilibrium at the transition point along the pressure–temperature transition line. The transition takes place below the melting point and there is a unique heat of transition associated with it. In *irreversible polymorphism* one form is never thermodynamically stable; that is, one of the polymorphs is unstable with respect to the other at all temperatures below the melting point. The transition temperature lies above the melting point and the metastable polymorph can be formed only by rapid

†In Sec. 4.2.1 a more exact form of Eq. (2.9) is presented and discussed.

‡In systems where the crystalline phase has been formed, T_M is known; in other systems it must be estimated.

§The term *allotropism* is usually reserved for polymorphism in elements.

quenching from a high temperature (white and red phosphorus are examples). A heat of transition is associated with the transformation from the metastable to the stable form, but the reaction is not reversible and there is no inversion curve. Such behavior is called *monotropism.*

The transformation or inversion curve gives the pressure dependence of the transition temperature for reversible polymorphism and it obeys the Clapeyron equation

$$\frac{dp}{dT} = \frac{\Delta H}{T \Delta V} \tag{2.13}$$

where ΔH and ΔV are the enthalpy and volume changes for the polymorphic transformation. Equation (2.13) can be integrated and the transformation curve can sometimes be generated from thermodynamic data. Figure 2.1 shows the diamond–graphite curve (Strong, 1961).

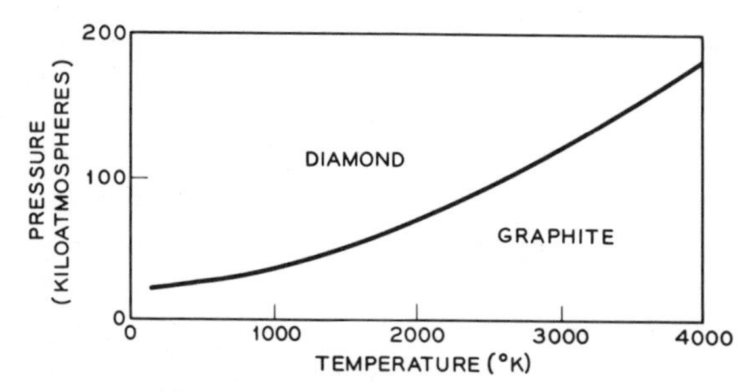

Fig. 2.1 Diamond–graphite phase diagram (after Bundy, Hibbard, and Strong 1963). (Courtesy of John Wiley & Sons, Inc.)

Polymorphic changes can be classified into three types (Smith, 1963):

1. *Minor symmetry* (*nonreconstructive or displacive*) *inversions.* The transformation keeps the general structure, coordination, and bonding the same, but details of structure, such as bond angles, change. The two forms usually belong to the same crystallographic system. It is often possible by carefully approaching the transition temperature to completely transform a single crystal of one polymorph to a single crystal of the other without cracking, twinning, or other gross crystal imperfection, but metastability and Dauphiné twinning often occur. The α-quartz/β-quartz transformation is an example.

2. *Major symmetry* (*reconstructive*) *inversions.* The coordination remains the same but the symmetry is altered. Bonds must be broken and reformed during the transformation, and polytypism and stacking faults often result. It is difficult but not impossible to completely transform a single crystal of one polymorph into another, and metastability is common. ZnS (sphalerite)–ZnS (wurtzite) is an example.

3. *Major structural (reconstructive) inversions.* The coordination and/or the bonding change. The structures are different and the energies and densities of the polymorphs differ much more than in types 1 and 2. Metastability can still take place but is less likely than in 1 and 2. An example is diamond–graphite.

It is easy to get transformations of types 1 and 2 to take place, but type 3 often requires the presence of a solvent and a *seed crystal.* Seed crystals often are the sine qua non of crystal growth. A seed crystal is a crystal of the material to be crystallized that is added to the system in which growth is to take place under conditions where the growth will take place on the seed. There is no infallible way to prove that a given material is the stable form under a given set of conditions. If phase α transforms to phase β at a specified pressure and temperature, we may assume that β is more stable than α; however, we have no proof that another phase, γ, is not more stable than β under these conditions. If crystals of γ are available we may test to see if the reaction phase $\gamma \rightarrow \beta$ proceeds to the right at the given pressure and temperature. However, if phase γ is unavailable or undiscovered (!) our test cannot be made. The usual convention is to assume that the most-stable known phase at a given pressure and temperature is the stable phase and has the lowest free energy. In many polymorphic-phase transitions involving minor symmetry changes (type 1), the high-temperature form has a higher volume and a higher symmetry than the low-temperature form, while dp/dT for the transformation–low-temperature form $\rightarrow$ high-temperature form, in agreement with Le Chatelier's principle is positive (e.g., α-quartz/β-quartz and α-cristobalite/β-cristobalite). Because

$$\frac{dp}{dT} = \frac{\Delta S}{\Delta V} \tag{2.14}$$

we may wonder how ΔS can be positive while the high-temperature form has a higher symmetry. This occurs because ΔS is composed of two parts, ΔS_c, the configurational entropy (or randomness), which is negative for the process because the order increases in the higher symmetry form, and ΔS_v, the vibrational entropy, which is positive because the volume increases in the process and the amplitude of vibration of the atoms about their equilibrium positions in the lattice increases. The net effect in transformations such as α-quartz/β-quartz is to make ΔS positive.

The structure of alloys is often strongly influenced by polymorphic transitions that occur in these polycomponent systems as they are cooled. The well-known *martensite* transformation is an important example. The hardness of quenched steel is caused by the polymorphic transformation of austenite to martensite. Martensite is not the equilibrium phase but is only a transitory intermediate formed before low-temperature equilibrium phases such as pearlite. However, pearlite is not a polymorph of martensite, and martensite structures often persist even under conditions where they are metastable. Martensite-like transformations occur in many other alloy systems.

2.4 Monocomponent Solid–Liquid Equilibria

In growth by liquid–solid reactions an important consideration in the equilibrium is the dependence of melting point on pressure. For melting, dp/dT is positive when ΔV is positive and negative when ΔV is negative. ΔS for a melting process is positive and thus, from Eq. (2.2), ΔH, the heat of fusion, is positive. That is, melting is an endothermic process.

At equilibrium $\Delta G = 0$ and from Eq. (2.2),

$$T_{eq} = \Delta H/\Delta S \tag{2.15}$$

where T_{eq} is the melting point. Because ΔH and ΔS are both single-valued functions of temperature, at constant pressure there will be only one temperature for melting at which Eq. (2.15) will be valid and thus the melting point (or equilibrium temperature for any other first-order phase change)† is unique for any given pressure.

If out of 1 g-atom of a substance a fraction α exists in phase 1 and $(1 - \alpha)$ in phase 2, then

$$G = \alpha G_1 + (1 - \alpha)G_2 \tag{2.16}$$

G (the total free energy) will reach its maximum value when $\alpha = 1$ if $G_1 > G_2$ (where G_1 is the specific free energy of phase 1 and G_2 the specific free energy of phase 2), or when $\alpha = 0$ if $G_2 > G_1$. If $G_1 = G_2$, G will be independent of α, which is another way of stating the well-known fact that the equilibrium does not depend on the amounts of the two phases present. The above will be true for any first-order phase change. In the case of solid–liquid equilibria the result is that the melting temperature remains constant so long as both phases are present.

A solid surface cannot be raised above its melting temperature (*superheating*) without melting, but it is rather easy to bring a clean liquid many degrees below its freezing temperature (*supercooling* or *undercooling*) without freezing. However, in the presence of a seed, very slight supercooling causes

†A first-order phase change is one with a discontinuity in the free energy vs. T or P plot at the temperature or pressure of the phase change.

That is, if there is a discontinuity in $(\delta G/\delta T)_P = -S$, $(\delta G/\delta p)_T = V$, or

$$\frac{\delta(G/T)}{\delta(1/T)} = H$$

the phase change is first-order. Common first-order phase changes such as melting, evaporation, and reversible polymorphic-phase changes show a discontinuity in ΔS, ΔV, and ΔH at the transition temperature. There are entropy and volume changes and a heat of transition associated with the transition. If a transition is discontinuous in the second derivative of its free-energy function it is *second order*. The transition from the paramagnetic to the ferromagnetic state is an example of a second-order transition.

growth and indeed supercooling is required for growth. The experimental determination of melting points is discussed in the literature (Weissberger, 1959).

2.5 The Distribution Coefficient

Another important equilibrium consideration in liquid–solid growth is the solid solubility of trace impurities or dopants in the solid. The important quantity to consider here is the *distribution coefficient* or *partition coefficient*. As stated before, in ideal monocomponent growth there would be no impurities or dopants present. We should perhaps reserve our discussion of distribution constant until we consider polycomponent growth. However, because the problems of growth when there are substantial quantities of other components present are quite different from those when only traces are present, it is useful to consider the "trace" situation as part of monocomponent growth. Where the border line between mono- and polycomponent growth lies with respect to concentration of a second component is not definite. Obviously growth from solution where the crystal is a solute and the mole fraction of solvent is greater than the mole fraction of solute is polycomponent growth. Even the terms *solvent* and *solute* are often used in an arbitrary sense, the lower-melting material sometimes being designated as solvent. We will adopt the more logical procedure of considering the material present in lower mole fraction as the solute. In many situations this will be the stable solid phase. When *constitutional supercooling*† is not a problem the concentration of other components is probably low enough so that we can consider our growth system monocomponent.

Under such conditions we can consider the solid solubility of a trace component to be given by

$$k_0 \equiv \frac{a_{s(\mathrm{eq})}}{a_{l(\mathrm{eq})}} \tag{2.17}$$

Where k_0 is the equilibrium distribution constant or partition constant for the material, $a_{l(\mathrm{eq})}$ is its equilibrium activity in the liquid and $a_{s(\mathrm{eq})}$ is its equilibrium activity in the solid. Thus k_0 is the same as the partition constant in solvent extraction or the ordinary equilibrium constant, where the equilibrium under consideration is

$$\text{Material in solution in liquid state } (a_l) \rightleftharpoons \text{Material in solid in solid state } (a_s) \tag{2.18}$$

In dilute solutions the approach to ideality is close enough so that k_0 may be defined as

$$k_0 = \frac{C_{s(\mathrm{eq})}}{C_{l(\mathrm{eq})}} \tag{2.19}$$

†See Sec. 3.12 for a discussion of constitutional supercooling.

where $C_{s(\text{eq})}$ and $C_{l(\text{eq})}$ are the respective equilibrium concentrations or the concentration at the solid–liquid interface.† The distribution constant, k_0, can also be defined in terms of $X_{s(\text{eq})}$ and $X_{l(\text{eq})}$, the respective atom fractions. For k_0 to be a true equilibrium value the crystal should neither be growing nor dissolving. Usually at slow rates of growth $k_{\text{eff}} \rightarrow k_0$, where k_{eff} is the effective-distribution coefficient given by

$$k_{\text{eff}} = \frac{C_{s(\text{act})}}{C_{l(\text{act})}} \tag{2.20}$$

where $C_{s(\text{act})}$ and $C_{l(\text{act})}$ are the actual concentrations in the two phases. $C_{l(\text{act})}$ is the "bulk" concentration in the liquid far enough from the growing interface that concentration gradients are negligible. The value of k_{eff} depends on growth rate, diffusion constant for the material in the melt, and width of the diffusion layer. If $k_0 > 1$, $C_{s(\text{eq})} > C_{l(\text{eq})}$ so that the impurity or dopant concentration will tend to be depleted close to the growing crystal. Thus the bulk of the melt or solution will be richer in the impurity or dopant than the region close to the growing crystal and diffusion of the impurity or dopant will be toward the crystal. Conversely if $k_0 < 1$, diffusion of the impurity or dopant will be away from the growing crystal. Equation (2.19) is obeyed when the solution is ideal or nearly ideal even if concentration gradients exist in the solution, if we substitute for $C_{l(\text{eq})}$ the concentration close to the growing crystal. Because the concentration close to the growing crystal cannot be determined easily it is often more convenient to use Eq. (2.20). It should be pointed out that $k_{\text{eff}} \rightarrow k_0$ when $C_{l(\text{act})} \rightarrow C_{l(\text{eq})}$ [because in most cases $C_{s(\text{act})} \approx C_{s(\text{eq})}$], that is, when the bulk- and interface-liquid concentrations are equal. The thickness of the diffusion layer will be a function of the viscosity of the melt and the rate of stirring, and thus k_{eff} will depend on

†We have chosen to assume that at low rates where the bulk-liquid concentration equals the concentration near the growing face, the situation is best described in terms of k_0, the equilibrium-distribution constant. The equilibrium-distribution constant may also be used at higher rates when the bulk-liquid concentration does not equal the concentration at the growing interface if $C_{l(\text{eq})}$ in Eq. (2.19) is replaced by $C_{l(\text{act-interface})}$, where $C_{l(\text{act-interface})}$ is the *actual* concentration in the liquid at the growing interface.

The difficulty is that $C_{l(\text{act-interface})}$ is hardly ever known. An additional problem is brought about because of adsorption and chemisorption of impurities at the growing interface. For instance, slight differences of impurity uptake on different crystallographic faces even at low rates (or when the rates are the same on different faces) are thought to be caused by different concentrations of adsorbed impurities on these faces. Hall (1953) has assumed that growth occurs by the rapid propagation of monoatomic layers that trap adsorbed impurity atoms and necessitate diffusion of these atoms back to the new interface. This viewpoint would dictate that, for true equilibrium to obtain, rates low enough that solid-state diffusion would erase concentration gradients in the solid would be required. Chernov (1962) performed a more rigorous analysis of ledge movements and impurity trapping by considering the diffusion of trapped impurities across a newly formed ledge toward both the liquid and the solid. Carruthers (1963) has taken into account sinusoidal-growth-rate fluctuations together with adsorption phenomena to derive a model that explains phosphorus striations in Czochralski-grown Si.

stirring speed as discussed in Chapter 3. In addition, the rate of crystallization will depend upon the crystallographic direction whenever the surface free energy is orientation dependent. (See Sec. 3.3 for a discussion of surface free energy.) When the growth rate is limited by diffusion processes (see Sec. 3.2 for a further discussion) the concentration profile across the diffusion layer and the thickness of the diffusion layer will be different on faces growing at different rates. Thus $C_{l(\text{act-interface})}$ will be different on faces of different orientation and k_{eff} will depend upon crystallographic orientation.† When a measured k is independent of stirring speed, crystallographic direction, and growth rate, the system may be taken to be close enough to equilibrium so that $k_{\text{eff}} = k_0$. Surprisingly, this situation obtains even at quite high growth rates in many systems.

The value of k_0 can be a strong function of the concentration of other impurities in the system. The added impurity or dopant may enter the lattice either interstitially or substitutionally. To enter substituationally, the ionic or atomic radius must approximate the radius of the element for which it substitutes. If the radius is within about $\pm 15\%$ of the radius of the lattice atom, the substitution is quite easy, and provided the charge of the substituting atom is the same as the lattice atom, k_0 has a value often approaching unity. If the charge is not the same, substitution is more difficult. Nassau and Loiacono's work (1963) on the distribution constant of rare earths in $CaWO_4$ is probably indicative of the effects that can be expected in general in ionic lattices. If a plus-two cation is substituted for Ca^{2+}, electroneutrality is preserved. However, the distribution constant for Nd^{3+} in $CaWO_4$ is 0.28 when the Nd^{3+} concentration in the melt is 1–2 atomic percent. The compound formed by Nd^{3+} addition is $Ca^{2+}_{1-3x/2}Nd^{3+}_x\Phi_{x/2}(WO_4)$, where Φ is a calcium vacancy. Calcium vacancies are energetically difficult to form and hence k_0 for neodymium is small, even though the radius of Nd^{3+} (1.04 Å) closely approximates the radius of Ca^{2+} (0.99 Å). If *charge compensation* is achieved by the addition of a singly-positively-charged atom of the appropriate size instead of by vacancy formation,‡ the distribution constant will increase markedly. For example, if both Nd^{3+} and Na^{1+} are added to the melt, the solid crystal can be represented by $Ca_{(1-2x)}Nd_xNa_xWO_4$. Under these conditions, when the concentration of Na^+ in the melt is 15 atomic percent, k_0 for Nd^{3+} is 0.8. The value of k_0 of Nd^{3+} is a strong function of the size and concentration of the unipositive compensating cation as Fig. 2.2 shows. As can be seen, ions whose ionic radii most closely approximate Ca^{2+} have the greatest effect. In a similar manner, k_0 for Na^+ or any other unipositive ion is affected by the size, charge, and concentration of the compensating

†Adsorption may also be important in making k_{eff} have a directional dependence. See footnote, p. 48.

‡There will still, of course, be a finite-vacancy concentration that can be calculated by the methods of Chapter 1, but it will be insignificant even at the melting point in comparison with Φ for reasonable concentrations of Nd^{3+} when plus-one cations are absent.

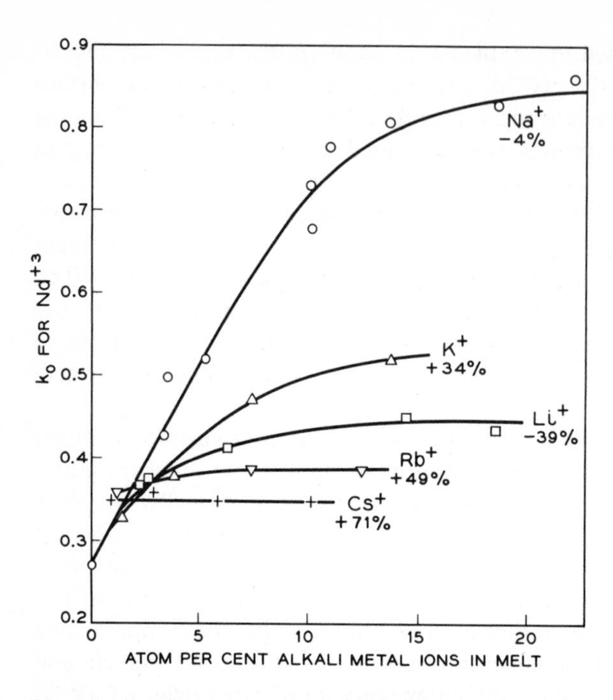

Fig. 2.2 Equilibrium distribution constant for Nd^{+3} in $CaWO_4$ as a function of alkali metal compensating ion concentration (Nd^{+3} concentration 1–2 atomic percent in melt). Percent difference ionic radius between ions and Ca^{++} shown (after Nassau and Loiacono, 1963). (Courtesy of *Journal of Physics and Chemistry of Solids.*)

cation such as Nd^{3+}. Charge compensation can be achieved in relatively complicated lattices such as $CaWO_4$ in a variety of ways. For instance, when doping with Nd^{3+}, Nb^{5+} may be added. Under these conditions the formula for the resultant crystal may be written as $Ca_{(1-x)}Nd_xW_{(1-x)}Nb_xO_4$ because Nb^{5+} substitutes for W^{6+} whenever Nd^{3+} substitutes for Ca^{2+}. The distribution constant for Nd^{3+} is again a function of the size, charge, and concentration of the pentavalent cation that charge-compensates it by substituting for W^{6+}. In a similar manner, k_0 for Nb^{5+} will depend on the size, charge, and concentration of the ion that charge-compensates for it by substituting for Ca^{2+}.

If dopant or impurities enter a lattice interstitially the situation is somewhat different. Alpha-quartz is an example of a structure that can accommodate several ions interstitially. The spiral channels parallel to the *c* axis formed by the silica tetrahedra comprising the structure have diameters of about 1 Å. Thus Li^+, Na^+, Mg^{2+}, and Ca^{2+} can enter the lattice interstitially comparatively easily. However, charge compensation is required for any interstitial atom. Thus when Li^+ enters the quartz lattice the formula can be represented as $Si_{(1-x/4)}Li_x\Phi_x/_4O_2$, where Φ is a silicon vacancy. Consequently, k_0 for Li^+ (radius 0.68 Å) is small. It is larger for Na^+ because the ionic radius (0.94 Å) of sodium more closely matches the interstitial "holes" in the lattice, while it is smaller for K^+ where the radius (1.33 Å) is a poor fit to the interstices. Aluminum will enter the quartz lattice, where it substitutes for Si^{4+} (under some conditions it will also take up interstitial positions). Thus when, for example, Al^{3+} and Li^{1+} are present in the quartz-growth solution, the k_0 for both is increased because they charge-compensate for each other. Laudise et al. (1965) have discussed distribution constants for quartz.

For an ideal system, A–B, close to the melting of A when the dopant or impurity concentration of B is low, the value of $k_{0(B)}$ is related to the melting point of A, T_A, and the melting point of B, T_B, by the equation

$$\ln k_{0(B)} = \frac{\Delta H_B}{R}\left(\frac{1}{T_A} - \frac{1}{T_B}\right) \tag{2.21}$$

where ΔH_B is the molar heat of solution of B in A. Equation (2.21) can be derived as a special case of the van't Hoff relation [Eq. (2.32)] and assumes that ΔH is independent of temperature. This is a valid assumption for temperatures close to the melting point. Over regions where Eq. (2.21) is valid, the logarithm of the distribution constant is linear with the reciprocal of the absolute temperature, and the slope of $\log k_0$ vs. $1/T$ can be used to calculate ΔH. Equation (2.21) can be put in a more simple form by means of several approximations (Thurmond, 1959):

$$\Delta T = (1 - k_0)\frac{RT_0^2}{\Delta H}x_{l(eq)} \tag{2.22}$$

where ΔT is the depression of the freezing point of the pure material and $x_{l(eq)}$ is the equilibrium mole fraction of solute in the liquid. Equation (2.22) assumes k_0 is not a function of concentration or temperature, but where it is, the extrapolated value at infinite dilution may be used.

Thurmond (1959) has shown that the above equations involving k_0 are valid not only in the case of dilute ideal liquid and solid solution, but also when the solute is not ideal, the distribution constant is small, and the enthalpy and excess entropy of the solute are independent of temperature.

A further discussion of the distribution coefficient under nonequilibrium conditions will be reserved until the kinetics of crystal growth are considered in Chapter 3.

2.6 Phase Diagrams and the Phase Rule

Distribution coefficients and indeed many other phase relationships are conveniently represented by means of phase diagrams. Phase diagrams and the phase rule, of course, apply only to systems at equilibrium. We will consider in this section the phase rule and phase diagrams in general. A *homogeneous system* is one in which all macroscopic parts of the system have the same chemical and physical properties. Examples are a crystalline solid, a solution, a liquid, or a gas or mixture of gases. A *heterogeneous system* is one in which the physical properties and composition vary in different macroscopic parts. A solid in equilibrium with its melt, a saturated solution in the presence of excess solute, two immiscible liquids, or a liquid in equilibrium with its own vapor are examples of heterogeneous systems.

A *phase* is defined as a part of a system that is homogeneous throughout and is physically separated from other phases by distinct boundaries. The number of *components* in a system is the minimum number of chemical constituents that must be specified to describe the composition of all phases present. Figure 2.1 is an example of a simple *phase diagram* or graphical representation of the equilibrium relationship between various phases and certain *intensive* variables such as composition, temperature, and pressure. An intensive variable is one that is independent of the amount of the phase present, while an *extensive* variable (e.g., mass) depends on amount. In Fig. 2.1 the number of phases is two (diamond and graphite) and the number of components is one (carbon). It will be noted that along the equilibrium line only one variable, either pressure or temperature, may be changed independently without causing the system to lose a phase (a mere decrease in the amount of a phase is not a loss of the phase). The system thus has one *degree of freedom.* Along the equilibrium line, specifying one variable defines the system. Willard Gibbs deduced a quantitative relationship, the phase rule, that expresses the relationship between F, the number of degrees of freedom in an equilibrium, C, the number of components, and P, the number of phases:

$$F = C - P + 2 \tag{2.23}$$

This relationship holds when the intensive variables considered are pressure, temperature, and composition. If other variables such as magnetic, electric, or gravitational field are considered, the value of 2 will be increased by the number of new variables considered.

Distribution coefficients can be understood in terms of the complete binary-phase diagram between the appropriate components. Figure 2.3 shows various phase relationships that may exist between two components A and B. In Fig. 2.3(a) the liquidus curve shows the temperature at which solidification of the melt begins as a function of composition, while the solidus curve shows the temperature at which melting of the solid begins as a function of composition. The horizontal lines 1–2, 3–4, are tie lines; that is, they connect liquid of a given composition with the solid of the composition that is in equilibrium with it. In Fig. 2.3(a), A and B are completely miscible with each other, both in the solid and in the liquid. Freezing solid tends to reject the lower-melting component B; that is, $k_{0(B)} < 1$, while $k_{0(A)} > 1$. The relation-

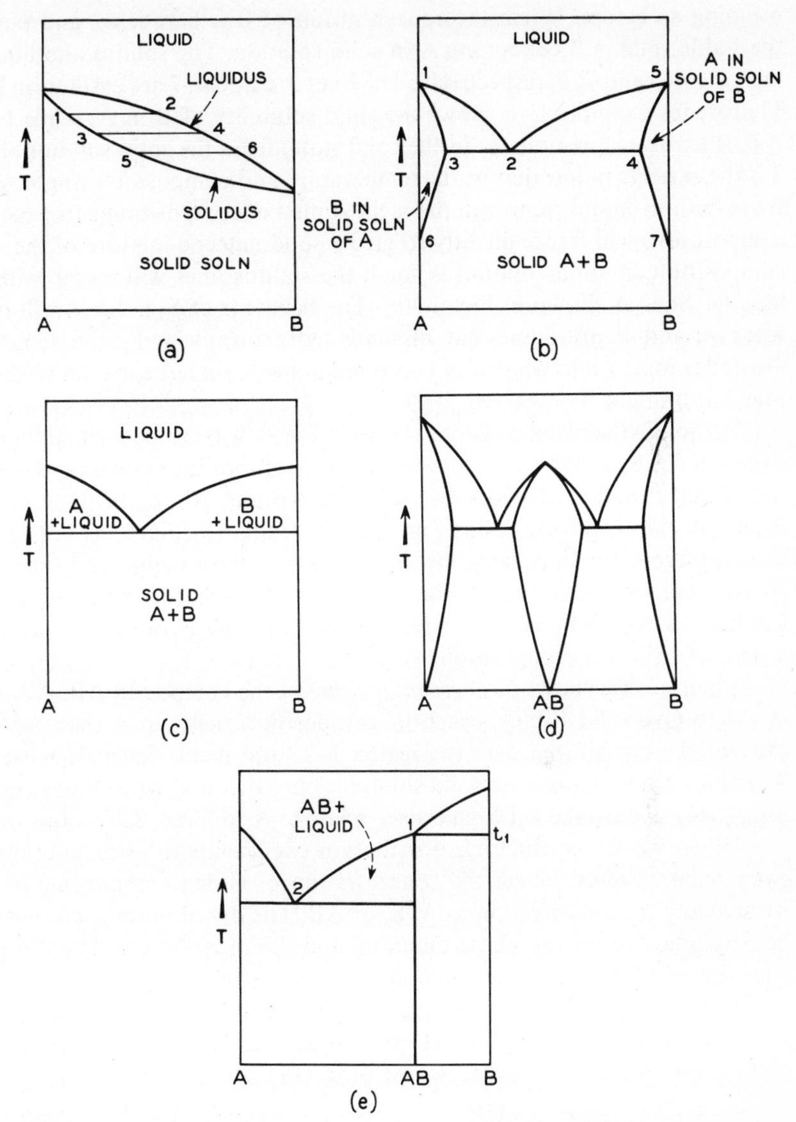

Fig. 2.3 Binary phase diagrams.

ship shown in Fig. 2.3(a) is common for isostructural compounds whose atoms have similar sizes, that is, for *isomorphous* compounds.

Figure 2.3(b) shows the phase relationships when A and B are not completely miscible in the solid state. For concentrations from pure A to the concentration corresponding to point 2, A containing B in solid solution is the stable phase. The liquidus line 1–2 and the solidus line 1–3 have the same meaning as before. When the concentration of B is between 2 and pure B, the stable solid is B containing A in solid solution. The solidus and liquidus lines are 4–5 and 2–5, respectively. The lines 3–6 and 4–7 are exsolution lines. That is, for example, 3–6 shows the solid solubility of B in A, while below 3–6, B exsolves (precipitates in the solid state) from the solid solution. Point 2 is the *eutectic* point, that is, the temperature and composition where A and B are both in equilibrium with the melt. A melt of composition 2 (the eutectic composition) will freeze directly to give a solid eutectic mixture of the same composition. If solid solution is small the solidus lines will merge with the liquidus lines as shown in Fig. 2.3(c). The behavior of Fig. 2.3(b) will occur when A and B are somewhat dissimilar structurally and have somewhat dissimilar ionic radii, while if A and B are quite dissimilar the relationship of Fig. 2.3(c) might be expected.

Although the phase relations shown in Fig. 2.3(d) are for formation of the compound AB, as can be seen under these conditions the system A–B may be considered in terms of two subsystems, A–AB and AB–B. In these subsystems, eutectics can form. Thus Fig. 2.3(d) is similar to Fig. 2.3(b) in each of its subsystems. In some cases the melting point of the compound formed in a diagram like that shown in Fig. 2.3(d) may be higher than either A or B. Compound AB melts congruently; that is, solid AB is in equilibrium with a melt of identical composition.

Figure 2.3(e) shows *incongruent melting* of the compound AB. AB melts at T_1† to give solid B plus a melt of composition richer in A than AB. AB can only be crystallized over the region 1–2 from melts richer than itself in A. Figure 2.3(e) assumes no solid solution, but obviously when it does occur, Fig. 2.3(e) could take on a character analogous to Figs. 2.3(b) and (d).

When we are considering distribution coefficients in essentially monocomponent systems, where the concentration of added components is low, we are close in concentration to A, B, or AB. The distribution coefficient can be determined from the phase diagram, and the diagram can sometimes be constructed from thermodynamic data, if available.

For further discussion of phase diagrams the reader should consult standard works such as Hansen (1958), Findlay et al. (1951), and Levin et al. (1964); for ternary diagrams, see Masing (1944).

2.7 Conservative Processes

The impurity or dopant concentration will depend strongly on the manner in which the freezing in a solid–liquid crystal-growth process is carried out. When no material is added to or removed from either the solid or liquid phase except by crystallization the process is called *conservative crystal growth*

†The invariant point whose temperature is T_1 and composition 1 is the *peritectic* point.

(Thurmond, 1959). Many *normal-freezing*† (Pfann, 1966) processes are examples. In normal freezing by a conservative process the entire volume of the material is liquid at the beginning of growth and the solid is grown from the liquid phase by solidification at a specific surface in the supercooled solution. The result is that a solid–liquid interface moves through the melt in a controlled manner. The distribution of impurity or dopant under such conditions can be rather easily predicted.

In normal freezing, growth is initiated ideally at a single site in the melt by causing this site to be cooler than any other region. Nucleation at this initial site provides a crystal on which all the subsequent growth is deposited. A typical arrangement is to contain the melt in a crucible with a conical bottom and to lower it through a temperature gradient so that freezing first takes place at the tip [Fig. 2.4(a)]. The solid–liquid interface then moves through the melt as the crucible is lowered. Usually a single crystal is nucleated at the tip or if several crystals are nucleated the fastest-growing one succeeds in dominating the interface. In some cases a seed may be placed in the tip before the growth is begun. When this is done care must be exercised not to melt the seed before growth is begun. This technique is generally called the Bridgman–Stockbarger method. In one variant of the technique a temperature gradient is imposed on the crucible so that the tip is coolest and the entire volume is above the melting temperature. The temperature of the whole crucible is then lowered while the gradient is maintained, and growth again starts in the conical tip and proceeds through the crucible. Another conservative method is the Czochralski or crystal-pulling technique, Fig. 2.4(b). The material is contained in a crucible and is entirely molten at the beginning of growth. Heating and heat leaks are arranged so that the melt is essentially isothermal‡ and there is a negative gradient above the melt. A seed is introduced into the melt so that it just makes contact and a small portion is melted to ensure that the crystal surface is clean and uncontaminated by spurious nuclei. Then it is slowly withdrawn. If the temperature profile and the withdrawal rate are proper, the solid–liquid interface will be established slightly above the level of the bulk of the melt and interfacial tension will support a small liquid column above the melt. In some cases the temperature

†We shall restrict normal freezing to mean growth from a solid–liquid system where a *single* solid–liquid interface moves through the melt. All normal-freezing techniques are not conservative. For instance, evaporation may take place while growth is proceeding. Conversely all conservative growth is not by normal freezing. A melt may be cooled in an irregular-temperature profile so that nucleation occurs in many sites. The resultant polycrystalline mass is grown under conservative conditions but not by normal freezing. Similarly, *nonconservative* growth may take place with or without normal freezing.

‡The temperature profiles discussed in connection with Fig. 2.4 are idealized. Many subtle and important effects will be later shown to depend on details of the temperature profile. The details of the growth techniques and pertinent literature references will be given in appropriate subsequent chapters.

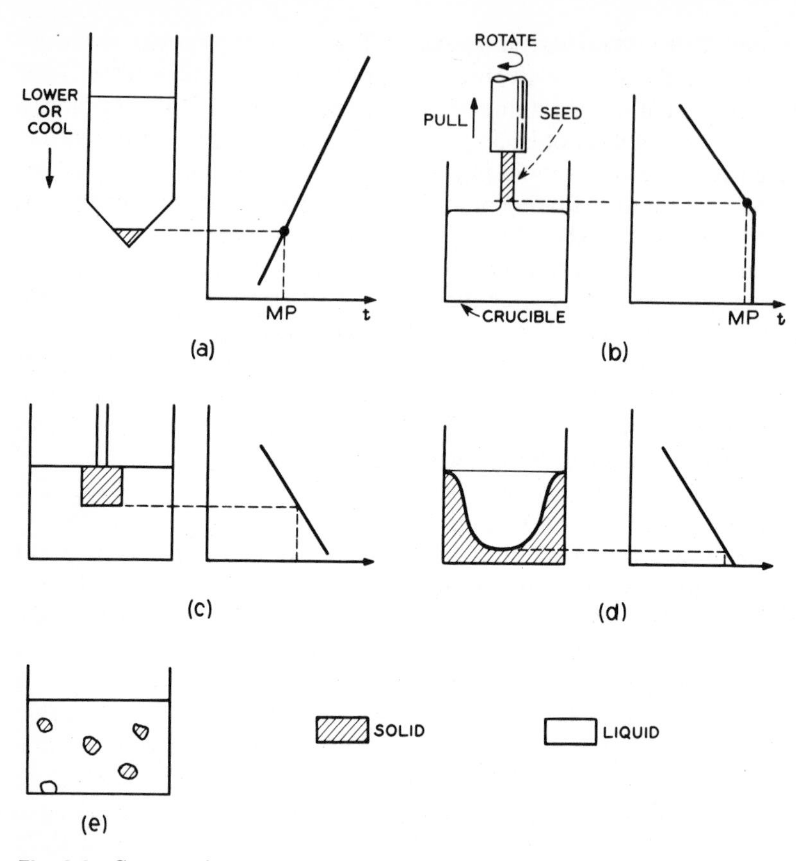

Fig. 2.4 Conservative crystal-growth processes.

profile is arranged so that the crystal grows right at the melt surface or even slightly below. The seed is slowly withdrawn (often while being rotated) as growth takes place.

In another variant of conservative growth by normal freezing shown in Fig. 2.4(c), a seed is immersed in a crucible. The temperature profile is such that growth occurs at the seed–melt interface only (it is often necessary to cool the seed through its holder to achieve this), and the entire melt is cooled while the gradient shown in Fig. 2.4(c) is maintained. In favorable circumstances almost the entire melt can be crystallized on the seed as a single crystal. This technique is called the Kyropoulos method.

A typical mode for the solidification of ingots, as shown in Fig. 2.4(d), can also be conservative growth by normal freezing. Heat loss is through the walls of the container so that nucleation begins at the walls. Although the growth is not usually single-crystal, a single solid–liquid interface does sweep through the melt in favorable cases.

Conservative growth by freezing that is not normal is illustrated by Fig. 2.4(e), where, because of an irregular-temperature profile, growth has been initiated at many sites with the result that many solid–liquid interfaces will be involved in the growth.

In all conservative crystal growth where the conditions are such that equilibrium impurity distribution occurs, the impurity distribution can be described by the differential equation (Thurmond, 1959)

$$\frac{d \ln X_l}{d \ln N_l} = k_0 - 1 \tag{2.24}$$

where X_l is the atom fraction of impurity or dopant in the liquid phase, N_l is the total grams of impurity in the liquid phase, and k_0 is the equilibrium distribution constant for the impurity or dopant when concentrations are expressed as atom fractions.

This equation holds for all conservative crystal growth but is difficult to apply when freezing is not normal. When freezing is not normal a knowledge of the fraction of melt still liquid when a given region solidified is generally not available. Even in normal freezing by Kyropoulos growth and in the freezing of ingots it is not usual to know the fraction solidified when a given region was crystallized. Equation (2.24) can most easily be used in Bridgman–Stockbarger and in Czochralski growth because the fraction solidified at any given time during the growth can be determined from simple geometric considerations. Figure 2.5 shows the impurity distribution in the solid in conservative freezing as a function of the fraction frozen for various values of k_{eff} (Pfann, 1952).

Figure 2.5 was constructed by the use of

$$C_s = k_0 C_{0l}(1 - g)^{k_0 - 1} \tag{2.25}$$

and is valid when $k_{eff} \simeq k_0$,† where C_s is the concentration of impurity in the solid, C_{0l} is the initial concentration in the liquid, and g is the fraction solidified. Equation (2.25) in one form or another has been derived by various workers (Pfann, 1966) and can be derived by solving the differential equation [Eq. (2.24)] with the assumption that the densities of solid and liquid are equal [or by introducing a "density ratio" (Pfann, 1966)] and by substituting concentrations for atom fractions. The impurity distribution is quite different from that obtained in nonconservative processes as we shall show.

In a conservative process, control of impurities can be obtained by cutting or *cropping* the impure part of the crystal, discarding it, remelting the pure part, and refreezing it. Successive cropping can be used to prepare quite pure materials or to secure some particular impurity distribution. The main disadvantage is that this procedure is wasteful of material because the impure

†This is equivalent to assuming the growth rate is such that mixing in the liquid is complete and solid-state diffusion is negligible (Pfann, 1966).

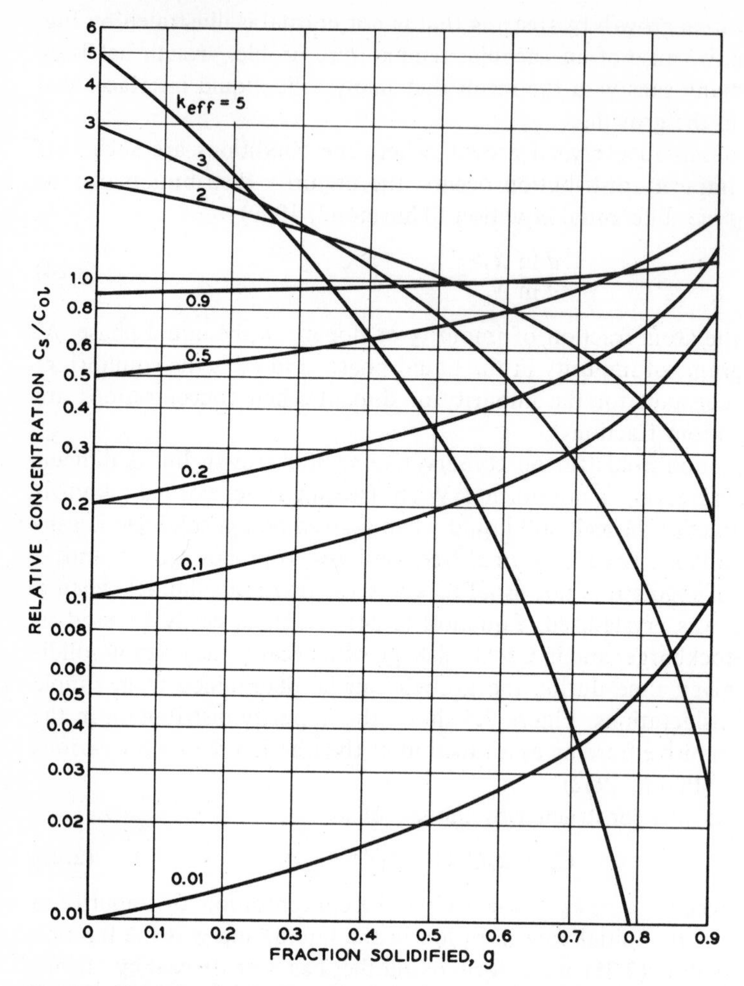

Fig. 2.5 Impurity distribution in solid during conservative freezing as a function of fraction frozen (after Pfann, 1966). (Courtesy of John Wiley & Sons, Inc.)

part of the crystal or the part with the undesired impurity distribution must be discarded in each cropping.

2.8 Nonconservative Processes

In nonconservative processes, material may be added to the molten region by any process or may be removed from the molten region by processes other than freezing (Thurmond, 1959). One way in which a system is nonconservative is by means of vapor loss. All solids and liquids have equilibrium-vapor pressures. However, if the vapor pressure is low or if the system is closed so that the volatilization of a relatively small amount of material will establish an equilibrium-vapor pressure, crystal growth will be conservative with respect to vapor loss. Growth can also be conservative with respect to vapor loss if some means is used to provide an equilibrium-vapor pressure from a source other than the melt from which the crystal is growing or the crystal itself. The common manner in which a growth process is nonconservative is that solid is melted into the molten zone while crystallization is in progress.

Zone melting, which was invented by Pfann (1952), is an example of an especially powerful nonconservative process that is used for purification and for the growth of single crystals. In horizontal zone melting, as Fig. 2.6(a) shows, the material is held in a boat and the temperature profile is arranged so that a narrow molten zone is produced. This zone is moved along the boat rather slowly in order to effect purification, or in some cases homogenization, of an impurity. Single crystals are also often grown by the technique. If further purification is required, the zone is reestablished at the front end of the boat and the process is repeated (another *pass* is made). Seeding may be effected

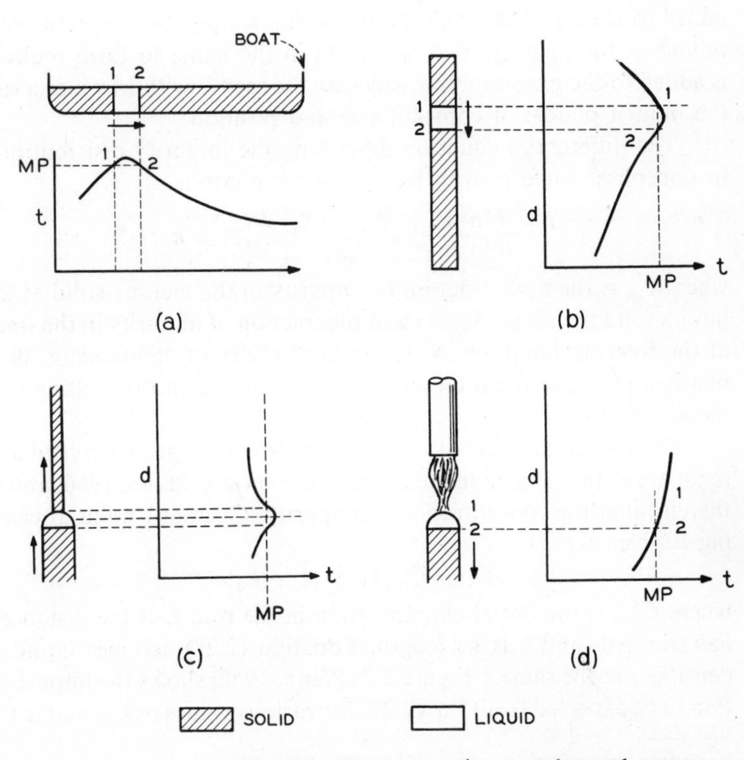

Fig. 2.6 Nonconservative crystal-growth processes.

by placing a seed at the front end of the boat and beginning the zone so as to avoid completely melting the seed.

The ingot to be melted can be arranged with its axis vertical, as shown in Fig. 2.6(b). This technique, called *float zoning* (Keck and Golay, 1953; Emeis, 1954; Theuerer, 1952, 1956), when no container is used for confining the melt and allows zone melting without crucible contamination. The molten zone, which is usually established by radio-frequency or radiant heating, is held in place by surface tension. The diameter of the melting ingot and the growing crystal need not be the same. If the molten zone is held stationary in space and the melting ingot is "pushed" into it while the growing crystal is pulled out of it, the crystal diameter will depend on the ratio of the pushing rate to the pulling rate. This method has been called *pedestal growth* or *crystal "pushing"* and can be thought of as a variant of crystal pulling or float zoning. Figure 2.6(c) shows one variant of this technique.

Figure 2.6(d) shows the *flame-fusion* or *verneuil* process. In this growth technique a molten "puddle," held in place by surface tension, is produced on the surface of the seed by a flame, by a plasma, or by radiant heating from a focused high-intensity light. Solid particles of the material or molten droplets are added to the puddle. When the source is a flame, powder is added to the flame through the tubes that supply the gas, or a volatile compound of the material that will react in the flame to form molten droplets is added to the gas supply. In any case the seed is withdrawn at a rate to keep the molten puddle of constant size and position.

The differential equation describing the impurity distribution obtained in nonconservative growth from a liquid phase is

$$\frac{d \ln X_l}{d \ln N_l} + \frac{X_{sm} - X_{sf}}{X_l}\frac{dN_{sm}}{dN_l} = k_0 - 1 \qquad (2.26)$$

when X_{sm} is the mole fraction of impurity in the melting solid at the melting liquid–solid interface, X_{sf} is the mole fraction of impurity in the freezing solid at the freezing interface, N_l is the total moles of components in the liquid phase, and N_{sm} is the total moles of components in the melting solid (Thurmond, 1959).

For the special case of zone melting when the grown crystal and melting ingot are of the same diameter, it has been shown (Pfann, 1966, and references therein) that the concentration of impurity, C_s, at any point after zone refining is given by

$$C_s = C_{0(s)}[1 - (1 - k_0)e^{-k_0(L/Z)}] \qquad (2.27)$$

where $C_{0(s)}$ is the initial concentration in the rod, L is the distance the zone has traveled, and z is its length. Equation (2.27) assumes liquid and solid densities are the same.† Figure 2.7 (Pfann, 1952) shows the impurity distribution to be expected from Eq. (2.27) for various values of k_{eff} and is valid when

†Pfann (1966) discusses the conditions of applicability of Eq. (2.27) when liquid and solid densities are unequal.

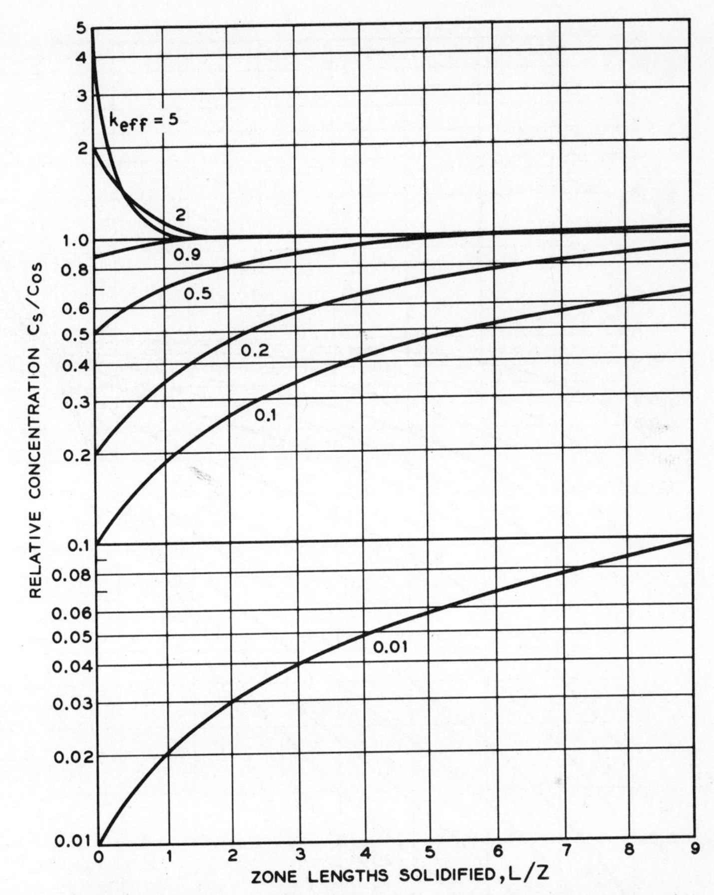

Fig. 2.7 Impurity distribution in a solid during nonconservative freezing by zone melting as a function of zone lengths solidified for various values of distribution constant (after Pfann, 1966). (Courtesy of John Wiley & Sons, Inc.)

$k_{eff} \simeq k_0$.† Figure 2.8 (Pfann, 1966) shows the effect of subsequent passes on the impurity distribution in zone refining and is valid when $k_{eff} = 0.5 \simeq k_0$.† The distribution obtained after one pass could be the same in any nonconservative process in which the volume of the melt is constant throughout the process, the cross section of the grown crystal is constant, and the concentration of impurity in the material that melts to form the molten zone is constant. The methods shown in Fig. 2.6(a)–(d) usually fulfill the above criteria. The impurity distribution after subsequent passes would be that given by Eq. (2.27) and by Fig. 2.8 when $k_{eff} = 0.5$, provided melt volume and crystal cross section are constant and the concentration of impurity in

---

†This is equivalent to assuming that the growth rate is such that mixing in the liquid is complete and solid-state diffusion is negligible (Pfann, 1966).

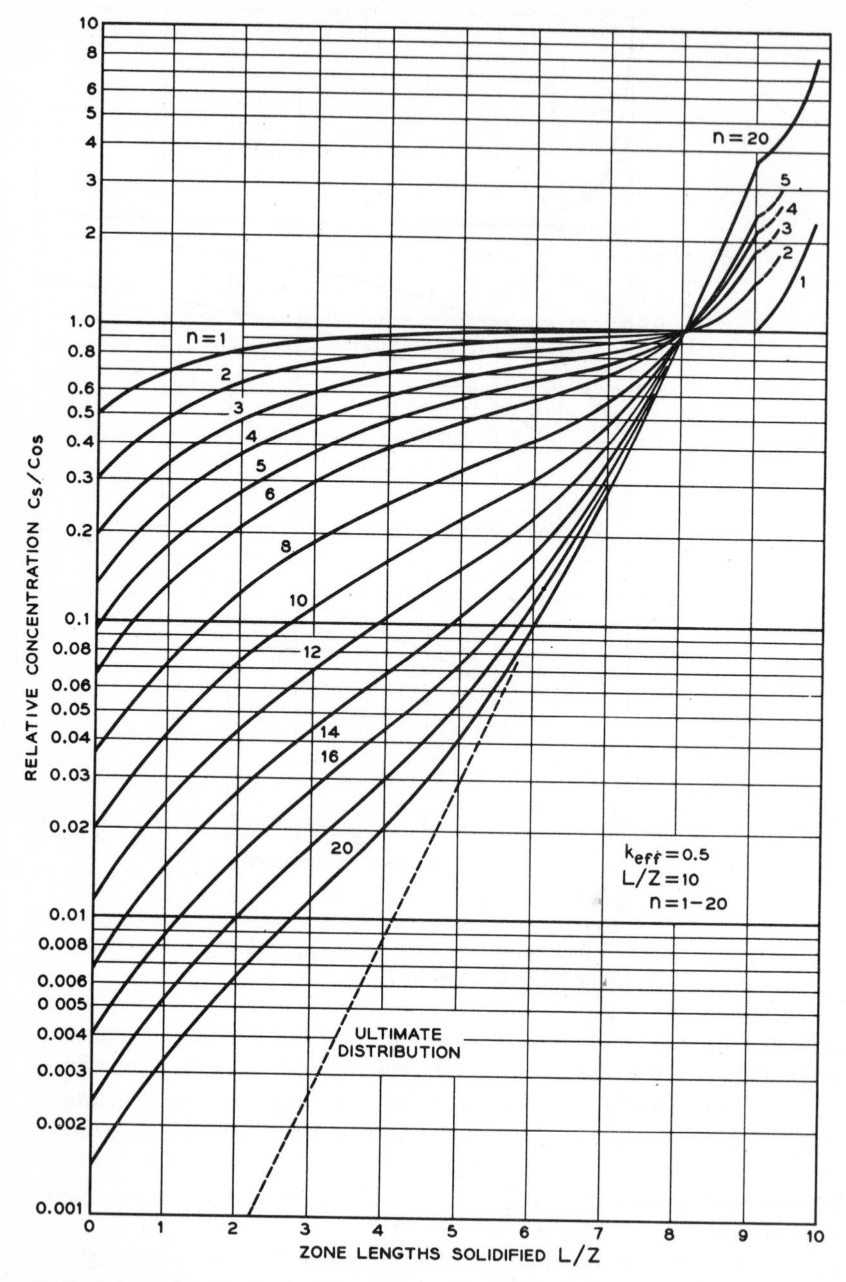

Fig. 2.8 Impurity distribution in a solid during nonconservative freezing by zone melting as a function of zone lengths solidified for various numbers of passes (n) when $k_{eff} = 0.5$ (after Pfann, 1966). (Courtesy of John Wiley & Sons, Inc.)

the solid used in the equation is that given by the previous recrystallization. All of the techniques of Fig. 2.6 would fulfill these criteria except Verneuil growth. Because a powdered feed is required in Verneuil growth, once the crystal is grown repowdering would destroy the impurity distribution. In practice such finely divided feed is required in Verneuil growth that only voluminous powders produced by calcination are suitable as feed materials.

The impurity distribution that results from conservative processes is quite different from that in nonconservative processes as a comparison of Figs. 2.5 and 2.7 will show. In a conservative process after a freezing, the impurity content is lower (when $k_{eff} < 1$) over a larger volume of the crystal than in a nonconservative process. However, if further purification is desired the impure region of the crystal that grew last (when $k_{eff} < 1$) must be cut off and discarded (cropped) before the second crystallization is begun. Growing pure crystals by successive croppings is cumbersome and wasteful because much larger sizes of crystal than are finally needed must be handled at the initial stages. In growth by nonconservative processes (when remelting is practical) the advantage is that subsequent recrystallizations (for instance, more passes in zone refining) are easy to make and markedly improve the purity. In nonconservative processes, techniques such as zone leveling, which is discussed in Sec. 5.5, make it relatively easy to get a uniform dopant concentration over a large volume of the crystal.

2.9 Monocomponent Gas–Solid Equilibria

For a material to be grown by a gas–solid equilibrium in a monocomponent system, it must have a sufficiently high vapor pressure. If the vapor pressure is not high enough, complexing agents or other reactants may be used to produce a volatile species that decomposes to form the material desired. Such growth is from a polycomponent system and will be discussed later. The vapor pressure need not equal 1 atm for practical growth by a gas–solid equilibrium below the melting point. If the material is vaporized from the solid in one part of the system and condensed from the gas in another part, the growth is by sublimation. The material need not be vaporized from the solid. For instance, vaporization from the liquid followed by rapid condensation from the vapor to the solid in another part of the system may take place close to the *triple point*.† In this case our concern insofar as the growth process is concerned would still be with a gas–solid equilibrium.

The vapor pressure is related to the heat of sublimation by the Clausius-Clapeyron equation:

$$\frac{d \ln p}{dT} = \frac{\Delta H}{RT^2} \tag{2.28}$$

†The triple point is an invariant point, the unique pressure and temperature in a monocomponent system where solid, liquid, and gas coexist.

where T is the absolute temperature, p is the pressure, ΔH is the heat of sublimation, and R is the gas constant. This equation is derived from the assumption that the vapor behaves as a perfect gas and that the molar volume of the gas is large in comparison with the molar volume of the solid. Equation (2.28) also describes the vapor pressure of a liquid where ΔH is the heat of vaporization, where the vapor obeys the perfect gas law, and where the molar volume of the vapor is large in comparison with the molar volume of the liquid.

2.10 Growth from Solution

Up to this point our discussion has followed the outline of Table 2.1 and we have described the equilibrium considerations involved in monocomponent crystal growth in terms of solid–solid, liquid–solid, and gas–solid equilibria. As Table 2.1 shows, polycomponent growth equilibria can be described in a similar manner. However, polycomponent growth systems also fall naturally into two other broad classes regardless of whether the medium is is solid, liquid, or gas. These classes are

1. Growth from solution
2. Growth by chemical reaction

Because equilibrium considerations are more dependent on which of these classes a system falls into than on the nature of the medium, we shall discuss polycomponent growth in terms of these classes.

In crystal growth from solution, the most important equilibrium consideration is solubility. When the concentration of solute is moderate, a single solvent is used, the concentration of impurities or dopants other than the solute is low,† and the solute solubility in the crystal is negligible, we are then concerned with the region of the binary-phase diagram of Fig. 2.9 labeled α–β. A represents the crystal to be grown. Figure 2.9 assumes no compound formation between A and B. As can be seen, Fig. 2.9 is similar to Fig. 2.3(c) except that the eutectic is closer to B, thus allowing the existence of dilute solutions where A is stable. The considerations that were pertinent to Fig. 2.3(b) and (d) would apply in the case of the formation of compounds or solid solutions.

It is usual to plot solubility vs. temperature as shown in Fig. 2.10 where α and β have the same meanings as before. The solubility line gives the concentration of a *saturated* solution, that is, a solution in equilibrium with a solid, as a function of temperature. Regions above the line are *supersaturated*; that is, they represent solutions containing more than an equilibrium concentration of solute. The supersaturation Δs is defined as $\Delta s = c - s_{eq}$, where c is the actual concentration and s_{eq} is the equilibrium solubility. Supersat-

†The solute itself is, of course, the principal "impurity" under these conditions.

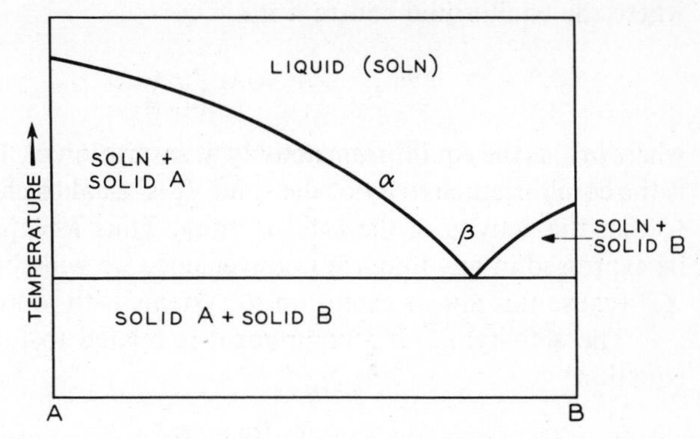

Fig. 2.9 Binary phase diagram showing region for solution growth.

urated solutions are not stable and will deposit solute or *spontaneously nucleate* in the solution or on the walls of the container if the supersaturation is great enough. If a fragment of solute crystal (a *seed crystal*) is introduced into the solution, excess solute will deposit on it until the concentration falls to the saturation concentration even if the supersaturation is small. However, in the absence of a seed, clean undisturbed supersaturated solutions may sometimes be kept for long periods of time in spite of their thermodynamic instability.

The region below the solubility curve of Fig. 2.10 represents an unsaturated solution, that is, one that contains less than the equilibrium concentration of solute.

The process of solution of a solid, A, is a special case of chemical equilibrium described by the reaction

$$A_{\text{solid}} \rightleftharpoons A_{\text{solution}} \tag{2.29}$$

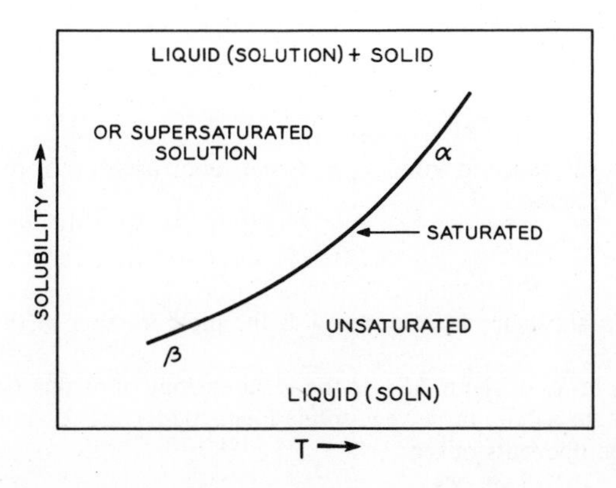

Fig. 2.10 Solubility vs. temperature curve.

where the equilibrium constant is

$$K = \frac{[a]_{\mathrm{eq}}}{[a]_{\mathrm{solid\ eq}}} \tag{2.30}$$

where $[a]_{\mathrm{eq}}$ is the equilibrium activity at saturation of the solution and $[a]_{\mathrm{solid\ eq}}$ is the equilibrium activity of the solid. It is usual to choose the standard state so that the activity of the solid is unity. Thus $K = [a]_{\mathrm{eq}}$. Concentration can be expressed in any units; for convenience we will choose the mole fraction, X, because this allows extension to systems with more than one component.

The activity, a_i, of a component is related to the mole fraction by the equation

$$a_i = \gamma_i X_i \tag{2.31}$$

where γ is the activity coefficient. If the activity coefficient is unity, the component obeys Raoult's law†; if the activity coefficient is not unity but is constant, the component obeys Henry's law. In an ideal solution both the solvent and the solute obey Raoult's law. The enthalpy and volume are the sum of the enthalpies and volumes of the solvent and solute. That is, the heat of mixing or of solution and the volume change during solution are zero. The entropy, however, is not the sum of the entropies of the components because the entropy is increased by the additional randomness possible in a solution.‡ In a nonideal dilute solution, the solute obeys Henry's law, while the solvent obeys Raoult's law.

The component whose mole fraction approaches unity is usually considered to be the solvent. In the intermediate region (the concentrated-solution region) neither Raoult's law nor Henry's law applies and experimental results are required to evaluate γ, which is usually a strong function of concentration. Unfortunately many solutions used in crystal growth are quite concentrated. Indeed it is often dangerous to make calculations predicated on weak solute–solute interactions. Because crystallization is taking place or is about to take place in a crystal-growth solution, solute–solute interactions are obviously of great importance.

---

†According to Raoult's law,

$$p = p_0 X$$

where p is the vapor pressure of a substance in solution, p_0 is the vapor pressure of the pure substance, and X is its mole fraction.

According to Henry's law,

$$p = kX$$

where p is the vapor pressure of a substance in solution, X is the mole fraction of the substance, and k is a constant.

‡$\Delta S_m = -R(X_A \ln X_A + X_B \ln X_B)$, where ΔS_m is the ideal entropy of mixing (in the absence of solvent–solvent, solute–solute, or solvent–solute interactions) and X_A and X_B are the mole fractions of the components mixed.

The temperature dependence of solubility is given by the van't Hoff equation

$$\frac{d \ln K}{dT} = \frac{-\Delta H}{RT^2} \tag{2.32}$$

which, when applied to solutions, becomes

$$\frac{d \ln [a]_{eq}}{dT} = -\frac{\overline{\Delta H_A}}{RT^2} \tag{2.33}$$

where $\overline{\Delta H_A}$ is the partial-molal-enthalpy change of the solute A in the solvent B. The partial-molal-enthalpy change is the enthalpy change that takes place in transferring one mole of pure A to the solution, where the solution volume is so large as not to change its concentration appreciably from saturation. In a solution obeying Raoult's law, X_A can be substituted for the activity of A.

The equilibrium constant for solution growth is related to the standard-free-energy change in the growth process ΔG^0 by the equation

$$\Delta G^0 = -RT \ln [a]_{eq} \tag{2.34}$$

because

$$\Delta G^0 = -RT \ln K \tag{2.35}$$

where K is the reciprocal of Eq. (2.30) for *growth* because

$$\Delta G = \Delta G^0 + RT \ln Q \tag{2.36}$$

where

$$Q = [a]_{act}^{-1}$$

ΔG is the free-energy change during growth, and $[a]_{act}$ is the actual activity of A in a supersaturated solution.

ΔG is the driving force for the growth and it may be expressed as

$$\Delta G = RT \ln \frac{[a]_{eq}}{[a]_{act}} \tag{2.37}$$

where $[a]_{eq}$ is the activity of A in a saturated solution and $[a]_{act}$ is the actual activity of A. Activities can be replaced with mole fractions or concentrations when $\gamma \simeq 1$ or γ is independent of concentration over the range of interest in Eqs. (2.33)–(2.37). Thus

$$\frac{[a]_{eq}}{[a]_{act}}$$

in Eq. (2.37) can often be replaced with s_{eq}/c_{act}, where s_{eq} is the equilibrium solubility at saturation and c_{act} is the concentration in the supersaturated solution. The ratio c_{act}/s_{eq} is often called the relative saturation or degree of supersaturation. Thus thermodynamic data may be obtained from, and may be used to generate, solubility relationships. Solution thermodynamics are treated in several standard works and will not be further discussed here

(Kubachewski and Evans, 1967; Swalin, 1962). The problems associated with various solvents used in solution–crystal growth will be treated in Chapter 7, where the growth of specific substances is discussed in detail.

In crystallization by a liquid–solid equilibrium the growth process may be conservative or nonconservative. Most of the conservative techniques in Fig. 2.4 can be adapted to solution growth. However, because the problem in growth from solution is to introduce selected portions of the system into a region where the solution is supersaturated, and because the temperature for saturation will change during the crystallization as the solute concentration changes, the temperature must often be lowered to maintain supersaturation.† Because pure solute is crystallizing from a solution, the composition of the solution and the temperature will change and thus the equations derived for impurity distribution in conservative growth will often not be applicable. Unless there is complete solid solubility between the solvent and the solute [phase diagram of Fig. 2.3(a)], a composition region where the eutectic or some other phase appears will be reached and the solute will no longer be crystallized alone. However, for the growth of a small mass of crystal from a large amount of solution the techniques of Fig. 2.4, where the regions labeled liquid will now be labeled solution, are applicable if the temperature at the freezing interface is slowly lowered. The equations described for impurity distribution in conservative growth are often valid approximations. It should be remembered that the problems of diffusion associated with growth from a polycomponent system give difficulties that are discussed in Chapter 3 and usually necessitate slower growth rates. Calculations of impurity distribution are, of course, more straightforward if growth is at a single solid–liquid interface, that is, if the process in a polycomponent system is analogous to normal freezing. Probably the most convenient geometry employed for conservative growth from solution is that shown in Fig. 2.11(a), which is essentially the same as that of Fig. 2.4(c). Here, supersaturation is obtained by slowly cooling the solution. It is helpful but not essential in preventing spontaneous nucleation to maintain the seed at a lower temperature than the solution. The impurity distribution can be directly calculated by the methods derived for conservative growth only if the fact that the temperature of crystallization is changed as solute deposits on the crystal is taken into account. Changing the temperature of crystallization changes the value of k_0 and consequently prevents direct application of the equations and figures for monocomponent conservative growth to polycomponent conservative growth. However, k_0 is often not a strong function of temperature, or the cooling interval is not large, and then the treatments of impurity distribution for monocomponent conditions can be applied with little or no modification.

---

†Isothermal methods of solution growth avoid this difficulty and are discussed later in this chapter and in Chapter 7.

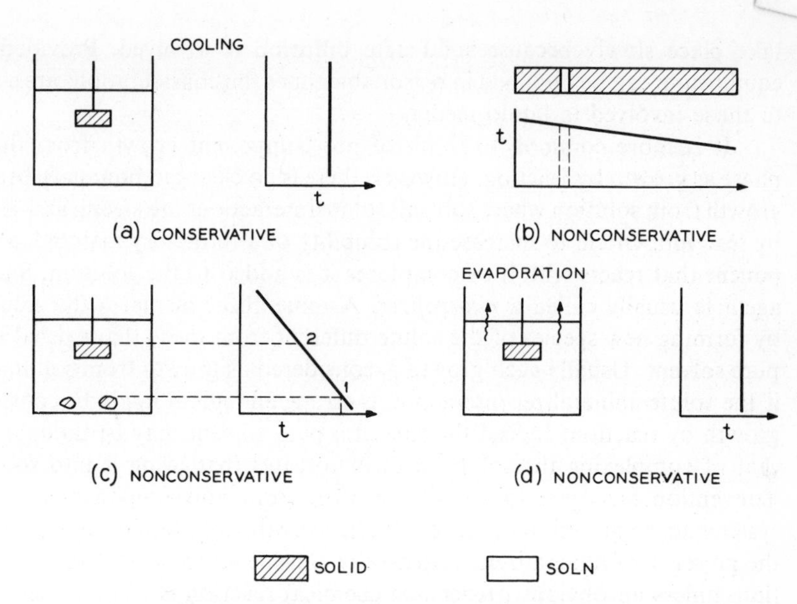

Fig. 2.11 Some conservative and nonconservative growth processes in solution.

If crystals are grown from a polycomponent system by a nonconservative solid–liquid process the techniques shown in Fig. 2.6 are directly applicable, with the change that the regions shown in Fig. 2.6 as liquid will instead be solution. In each case, solute dissolves from a source at a temperature above that where it crystallizes on a seed. The point where dissolving takes place is labeled 1, while the growth region is labeled 2, in Fig. 2.6. It might be mentioned that if in zone melting in a polycomponent system the gradient is linear across the whole ingot, as shown in Fig. 2.11(b), the zone can move of its own accord. This technique is *temperature gradient zone melting* (Pfann, 1955) and will be discussed further in Sec. 5.5. Two other methods of nonconservative solution growth that are often used are shown in Fig. 2.11(c) and (d). In Fig. 2.11(c) the source of solute (sometimes called the *nutrient*) is separated by a considerable distance from the seeds, which are at a lower temperature. In Fig. 2.11(d) saturation is produced by evaporation. Nonconservative solution growth often takes place at constant temperature and then equations and curves presented for monocomponent nonconservative growth are all directly applicable to the solution-growth techniques discussed here, except for growth by evaporation. In growth by evaporation the solution concentration is changing because of solvent loss, and corrections for this effect must be made in calculating impurity distributions.

Growth from solution can take place from the solid and gaseous states as well as from the liquid. Solid-state precipitation or *exsolution* often occurs in alloys and is responsible for their structure. In general, such processes take place slowly because solid-state diffusion is involved. Provided that equilibrium can be obtained in reasonable times the considerations are similar to those involved in liquid media.

It is more common to think of polycomponent growth from the gas phase as growth by reaction. However, there is no clear-cut boundary between growth from solution where solvent–solute interactions are strong and growth by reaction. Often, to increase the solubility of a refractory material, a component that reacts with it or complexes it is added to the solution. Such an agent is usually called a *mineralizer*. A mineralizer increases the solubility by forming new species of the solute different from those that existed in the pure solvent. Usually such growth is considered as growth from solution but if the solute–mineralizer interaction is strong enough, it might be considered growth by reaction. Indeed the role of a pure solvent may be thought of as that of complexing the solute to form solvates that bring it into solution. Convention has been to consider growth from polycomponent gas-phase systems to be growth by reaction, while growth from liquid phases, even in the presence of mineralizers, is generally considered to be growth from solution, unless an obvious irreversible chemical reaction is responsible for the growth.

2.11 Growth by Chemical Reaction

Consider a generalized chemical reaction in the gas phase.

$$aA + bB \cdots \rightleftharpoons gG(s) + hH \cdots \tag{2.38}$$

where G is a solid crystalline product of which single crystals are desired and the other products and reactants are gases. The equilibrium constant is

$$K = \frac{[G]_{eq}^{g}[H]_{eq}^{h}\cdots}{[A]_{eq}^{a}[B]_{eq}^{b}\cdots} \tag{2.39}$$

where $[\ ]_{eq}$ indicates equilibrium activity. Because the activity of solid G may be taken as unity and because we may approximate activities in a gas-phase system with pressures,

$$K \simeq \frac{(p_H)_{eq}^{h}\cdots}{(p_A)_{eq}^{a}(p_B)_{eq}^{b}\cdots} \tag{2.40}$$

where $p_{I_{eq}}$ are the equilibrium pressures of the reactants and products. It is usually desired that the concentration of the volatile species be fairly high so that the transport of matter to the growing seed crystal can be rapid. This would require a small value of K. However, one would very often, like to volatilize G in one part of the system and crystallize it in another part. For growth by this method differences in temperature or in concentration of reactants are used to shift the equilibrium. For ease in volatilizing G one would like the concentration of A and B in the equilibrium to be large, which would require a small value of K. For a reaction that would be easily reversible K should be near to unity. Thus because

$$\Delta G^0 = -RT \ln K \tag{2.41}$$

where ΔG^0 is the standard free energy of formation of the products in their standard states from the reactants in their standard states, a useful reaction will usually have small values of ΔG^0. Because ΔG^0 values are tabulated extensively and because pressures are a fair approximation to activities, a choice of systems beforehand on thermodynamic grounds is often possible in growth from the gas phase by reaction.

Equation (2.36) is valid in growth by chemical reaction, where Q is defined by the equation

$$Q = \frac{[\mathrm{G}]^g_{\mathrm{act}}[\mathrm{H}]^h_{\mathrm{act}}\cdots}{[\mathrm{A}]^a_{\mathrm{act}}[\mathrm{B}]^b_{\mathrm{act}}\cdots} \tag{2.42}$$

where $[\ \ ]_{\mathrm{act}}$ indicates actual activity. A good approximation to Eq. (2.42) is

$$Q = \frac{(p_{\mathrm{H}})^h_{\mathrm{act}}\cdots}{(p_{\mathrm{A}})^a_{\mathrm{act}}(p_{\mathrm{B}})^b_{\mathrm{act}}\cdots} \tag{2.43}$$

where $p_{\mathrm{I\,act}}$ are the actual pressures. Equation (2.37) becomes in growth by reaction

$$\Delta G = -RT \ln \frac{K}{Q} \tag{2.44}$$

where ΔG is the driving force for the reaction and K/Q is the analogue of relative saturation or degree of supersaturation. The temperature dependence of K is given by Eq. (2.32). If $\Delta H > 0$ (endothermic reactions) volatilization can be made to occur in a hot region of a system and crystallization in a cooler region. If $\Delta H < 0$ the reaction will transport from cold to hot. The magnitude of ΔH will determine the change in K with temperature and will dictate the temperature difference between volatilization and crystallization regions that will be required for growth. Appreciable rates can be obtained with small values of $|\Delta H|$ by using large temperature differences, but if $|\Delta H|$ is too large such small temperature differences are required to prevent excess nucleation that temperature control is difficult. If K is quite large the growth reaction is essentially irreversible and transport processes are not practical, but growth by *reaction* in the vapor phase, as will be discussed in Chapter 6, often is practiced.

When the reaction takes place in the liquid or solid phase Eq. (2.39) will become

$$K \simeq \frac{(X_{\mathrm{H}})^h\cdots}{(X_{\mathrm{A}})^a(X_{\mathrm{B}})^b\cdots} \tag{2.45}$$

provided the system is ideal enough that the mole fractions, X_{I}, approximate the activities. Few liquid-phase reactions with a K near to unity that are useful for crystal growth exist so that most growth by reaction is in the gas phase.† Several suggestions for utilizing reactions with large K's have been made and these will be discussed in Chapter 7. Equilibrium times for solid-state reactions will be long because solid-state diffusion will be involved.

Electrochemical reactions provide a specialized reaction type that is sometimes useful in crystal growth. The form of the overall equilibrium constant will be the same as Eq. (2.45). However, a consideration of the electrode processes will show that certain inherent advantages often occur in growth by electrochemical reaction (see Sec. 7.2.4-3). If a metal is to be deposited, a typical cathode reaction will be

$$M^{+n} + ne^- \longrightarrow M \tag{2.46}$$

The sum of the potential of the electrode reactions when the reactants and the products are in their standard state is the standard electrode potential, E^0, which is related to the equilibrium constant by the equation

$$E^0 = \frac{RT}{n\mathscr{F}} \ln K = \frac{0.05914}{n} \log K \tag{2.47}$$

where n is the number of electrons in the reaction, $\mathscr{F}$ is the Faraday constant, R is the gas constant, and T is the absolute temperature.

It can be shown that

$$\Delta G^0 = -n\mathscr{F}E^0 \tag{2.48}$$

Ordinarily, one of the chief difficulties in growth by chemical reaction is controlling the supersaturation to avoid spontaneous nucleation. If transport reactions are not used, reactants must be brought together without exceeding the supersaturation by enough to cause spontaneous nucleation. If K is not close to unity this is difficult to do, especially in liquid systems, where diffusion is slow and mixing times are long. However, in growth by electrochemical reaction, electrons can be thought of as essential reactants whose concentration is high only on the electrode where deposition and growth are desired, and thus electrochemical reactions provide some advantages with respect to control of nucleation.

If one desires a large concentration gradient of some dopant (for instance, in semiconductor diodes) low-temperature processes provide advantages because at low temperatures solid-state diffusion is negligible. If a seed (*substrate*) containing impurity is used and growth takes place at low-temperature conditions, diffusion of the dopant will be small and a large decrease in dopant concentration will occur where growth began. Large concentration gradients may also be obtained by introducing dopant into a vapor growth system during growth. Vapor-phase processes are thus especially useful because

†It would perhaps be more accurate to say that where $K \rightarrow 1$ in liquid-phase systems the growth is more often considered to be growth from solution.

controlled-property thin layers can often be easily produced. This technique is called *epitaxial*† crystal growth.

Growth by reaction may be conservative or nonconservative but where large crystals are desired nonconservative processes are generally used.

2.12 Growth of Metastable Phases

It is sometimes possible to form crystals of a material under conditions where it is thermodynamically unstable. Little success has so far been obtained in the growth of very large crystals of materials under metastable conditions but the technique if made practical would offer so many advantages that it is worth some discussion. If one could directly grow a high-temperature polymorph under conditions where it was metastable, then all the advantages of low-temperature growth mentioned in Sec. 2.1 would accrue. If one could grow a high-pressure polymorph at low pressures where it was metastable the process would obviously be considerably easier experimentally. If the rate of the phase transformation

$$\text{Metastable phase} \longrightarrow \text{Stable phase} \tag{2.49}$$

is slow at the growth temperature it is sometimes possible to form the unstable phase during the growth and rapidly quench to ambient conditions where the rate of Eq. (2.49) is essentially zero. Phase transformations sometimes proceed stepwise through increasingly less metastable phases on their way to the stable form (Ostwald's law of stages) and thus "freezing out" a metastable intermediate is possible. If a seed of a metastable form is placed in a saturated solution the seed may exert such a strong orientational influence that growth of the metastable form will result. The solubility relationships of polymorphs at a transition are shown in Fig. 2.12. *ABC* is the solubility of one polymorph, α, while *DBE* is the solubility of the other, β. *B* at temperature t_1 and solubility s_1 is the transition point at a given pressure. The solubility of β in the region where it is unstable (*DB*) is greater than that of α over the same temperature region. Similarly the solubility of α in the region *BC* is greater than that of β. Thus if a solution is saturated with respect to α in the α-stable region it will not be saturated with respect to β. However, a solution having the composition and temperature of *F* will be supersaturated with respect to α and β. If the supersaturation is sufficient to cause growth on a β seed of β without causing spontaneous nucleation of α, it may be possible to grow β on an β seed in the α region.‡ If β precipitates first because of Ostwald's law of stages it may even be possible to nucleate β in the α region without a

†See Sec. 3.9 for a more complete definition of epitaxy.

‡It will be shown in Chapter 3 that higher supersaturation is required for spontaneous nucleation than for seed growth.

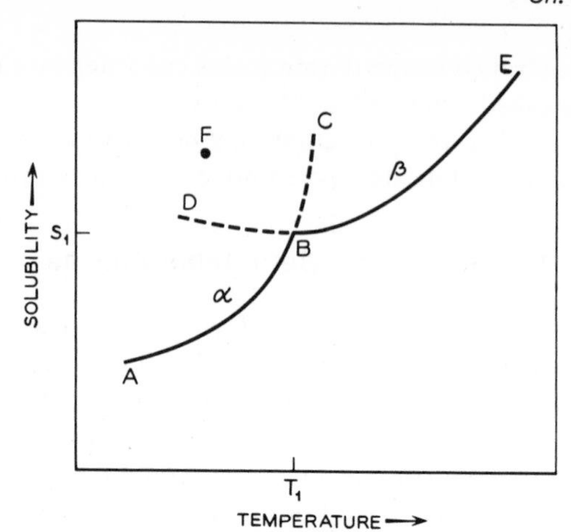

Fig. 2.12 Solubility of polymorphs near their transition temperature.

seed. An interesting example of the growth of a metastable phase, ethylene diamine tartrate, is given in Sec. 7.2.3.

REFERENCES

Carruthers, J. R., *Canad. Met. Quart.* **5** (1), 55 (1963).

Chernov, A. A., *Growth of Crystals*, Ed. by A. V. Shubnikov and N. N. Sheftal, Consultants Bureau, New York, 1962.

Emeis, Von R., *Z. Naturforsch.* **9a**, 67 (1954).

Findlay, A., A. N. Campbell, and N. O. Smith, *The Phase Rule and Its Applications*, Dover, New York, 1951.

Hall, R. N., *J. Phys. Chem.* **57**, 836 (1953).

Hansen, M., *Constitution of Binary Alloys*, 2nd ed., McGraw-Hill, New York, 1958.

Keck, P. H., and M. J. E. Golay, *Phys. Rev.* **89**, 1297 (1953).

Kubachewski, O., and E. Evans, *Metallurgical Thermochemistry*, 4th ed., Pergamon, New York, 1967.

Laudise, R. A., A. A. Ballman, and J. C. King, *J. Phys. Chem. Solids* **26**, 1305 (1965).

Levin, E. M., C. R. Robbins, and H. F. McMurdie, *Phase Diagrams for Ceramists*, American Ceramic Society, Columbus, Ohio, 1964.

Masing, G., *Ternary Systems*, Dover, New York, 1944.

Nassau, K., and G. M. Loiacono, *J. Phys. Chem. Solids* **24**, 1503 (1963).

Pfann, W. G., *Trans. AIME* **194**, 747 (1952).

Pfann, W. G., *Trans. AIME* **203**, 961 (1955).

Pfann, W. G., *Zone Melting*, 2nd ed., Wiley, New York, 1966, pp. 8ff, 31ff, 290.

Smith, F. Gordon, *Physical Geochemistry*, Addison-Wesley, Reading, Mass., 1963, pp. 74ff.

Strong, H. M., in *Progress in Very High Pressure Research*, Ed. by F. P. Bundy, W. R. Hibbard, Jr., and H. M. Strong, Wiley, New York, 1961, p. 192.

Swalin, Richard A., *Thermodynamics of Solids*, Wiley, New York, 1962, pp. 91ff, 200ff.

Theuerer, H., U.S. Pat. 3,060,123, filed December 17, 1952, issued October 23, 1962.

Theuerer, H., *Trans. AIME* **206**, 1316 (1956).

Thurmond, C. D., in *Semiconductors*, Ed. by N. B. Hannay, Reinhold, New York, 1959, pp. 145ff.

Weissberger, Arnold, *Physical Methods of Organic Chemistry*, Part I, Wiley-Interscience, New York, 3rd ed., 1959.

3

KINETICS OF CRYSTAL GROWTH

3.1 The Rate-Determining Process

In Chapter 2, we have dealt with crystal growth equilibria. In this chapter we shall discuss the rate at which crystal growth takes place. Kinetics deals with the rate of a process or reaction, with all the factors that influence the rate of the process, and with an explanation of the rate in terms of the reaction path or reaction mechanism. For a chemical reaction to take place or for a crystal to grow, a number of steps must occur. If one of these steps is much slower than the others, then the observed rate will be determined by the velocity of this *rate-limiting step*. If several of the steps are of comparable velocity, then the observed rate will depend on the velocities of all these steps and the reaction is a *consecutive reaction*. The rate-limiting step and mechanism will depend on the process employed for growth and on the conditions.

3.2 Diffusion

Diffusion is often an important step in crystal growth from polycomponent systems. At the growing crystal face, material can be accepted or rejected.

Obviously, the interface is a sink for the component that forms the crystal. It may either accept or reject some impurity or dopant in the system depending on whether k_{eff}, the effective distribution coefficient, is greater or less than unity. If the growth is from solution, the solvent is rejected and if the growth is by reaction, the reaction products that do not form the crystal are rejected. In both of these cases, as shown in Fig. 3.1, there will be a gradient in the concentration such that the concentration increases outward from the interface for a material that is accepted in the growing crystal and decreases outward from the interface for a material that is rejected. The

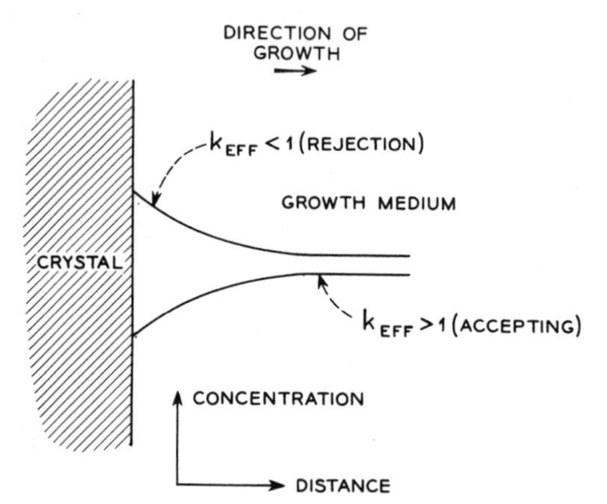

Fig. 3.1 Concentration gradients near a growing crystal.

details of the impurity- or dopant-concentration profile very close to the growing interface depend on whether an adsorbed layer is present and on other factors that need not concern us here. The diffusion rate will be given by Fick's laws. Fick's first law can be expressed as

$$J_i = -D_i \frac{\partial c_i}{\partial x} \tag{3.1}$$

where J_i is the flux† or flow-per-unit time per unit cross section of the ith component in a particular direction, D is the diffusion constant (which is here assumed to be independent of direction) and $\partial c_i/\partial x$ is the concentration gradient parallel to the direction of diffusion, which is here taken to be the x direction. Analogous equations can be written for y and z diffusion. The negative sign indicates that diffusion is toward the region of lower concentration. The area considered in establishing the flux is normal to the diffusion direction. J is a vector quantity so that the laws of vector addition can be used to calculate J in some direction not parallel to the reference axis. Fick's first law is most useful for steady-state conditions, that is, where the concentration of i is not changing with time.

†The present use of *flux* should not be confused with the use of the term to mean *molten-salt solvent* elsewhere.

If steady-state conditions do not exist, diffusion problems are more conveniently dealt with by means of Fick's second law, which is

$$\frac{\partial c_i}{\partial t} = D_i\left(\frac{\partial^2 c_i}{\partial x^2} + \frac{\partial^2 c_i}{\partial y^2} + \frac{\partial^2 c_i}{\partial z^2}\right) \tag{3.2}$$

where D is independent of concentration and t is time. For diffusion along only one direction, Eq. (3.2) reduces to

$$\frac{\partial c_i}{\partial t} = D_i\frac{\partial^2 c}{\partial x^2} \tag{3.3}$$

Solutions to Eqs. (3.1)–(3.3) have been reported in the literature (Jost, 1960; Crank, 1956) for most common geometries. The diffusion constant is usually independent of direction in most crystal-growth applications except when diffusion is in the solid state where the medium is often anisotropic. The two most common geometries of importance in crystal growth are diffusion toward a sphere, which takes place when a small crystal or nucleus grows, and diffusion toward a planar surface, which takes place when growth occurs on a planar seed. Very often the seed can be considered an infinite plane and diffusion from directions that are not normal to the plane which would occur near the edges can be ignored.

3.3 Crystal Surfaces

An important step in both mono- and polycomponent crystal growth processes is the *interfacial* reaction. This reaction may take place in a number of subreactions, but regardless of the details of these subreactions, the geometry of the growing interface will be of importance.

First, let us consider a partly completed atomically smooth surface such as that shown in Fig. 3.2. Figure 3.2 is idealized in that the crystal has no bulk imperfections and the atoms are represented as cubes. The atoms are identical and are closely packed so that each atom has 6 nearest neighbors. The 6 nearest neighbors touch each atom on the faces of the cube. In addition, each atom has 12 second-nearest neighbors that touch it on the cube edges and 8 third-nearest neighbors that touch it at the cube apexes. When a new atom is added to the crystal, the most probable position for addition will be that which is energetically most favorable. When energy is liberated during "bond" formation in this idealized model, the energetically most favorable position will be that where the greatest number of bonds can be formed. Bond formation is used here to mean covalent, or van der Waals' bonds. When ionic heterpolar bonds are formed, as we shall see below, repulsive terms must be considered. The row of atoms where faces form the plane $ABCD$ in Fig. 3.2 is a *step* that results when an atomically smooth surface is partly completed. The atoms of *ABCDEFGH* form a *kink* in a step that results when a new row of atoms is partially complete. Based on nearest-neighbor considerations, the most favorable position for the addition of a new atom is at a kink in a step [atom (1) in Fig. 3.2], because bonds may be formed to three nearest neighbors (at the cube faces at a distance from the center of the cube of a, where a is the cube edge). The next most favorable position for addition is at the front of a step [atom (2)], because bonds may be formed to 2 nearest neighbors. The least favorable position for addition based on nearest-neighbor considerations is as a lone atom on the surface [atom (3)]. If we consider second-nearest neighbors, certain energy differences in otherwise equivalent atoms will be apparent. In the lattice of Fig. 3.2, atom (3) has 4 second-nearest neighbors (at the cube edges at $a\sqrt{2}$ from the center of the cube), atom (4)

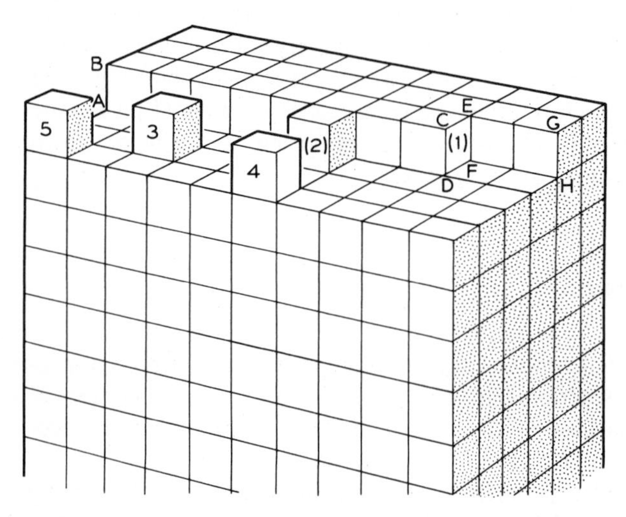

Fig. 3.2 Partly completed atomically smooth crystal surface.

has 3 second-nearest neighbors, and atom (5) has 2 second-nearest neighbors. Thus atoms (3), (4), and (5) all have 1 nearest neighbor, although the number of bonds formed is in the order atom (3) > atom (4) > (5). When energy is liberated by bond formation, the ease of attachment is atom (3) > atom (4) > atom (5). Similar reasoning would apply with regard to the effect of second-nearest neighbors on atoms, such as atom (1), which has 3 nearest neighbors, or atom (2), which has 2 nearest neighbors. The effect of third-nearest neighbors (at the cube apexes at $a\sqrt{3}$ from the center of the cube) and more-distant nearest neighbors is not so marked but is analogous. If one assumes that the bonding forces are inversely proportional to the square of the distance to the center of the cube then if the cube edge is a the bonding forces are

Nearest neighbor $\propto 1/a$

Second-nearest neighbor $\propto 1/2a$

Third-nearest neighbot $\propto 1/3a$

All that has been said above applies to crystals where the bonding energy is positive. Complications, however, arise in the case of heteropolar or ionic crystals, such as A^+B^-, because when an ion of a given charge approaches the face it will be attracted by ions of opposite charge, while at the same time it will be repelled by ions of like charge. Consider the attachment of an A^+ atom at a smooth A^+B^- surface. It will attach above a B^- atom so that the energy gain will be proportional to $1/a$. However, it will have as all of its second-nearest neighbors A^+ atoms so that it will be repelled from them by a quantity proportional to $(-)n/2a$, where n is the number of second-nearest neighbors. Similarly it will have as third-nearest neighbors all B^- atoms so that it will be attracted to them by a quantity proportional to $m/3a$, where m is the number of third-nearest neighbors. Thus for the addition of an A^+ atom on a smooth A^+B^- face neglecting greater-than-third-nearest neighbors, the energy released, E, is

$$E = \frac{\alpha}{a}\left(1 - \frac{n}{2} + \frac{m}{3}\right) \tag{3.4}$$

where α is a proportionality constant.

Thus if we consider atoms (3), (4), and (5) of Fig. 3.2,

$$E_3 = \frac{\alpha}{a}\left(1 - \frac{4}{2} + \frac{4}{3}\right) = \frac{\alpha}{3a} \tag{3.5}$$

$$E_4 = \frac{\alpha}{a}\left(1 - \frac{3}{2} + \frac{2}{3}\right) = \frac{\alpha}{6a} \tag{3.6}$$

$$E_5 = \frac{\alpha}{a}\left(1 - \frac{2}{2} + \frac{1}{3}\right) = \frac{\alpha}{3a} \tag{3.7}$$

Therefore, attachment at atom (3) or (5) is favored over attachment at atom (4). Kossel (1927, 1934, 1938) and Van Hook (1961) have made calculations based on the assumption that interatomic forces fall off more rapidly than the square of the distance and conclude that in a heteropolar crystal the energy relations are $E_5 > E_4 > E_3$. Analogous calculations for atoms (1) and (2) can be made for varying numbers of second- and third-nearest neighbors. However, atom (1) will still be the most-favorable site energetically and atom (2) will be the next most favorable. The important result that comes from these calculations is that in beginning a new layer, atom (5) is the most favorable site.

It should be obvious that for a structure different from Fig. 3.2 new calculations would have to be carried out.

The above considerations can be best applied to atomically smooth planes. It turns out that rough surfaces or stepped surfaces are present in many crystals, particularly for the higher-index faces. A qualitative idea of the need for rough faces can be gleaned from Fig. 3.3. The plane of atoms shown in Fig. 3.3 is an (001) plane in a cubic crystal, AB. 1–3–5–7–9–11–13 is the intersection of (111) with (001) and is bounded entirely by B atoms, as is the entire (111) plane. Such a plane will have a high surface energy because there are two unsatisfied bonds at each B atom. However, if the crystal is bounded by the rough surface whose intersection with (001) is 1–2–3–4–5–6–7–8–9–10–11–12–13 the surface energy is sometimes reduced. Thus by extension in three dimensions, in our illustration (111) would tend to be rough. Such an atomically rough surface would provide many easy steps for nucleation. In a real crystal the step-height regularity would probably not be maintained, but the roughness would still provide many easy sites for nucleation.

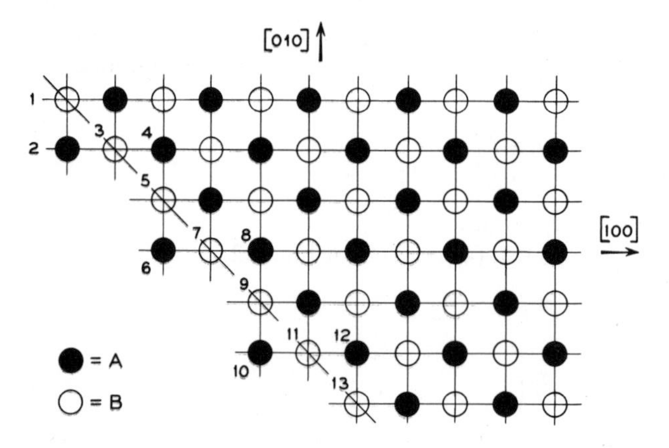

Fig. 3.3 The (001) plane in a cubic crystal, AB.

The problem of how nucleation of a new step can begin on an atomically smooth surface is one that has received great attention. As we have seen, the nucleation of a new step is energetically less favorable than growth at a step or kink. Once a step has grown across the surface, nucleation of a new step is required. Quantitative calculations show that large supercooling or supersaturation is required for the nucleation of a new step on an atomically smooth surface. However, crystals are observed to grow at quite low supercooling or supersaturation (Volmer and Schultze, 1931). What is required is a step that does not grow out of existence or a repeatable step (in the German literature a *wiederholbarer Schritt*). On a rough surface repeatable steps are always present. However, for many years the problem of growth at low supersaturations on atomically smooth surfaces was baffling. The realization that the spiral or screw dislocation (Frank, 1949), Fig.

3.4, supplies a step that would not grow out of existence and the observation of such steps on grown crystals provided an understanding of such growth. (See Sec. 3.7 for a quantitative discussion of the dependence of rate on supersaturation when growth is on spiral dislocation.) Step heights on most materials are many lattice parameters high and are often visible to the naked eye. As we shall see below, the bunching theory, as put forward by Cabrera and Frank, explains the genesis of macroscopic steps.

The first quantitative treatment of surface free energy of crystal faces is attributed to Wulff (1901) and has been amplified and extended by Herring (1951).

Fig. 3.4 Spiral dislocation.

Gibbs in 1878 (Gibbs, 1928) pointed out that the form of a small crystal in equilibrium with a solvent had to satisfy the condition that $\sum \sigma_i A_i$ be a minimum for the crystal, where σ is the specific interfacial surface free energy of surface i (the surface free energy per unit area) and A is the area of that surface on the crystal. When growth (or dissolving) takes place very far from equilibrium the form of a crystal may depart from the equilibrium form. The specific interfacial surface free energy is sometimes called the capillary constant or more commonly the interfacial surface tension. A convenient way to plot the directional dependence of σ in a crystal was suggested by Wulff and the plot of Fig. 3.5 is a so-called Wulff plot of a hypothetical crystal. It is a polar plot of the surface free energy along a great circle through the approximately spherical surface representing the complete directional dependence of σ in a crystal. The distance from the origin is proportional to the magnitude of σ. If planes are constructed at each point on the surface normal to the radius vector at that point, then the volume that can be reached from the origin without crossing any planes is geometrically similar to the equilibrium form of the crystal. This can be made plausible by imagining that a small spherical nucleus grows for a time long enough to produce the

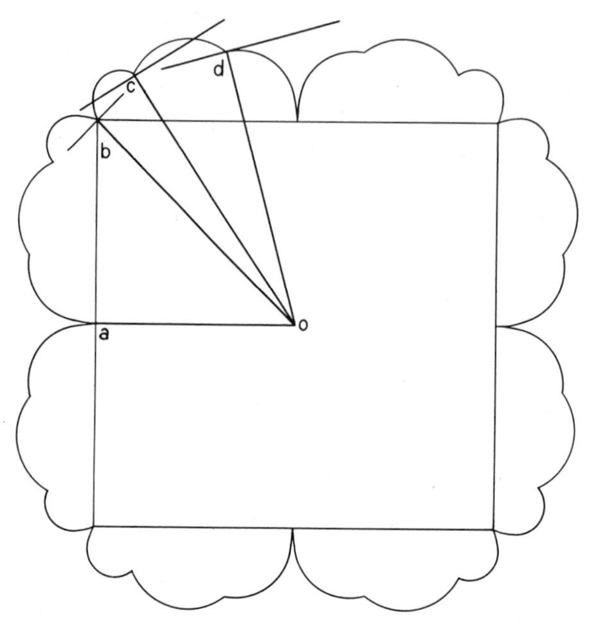

Fig. 3.5 Polar plot of crystal surface free energy.

equilibrium form. If a particular face forms, such as AB in Fig. 3.6(a), and the rate on this face (proportional to the distance 1–2) is greater than the rate on another face, such as BC (where rate is proportional to the distance 3–4), then the area of the faster-growing face will decrease with time ($A'B' < AB$), while the area of the slower-growing face will increase with time ($B'C' > BC$). Eventually the faster-growing face will "grow out of existence." Figure 3.6 is a two-dimensional schematic representation and on real crystals the geometric relationship of faces in addition to their relative rates will deter-

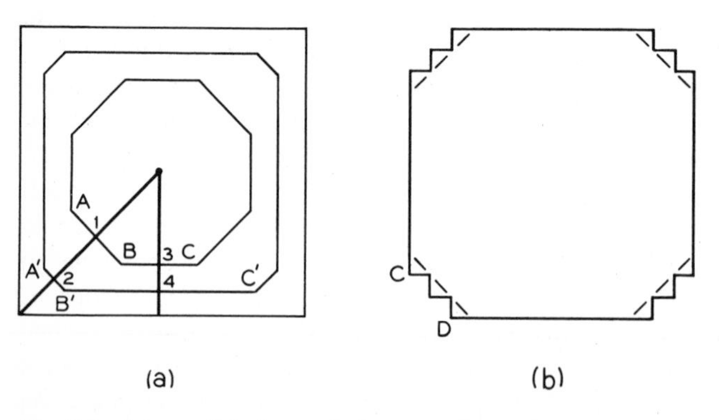

Fig. 3.6 Intermediate forms during crystal growth.

mine whether a given face will grow out of existence. However, a face whose area is increasing with time will always have grown slower than a face whose area is decreasing. Thus the distance from the center in Figs. 3.5 and 3.6(a) is proportional to both σ and the growth rate if growth takes place near equilibrium.† The closed figure of smallest volume that can be constructed by planes normal to the radius vectors at points where the faces interrupt the surface free-energy surface of Fig. 3.5 will be the equilibrium form of the crystal and will involve faces with rates slower than any other form. Because cusps in the surface free-energy surface will generally have the shortest radius vectors (i.e., be associated with surfaces of low σ), crystal faces will generally occur where the radius vector intercepts a cusp or saddle point on the Wulff plot and the volume that can be reached from the origin without crossing any planes is the shape that minimizes surface free energy.

The equilibrium shape is more often realized in small crystals. Because a significant change in shape can occur only by the transport of considerable crystal mass over considerable distance, the energy required for such transport is large in comparison with the decrease in surface free energy obtained.

Herring (1951) pointed out that for large crystals it is important to inquire how the free energies of neighboring configurations in a crystal compare even when none of the configurations is close to the minimum free energy given by the Wulff construction. Herring has analyzed the Wulff plot for neighboring configurations and his most important conclusions can be summarized as follows:

1. If a given macroscopic surface of a crystal does not coincide in orientation with some portion of the boundary of the equilibrium shape, there will always exist a hill-and-valley structure, such as *CD* in Fig. 3.6(b), which has a lower free energy than the flat surface. Conversely, if the given surface does occur in the equilibrium shape no hill-and-valley surface can be more stable.

2. When the free-energy surface in the Wulff plot is outside the sphere drawn tangent to the surface at the point where the radius vector interrupts the surface, that surface or "face" will be curved in the crystal. If the free-energy surface lies inside the sphere anywhere, the surface in the crystal will be bounded by a hill-and-valley structure of the sort described in conclusion 1.

3. Where planar edges intersect, surface free energy can be minimized by the rounding of edges. This rounding almost always will be so small as to be imperceptible but will be greater the smaller the areas (or volume in three dimensions) between the Wulff surface and the sphere drawn tangent to the intersection of the planar edges.

---

†The question of whether growth at *any* finite rate implies such a great departure from equilibrium as to invalidate equilibrium thermodynamics has not been settled.

It should be pointed out that as the temperature of a crystal approaches its melting point the surface becomes more and more rough, so assumptions about preferential attachment sites or the detailed nature of Wulff surfaces in growth by any equilibrium at or close to the melting point are likely to be unreliable. Thus Wulff-plot-type reasoning is considerably safer when applied to solution growth.

Attempts have been made to predict which faces will have the lowest specific surface free energy by direct calculation of the reticular density of the atoms in a given plane (Donnay and Harker, 1937). While high-density close-packed faces do have low specific surface free energy one must take into account the directional nature of bond formation and the details of the spatial arrangement of atoms, particularly when bonding is not homopolar in a compound composed of more than one element. Some improvement in compensating for certain of these factors is obtained in the periodic bond chain (P.B.C.) method of Hartman and Perdok (1955). In this method every crystal is divided into periodic bond chains and growth is predicted to be fastest in the direction of strongest chains.

3.4 Processes That Take Place During Growth from Solution

As we have stated, crystal growth involves a number of processes one or several of which can establish the overall rate. The nature of these steps will depend on the system from which growth takes place. The greatest number of steps is possible in a polycomponent system where growth is from solution and dissolving of solute takes place in one part of the system and growth takes place in another. Under such conditions the process may include (in the order in which the steps might take place):

G1. Dissolving.

G2. Transport of dissolved species.

†G3. Diffusion across a locally solute-depleted zone at the growth interface.

‡G4. Partial or complete desolvation or the reversal of the reaction with the mineralizer (see below).

†G5. Diffusion away from the growth interface of the solvent or mineralizer and of impurities.

G6. Desorption of impurity, solvent, or mineralizer species from the growth interface.

---

†Diffusion or diffusion-related steps.

‡Desolvation or desolvation-related steps.

G7. Adsorption or chemisorption on surface of seed of the species that forms the crystal.

†G8. Two-dimensional diffusion of adsorbed or chemisorbed species to a step.

‡G9. Partial desolvation or "dereaction" with the mineralizer.

G10. Attachment at the step.

†G11. Diffusion away from the growth interface of the solvent or mineralizer (newly released by step G9).

†G12. One-dimensional diffusion along the step to a kink.

‡G13. Partial desolvation or "dereaction" with mineralizer.

G14. Attachment at the kink.

†G15. Diffusion away from the growth interface of the solvent or mineralizer.

G16. Dissipation of the heat of crystallization, which may occur stepwise after each of the attachment steps. If any of the processes are endothermic, absorption of heat will take place.

The dissolving step itself can be broken into the following series of substeps:

D1. Detachment from a kink.

D2. One-dimensional diffusion away from the kink.

D3. Partial solvation or reaction with the mineralizer.

D4. Diffusion toward the dissolving interface of the solvent or mineralizer.

D5. Detachment from a step.

D6. Two-dimensional diffusion away from the step.

D7. Partial solvation or reaction with the mineralizer.

D8. Diffusion toward the dissolving interface of the solvent or mineralizer.

D9. Desorption from surface.

D10. Bulk diffusion across a locally solute-enriched zone at the dissolving surface.

D11. Partial or complete solvation or reaction with mineralizer.

D12. Diffusion toward the dissolving interface of solvent or mineralizer.

D13. Absorption of the heat of solution, which may occur stepwise after each of the detachment steps.

The exact order with which the steps take place in growth and dissolving is generally not known. Indeed, whether certain steps such as G4, G9, and G13 take place in a stepwise manner is not known for most systems.

Some steps are not possible in certain systems but it is generally easy, by the use of steps Gl–G16 and Dl–D13 as a guide, to list the possible steps in a given system. We shall now discuss some of the steps in detail and we shall show in the next section how they apply to growth by various types of equilibria.

If growth occurs without need of dissolving, as in growth by slowly cooling a saturated solution, dissolving (G1) and transport (G2) will not be present. By control of geometry, dissolving can usually be made to be of minor importance in establishing the rate. For instance, the surface area of the solid solute that it is desired to dissolve (the *nutrient*) can be made quite large. Transport can be by means of diffusion, convection, mechanical agitation, and flow caused by pressure differentials. We have discussed bulk diffusion (G3) in Sec. 3.2.

In a solvent–solute system it is the solvent–solute interaction that is responsible for dissolving. At some stage during growth this solvent–solute reaction must be reversed and the solute must become desolvated. In many solvent–solute systems additional components are added, such as complexing agents, that interact even more strongly with the solute than with the solvent. These additives that increase solubility by forming new species different from those existing in the pure solvent are usually called *mineralizers*. Obviously, the solute–mineralizer reaction must be reversed. It is probable that the solute need not completely desolvate or shuck off the mineralizer in one step. Similarly, it is probable that the solute species does not attach itself to the growing interface in only a single step. As we have seen in Sec. 3.3, atoms may be attached to a crystal surface by a single bond (adsorbed or chemisorbed), by two bonds (attachment at a step), or by three bonds (attachment at a kink in a step).† It is probable that, as the number of bonds made to the crystal surface increases, the number of solute–mineralizer or solute–solvent bonds decreases, thereby tending to keep the coordination number of the solute species constant during crystallization. Thus we have listed desolvation or reversal of reaction with mineralizer as steps G4, G9, and G13 in the generalized growth sequence. In each case a desolvation step is followed by the formation of an additional bond to the crystal surface (steps G7, G10, and G14). It seems reasonable to assume that some sort of diffusion step in which the solute species moves to the proper site always precedes the formation of an additional bond to the crystal surface (steps G3, G8, and G12). Similarly, bulk diffusion away from the growing crystal of the solvent or mineralizer that has been shucked off always follows the desolvation process (steps G5, G11, and G15). If the site where the solute particle will form a bond to the crystal surface is blocked by an adsorbed or chemisorbed atom, then the atom must be displaced for bond formation to occur (step G6). A step similar to G6 might take place after step G9 and also after step G13.

Whether the first bond the solute species makes to the crystal surface (step G7) should be considered adsorption or chemisorption is an interesting

†The discussion of Sec. 3.3 was based on a cube model of the atoms forming the crystal. In real systems the number of nearest neighbors is often not six and hence attachment at a surface might involve more than one bond, attachment at a kink more than two bonds, etc.

question. It is likely that, if the solute closely approaches the crystal surface, the bond formed would be similar to the bonds in the bulk of the crystal and hence the process would be chemisorption. However, if the solute particle's approach to the crystal surface were not close enough to allow a chemical bond (the solvation or mineralizer shell might sterically prevent close approach), van der Waals' forces would hold the particle to the crystal surface and the process would be described as adsorption. Eventually, however, the bond would become the normal bond that holds the crystal together. This may take place later in the growth sequence, perhaps when kink attachment (G12) occurs.

The dissolving process has been broken down into a number of steps (D1–D13). In many cases, these steps are the reverse of the steps of the growth process. The exact order of steps listed in (D1–D13) and the number of steps are not the same as in (G1–G15). We have done this to emphasize that the exact order and number of steps both in growth and dissolving are clearly still a matter of speculation. However, for all the steps of growth listed in G1–G15, a reverse step in dissolving is at least conceptually possible. If in practice the reverse steps indeed take place, the mechanism and rate-limiting steps are the same for growth and dissolving, and the rate equation is identical except for a change of sign. Under such conditions growth and dissolving would be *reciprocal* processes. Most experimental evidence indicates that growth and dissolving are not reciprocal. One reason for this is that adsorbed impurities, dopants, mineralizers, or solvents will generally tend to block growth but will often tend to accelerate dissolving.

It should be pointed out that growth at high supersaturation or dissolving at high undersaturation do not require processes involving steps or kinks.

3.5 Processes That Take Place During Growth by Various Types of Equilibria

In growth from a monocomponent system by a solid–solid equilibrium, because the solids are of identical chemical composition, diffusion in the ordinary sense will not be rate limiting. However, if the structural change is large, atoms may have to move appreciable distances for the new structure to be formed and this "microdiffusion" may be involved in the rate-limiting step. Polymorphic-phase changes are usually first-order reactions; that is, the rate depends on the first power of the instantaneous concentration of the phase transforming. The rate constant usually shows an Arrhenius dependence on temperature. That is,

$$k = ae^{-b/RT} \tag{3.8}$$

where k is the rate constant, a and b are, to a first approximation, constants, R is the gas constant, and T is the absolute temperature. The value of b is the energy of activation. An abrupt change in slope in a log k vs. $1/T$ plot is suggestive of two different mechanisms of reaction operative in different temperature ranges. The frequency factor a may itself be temperature-dependent. A more complete discussion of the dependence of the rate constant on temperature is given in standard works on chemical kinetics (Glasstone et al., 1941; Slater, 1959; Frost and Pearson, 1953). Figure 3.7 shows a schematic representation of the energy content of a stable and an unstable phase. The higher the potential-energy barrier between the stable and unstable phases, the larger the energy of activation, ΔE^*, and the slower the rate of transformation at a given temperature. Raising the temperature will increase the value of k, but, of course, if the temperature increase takes the system out of the range of thermodynamic stability of β, it cannot be formed. If the potential-energy barrier is so high that α transforms to β at a slow rate and α persists for appreciable times under unstable conditions, α is said to be *metastable*.

Nucleation of the stable phase in the solid is often not difficult because nucleation by epitaxial growth on the unstable phase is usually easy. Indeed multiple nucleation in solid–solid transformation makes it difficult to grow single crystals.

Diffusion cannot be involved in growth from a true monocomponent system by a liquid–solid equilibrium, because the liquid and solid are identical in composition. In real systems, however, some impurity diffusion is usually present. The growing interface is probably always atomically rough because growth is at the melting point and therefore steps and kinks are readily available. Attachment on the face and two-dimensional diffusion to the site

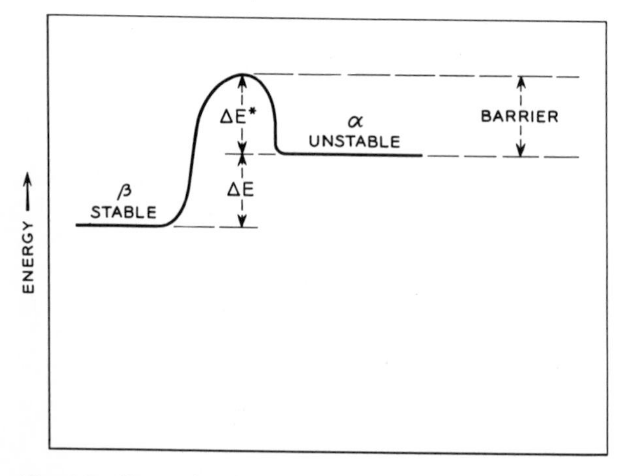

Fig. 3.7 Energy barrier for transformation.

of inclusion in the crystal are possible rate-limiting steps. However, in most systems these steps are probably rapid in comparison with dissipation of the heat of fusion. The heat of fusion is transported away from the growing interface by conduction, convection, and radiation. The principal means of heat transport may involve any one or any combination of these, depending on the condition, physical properties, and geometry of the system. For instance, radiation is generally not important for growth below about 800–1000°C. For the growth of metals and semimetals, conduction along the grown crystal is important in the geometries used in crystal pulling.

For temperatures above about 800–1000°C, radiation is important and its importance increases when the growing crystal is transparent in the spectral range in the vicinity of the peak of the blackbody radiation curve for the growth temperature. Then the crystal can act as a "light pipe" to "leak" heat from the growing interface. In growth from the melt, perfection is often a function of the shape of the growing interface. The interface shape is controlled by the heat-flow pattern in the system, so some knowledge of the heat-transport processes is usually necessary in order to optimize conditions in a given system.

Strictly speaking, we should consider growth in the presence of air or an inert gas under polycomponent growth, but, practically speaking, if an added gas does not interact with the material grown, the growth may be considered monocomponent. In the case of growth by a gas–solid equilibrium in a monocomponent system, diffusion can be an important step. The gas phase is so dilute in comparison with the solid phase that the gas species must move toward the growing crystal from appreciable distances in the gas phase. The mean free path, or average distance a molecule in the gas moves before colliding with another molecule, will be a measure of the rate with which the gas species can move toward the growing interface. The mean free path is given, to a good approximation, by the equation

$$\lambda = \frac{\alpha}{Pd^2} \tag{3.9}$$

where λ is the mean free path, P is the total pressure, α is a constant, and d is the atomic or molecular diameter. Thus the higher the pressure, the shorter the λ. The flow-per-unit area or flux of material toward the growing crystal interface, φ, will be a function of λ and of the partial pressure of the material. Thus in gas–solid growth it will often be difficult to have as high a rate as in other monocomponent growth because diffusion is involved and large φ's are not obtainable. Increasing pressure will not increase φ because it will decrease λ. The rate-limiting process under such conditions can thus involve only those steps listed in G1–G16 that do not involve the solvent or mineralizer. The dissolving steps will, of course, become volatilization steps. If a nonreactive gas is present, its diffusion away from the growing crystal could be involved in the rate-limiting process. Actually, as we shall see, the best fit between theories of crystal growth and experimental results has been obtained in vapor–solid crystal growth.

In growth from a polycomponent system the diffusion steps of G1–G16 can be important in solid–solid, liquid–solid, and gas–solid growth. Thus in all cases all of the steps of G1–G16 can be important. In solid–solid growth, diffusion will be slow and is often of overriding importance. Controlled growth will be difficult because the solid will provide many sites for nucleation.

In growth by polycomponent liquid–solid and gas–solid equilibria, diffusion is not as slow but still is generally involved in the rate-limiting step. In addition, interfacial reactions are almost always of importance. Indeed, if bulk diffusion alone were rate limiting, we should not expect a directional dependence of rate and the "habit" of the grown crystal would be spherical.

It should be apparent that the details of steps G1–G16 and D1–D13 will depend on the system where growth takes place. For instance, those steps involving the solvent or mineralizer in growth from solution will be replaced by steps involving reactants in growth by chemical reaction. In monocomponent growth, steps involving the heat of crystallization or solution in solution growth will be replaced by steps involving the heat of transition or devitrification in solid–solid equilibria, the heat of fusion in growth from the melt, and the heat of sublimation in growth from the vapor. In polycomponent growth the heat of reaction will replace the heat of crystallization if growth is by means of a reaction.

It is usual to speak of supercooling in liquid systems that are monocomponent or essentially so and supersaturation in polycomponent systems. The supercooling ΔT is defined by the relation $\Delta T = t_{act} - t_{eq}$, where t_{eq} is the equilibrium temperature and t_{act} the actual temperature, while supersaturation Δs is defined as $\Delta s = c_{act} - s_{eq}$, where c_{act} is the actual concentration and s_{eq} is the equilibrium solubility. In gas-phase systems, $\Delta s = p_{act} - p_{eq}$, where p_{act} is the actual pressure and p_{eq} is the equilibrium pressure. Defining supersaturation in chemical reactions is not as straightforward and depends on the purpose at hand. Relative supersaturation as defined in Secs. 2.10 and 2.11 has particular use in unambiguously defining the driving force in growth by reaction and has importance in crystal-growth theories.

3.6 Crystal Growth as a Consecutive Reaction

Crystal growth involves many steps and almost always several of the steps are of comparable velocity and will thus be involved in determining the rate. If we lump all the interfacial steps together and assume that the observed rate depends on two steps, bulk diffusion and one interfacial step, it is possible to derive a rate equation for crystallization.

If the equilibrium at the interface is so rapid as not to be involved in the rate-limiting step, so that bulk diffusion alone establishes the rate, Fick's first law expresses the rate of diffusion:

$$\frac{dm}{dt} = DA\frac{dc}{dx} \tag{3.10}$$

where dm/dt is the mass added to a growing surface per unit time, dc/dx is the concentration gradient normal to the surface, D is the diffusion constant, and A is the area of the growing surface.

If the width of the diffusion layer is σ, c_B is the concentration in the bulk of the solution, and c_{su} is the concentration at the surface, Eq. (3.10) reduces to

$$\frac{dm}{dt} = \frac{DA}{\sigma}(c_B - c_{su}) \tag{3.11}$$

If the $\mathscr{R}$ is the linear rate of crystallization on the surface, because

$$\frac{dm}{dt} = \mathscr{R}A\rho \tag{3.12}$$

where ρ is the density of the crystal, Eq. (3.11) becomes

$$\mathscr{R} = \frac{D}{\rho\sigma}(c_B - c_{su}) \tag{3.13}$$

If σ is a constant it is convenient to replace D/σ with k_d, the "diffusion velocity constant," so that Eq. (3.13) becomes

$$\mathscr{R} = \frac{k_d}{\rho}(c_B - c_{su}) \tag{3.14}$$

Eq. (3.14) is valid even if the rate of the interfacial reaction is not fast compared to diffusion. If the interfacial reaction is fast compared to diffusion, then $c_{su} = c_{eq}$, where c_{eq} is the equilibrium concentration.

If the interfacial reaction is overriding for a first-order heterogeneous reaction,

$$\frac{dm}{dt} = k_i A(c_{su} - c_{eq}) \tag{3.15}$$

where k_i is the "interfacial velocity constant." Thus by appropriate substitution,

$$\mathscr{R} = \frac{k_i}{\rho}(c_{su} - c_{eq}) \tag{3.16}$$

The value of k_i will generally depend on the crystallographic direction in which growth takes place. Equation (3.16) is valid even if the diffusion reaction is not rapid compared to the interfacial reaction. If the diffusion reaction is fast compared to the interfacial reaction, then $c_{su} = c_B$.

For consecutive reactions, we may eliminate c_{su} between Eqs. (3.12) and (3.14), so that

$$\mathscr{R} = \frac{1}{\rho[(1/k_d) + (1/k_i)]}(c_B - c_{eq}) \tag{3.17}$$

which describes crystal growth when two consecutive reactions, diffusion and an interfacial step, are of comparable velocity. In Eq. (3.17), $\mathscr{R}$ will, of course, be directionally dependent and Eq. (3.17) will reduce to Eq. (3.14) for large values of k_i and to Eq. (3.16) for large values of k_d. If k_d and k_i are constants over the conditions of a given set of experiments, then

$$\frac{1}{k_d} + \frac{1}{k_i} = \frac{1}{\bar{k}} \tag{3.18}$$

$$\mathscr{R} = \frac{\bar{k}}{\rho}(c_B - c_{eq}) \tag{3.19}$$

The values of k_d and k_i can be expected to be concentration and pressure independent over fairly broad ranges but will usually depend on temperature. The temperature dependence, at least to a first approximation, will be, according to the Arrhenius equation,

$$k_d = a_d e^{-\Delta E_D^*/RT} \tag{3.20}$$

$$k_i = a_i e^{-\Delta E_i^*/RT} \tag{3.21}$$

where a_d and a_i are, to a first approximation, temperature-independent constants and ΔE_d^* and ΔE_i^* are the energies of activation of the diffusion and interfacial steps, respectively. Because ΔE_d^* is generally a few kilocalories per mole, while ΔE_i^* is considerably larger, k_i will be more strongly temperature dependent. In many cases $\bar{k}$ will obey the Arrhenius equation over a fairly wide range of conditions.

3.7 Reaction Steps in the Interfacial Process

A basic problem in crystal-growth kinetics is to solve the simultaneous differential equations describing consecutive growth. This is especially difficult where more than two steps are involved and particularly complex when we consider the interfacial process, where the reaction steps are experimentally difficult to study.

A complete historical treatment of the development of the ideas involved in reaction kinetics is not appropriate here. N. Cabrera and F. C. Frank are responsible for the first synthesis of a theory that takes into account the persistence of growth even at low supersaturation and that makes use of the spiral dislocation as a repeatable step. Their work was in particular based on the earlier theoretical analysis of I. N. Stranski and M. Volmer (Frank,

1949, 1958; Burton and Cabrera, 1949; Frank, 1952; Cabrera and Coleman, 1963; Cabrera and Vermilyea, 1958). This subject is best approached from the point of view of growth from the vapor in a monocomponent system on a low-index face at a reasonable temperature below the melting point under conditions where diffusion can be neglected. Many of the conclusions can be extended to more complex systems but the basic concepts are best developed under the above conditions. A low-index face at a reasonable temperature below the melting point will remain atomically smooth and will grow by means of the following steps in a monocomponent vapor system:

1. Attachment to the surface (chemisorption or adsorption)
2. Two-dimensional diffusion to a step
3. Attachment at a step
4. One-dimensional diffusion to a kink
5. Attachment at a kink

The reverse of steps 1–5 is evaporation. At equilibrium, the rates of evaporation and condensation are equal. For growth to occur, the rate of condensation must exceed the rate of evaporation.

If the kink density is high compared to the step density and if two-dimensiona diffusion and one-dimensional diffusion are of comparable velocity, then steps 4 and 5 cannot be rate limiting. By assuming that the energy of formation of a kink on a step is about $\frac{1}{10}$† the evaporation energy, Frank has calculated that, at the lowest temperature at which the vapor pressure would be high enough even to contemplate growing a crystal, kinks occur at intervals of every three or four atoms along a step.

Two-dimensional diffusion and probably one-dimensional diffusion are rapid and of comparable velocity. Frank estimates that the energy of activation for two-dimensional diffusion on a planar surface is about $\frac{1}{20}$ of the energy required to transfer an atom attached at a kink along the surface to a point where it exists as an isolated adsorbed atom. Thus one-dimensional diffusion and attachment at a kink are ordinarily not rate limiting. At vapor pressures as low as 10^{-10} atm, it can be shown that an atom will, on the average, diffuse two-dimensionally several hundred atom diameters before evaporating. Thus kinks will be fed principally by surface diffusion and not by the direct arrival of atoms from the vapor.

We may thus conclude that the adsorption of atoms and the movement of atoms to steps will be rate limiting, provided a source of steps is present. If a step is initially present when it grows to the edge of the crystal a new step must be nucleated. A step will begin by the adsorption or chemisorption of

†This estimate can be made by considering the difference in the total bond energies of an atom in a step and an atom at a kink compared to an atom in the bulk of the crystal. The bond energies are calculated by considering the number and distance of first-, second-, and third-nearest neighbors in a manner analogous to that described in Sec. 3.3.

a single atom and will grow by the addition of atoms adjacent to this atom as a sort of "island" on the surface. The average "radius," r, for an island that will grow, that is, an island for which the probability of attachment of an atom is equal to the probability of detachment, is

$$r = s_0\gamma/kT \ln \alpha \tag{3.22}$$

where s_0 is the area per atom in the layer, γ is the specific free energy of the boundary of the island (which will be a function of its orientation), k is the Boltzmann constant, and α, the relative supersaturation, is the ratio of the vapor pressure p to the equilibrium vapor pressure $p_0(\alpha = p/p_0)$. An island with a radius larger than r will grow, while one with a smaller radius will shrink.

Equation (3.22) is a modification of the equations developed for critical droplet size in three-dimensional vapor–liquid nucleation. The formation of a new layer thus depends on fluctuations of adsorbed molecules producing a critical nucleus. The rate of formation of critical nuclei, Γ, is given by the formula

$$\Gamma = [ZS/s_0]e^{-A_0/kT} \tag{3.23}$$

where Z is the rate of arrival of molecules at single surface-lattice sites, S is the surface area of the crystal under consideration, and A_0 is half the total edge free energy of the nucleus, which is a sort of activation energy for island nucleation. A_0 is given by the formula

$$A_0 = \varphi^2/kT \ln \alpha \tag{3.24}$$

where $\frac{1}{2}\varphi$ is the energy of the edge per atom. On simple nearest-neighbor considerations, φ is about $\frac{1}{6}$ the evaporation energy per atom.

For Γ to be a few microns per month, we can estimate from Eqs. (3.23) and (3.24) what value of α would be required. Kinetic theory of gases shows that, even at appreciable pressures, Z cannot be much greater than 10^{13}/sec. On a crystal of millimeter dimensions, S/s_0 will be no greater than 10^{14}, and will generally be more like 10^9, so that the preexponential factor in Eq. (3.23) will be 10^{27}–10^{22}. Thus $\ln \alpha$ must be at least $(\varphi/kT)^2/90$ for a growth of about 1 μ/mo. Using reasonable values of kT this requires p/p_0 to be not less than 25–50%. Volmer and Schultze (1931) showed that, for supersaturations above 1%, the growth rate of iodine was linear with supersaturation and growth proceeded at tenths of millimeters per hour. Similar high rates at low supersaturation and linear dependence of rate on supersaturation have been observed in a variety of materials. Analysis similar to the above was made as early as 1935 by Becker and Döring (1935), and by others. However, it was Cabrera and Frank who pointed out that, if a spiral dislocation were present on the low-index face, island nucleation would not be required and finite rates at low supersaturation together with linear dependence of rate on supersaturation would be expected.

When a screw dislocation is present the growth process will consist of the rotation of the step about the point where it joins the dislocation. The crystal will thus consist of a single helicoidal layer, as shown in Fig. 3.4. Of course, a single-crystal surface may have several screw dislocations terminating on it and they may grow independently or interact depending on their Burgers vectors and separation.

The step in a crystal face with one dislocation will be straight when the crystal is at equilibrium saturation and essentially straight at low supersaturation. If the environment is supersaturated the step will advance with a velocity v_0. Provided it remains an essentially straight step, v_0 will be given by

$$v_0 = 2(\alpha - 1)X_S Z \beta \tag{3.25}$$

where X_S is the mean distance an atom diffuses on the surface before evaporation, β is a factor usually ≈ 1, and Z is the collision frequency with which molecules from the equilibrium vapor strike a surface-lattice site.

Equation (3.25) can be justified by the following argument. One half of the atoms that strike the surface in the diffusion zone of length "X_S" closest to the step will still remain by the time they have diffused to the step. Similarly, one half of those in the zone $2X_S$ removed from the step will survive to enter the region X_S removed from the step and $\frac{1}{2} \times \frac{1}{2}$ of these will reach the step. Thus the number reaching the step from one side will be

$$\sum_{n=1}^{\infty} \frac{ZX_S}{2^n} = ZX_S \tag{3.26}$$

but because the step is "fed" from both sides it will receive $2ZX_S$ at equilibrium. At equilibrium it will also lose $2ZX_S$ atoms. If the vapor pressure exceeds the equilibrium value the rate of advance will be that given by the excess of atoms arriving over those departing, which introduces the factor $(\alpha - 1)$ in Eq. (3.25). The factor β will depart from unity if the kinks are not close together or if the sticking probability at kinks is not close to unity.

A straight step will in time tend to become curved, as shown in Fig. 3.8. Figure 3.8(b) is a view from above of the dislocation at several times during growth. The line labeled (1) in Fig. 3.8(b) corresponds to the configuration in Fig. 3.8(a) at an early stage of growth, where a is a kink. After growth the step moves in the direction of the arrow and assumes the configuration shown by the line labeled (2). At later stages the configuration is (3), (4), and (5). The final result is a helical ramp whose configuration from above is an Archimedean spiral. The atoms attach themselves all along the ramp at the same rate but the angular velocity will be greater near the center. The rate of advance of a step with a radius of curvature ρ_A, according to Frank, is

$$v = v_0(1 - \rho_c/\rho_A) \tag{3.27}$$

where ρ_c is the radius of curvature of the critical nucleus.

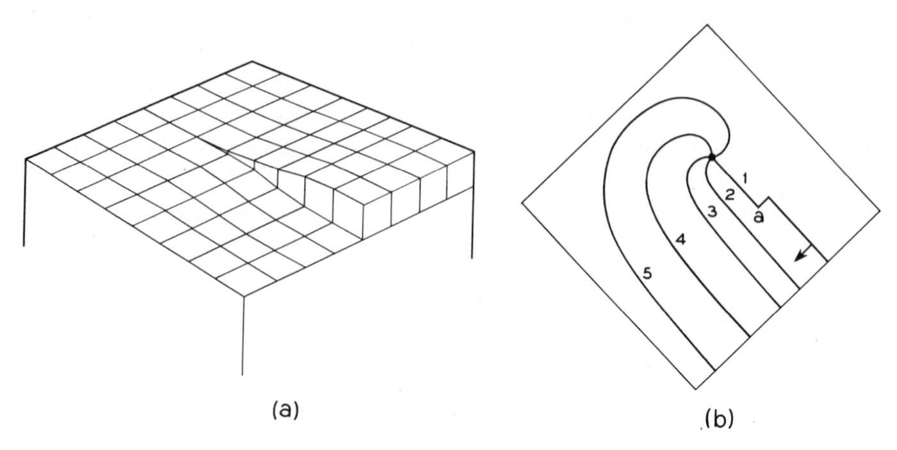

Fig. 3.8 Genesis of spiral.

The number of turns of the spiral passing a fixed point per unit time multiplied by the step height, d, will be the rate of growth, $\mathscr{R}$, for the surface under consideration.

Frank shows that the rate at low supersaturation should follow the equation

$$\mathscr{R} = \beta d Z \sigma^2/\sigma_1 \tag{3.28}$$

where

$$\sigma = \alpha - 1 \tag{3.29}$$

and

$$\sigma_1 = 10\gamma s_0/X_S kT \tag{3.30}$$

At high supersaturation,

$$\mathscr{R} = \beta d Z \sigma \tag{3.31}$$

Thus at low supersaturation the growth-rate dependence on supersaturation is parabolic, while at high supersaturation the dependence is linear.

The crossover from parabolic to linear dependence will occur when the turns of the spiral are close enough to compete for molecules from the vapor, that is, when they are closer than $2X_S$. In the case of iodine the expected crossover is calculated to take place at about 10% supersaturation and this calculation is in agreement with Volmer and Schultze's experimental results.

3.8 Movement of Steps

The foregoing has applied to growth at steady state. The theory of growth before steady state is achieved and the stability of growth rate to minor perturbations is treated in the so-called "kinematic" theory of crystal growth (Cabrera and Coleman, 1963).

Cabrera has developed a kinematic theory of step motion, adopting the ideas advanced by Lighthill and Whitham (1955) for the study of traffic flow on highways. When the rate of growth depends on step density, the theory describes the crystal profile and rate of change of step spacings as growth proceeds. In much the same manner in which traffic bunches on a busy highway, steps will bunch on a growing face. Such bunches account for the macroscopically observed step heights on grown crystals, as shown in Fig. 3.9. Attempts have been made to treat cases where the rate depends on factors other than step density, such as impurity adsorption, but such cases have been difficult to analyze quantitatively. Figure 3.10(a) shows a cross section of a given crystal containing steps and bunches. The step height will usually be comparable to the lattice constant for the material, while the bunch height will be of macroscopic dimensions. It is the bunch height that we observe in growth "steps" and spirals on crystals. The step and bunch velocities are normal to the rate of growth in the usual sense and thus the rate of growth will be the average step velocity times the step height or the average bunch velocity times the average bunch height divided by either

Fig. 3.9 Phase contrast micrograph of growth spirals with circular symmetry on SiC crystal polytype 6 H. The spiral height as measured by multiple beam interference fringes is 15 ± 1A. The left half of the figure illustrates the formation of closed loops (nearly circular) due to two dislocations of opposite sign. The right half of the figure shows the hyperbolic curve of intersection of two circular spirals unequally developed (500 ×). (Courtesy of Dr. A. R. Verma.)

the average step or bunch spacing. Bunching on a surface where steps are uniformly distributed will occur only when for some reason the velocity of a step slows so that steps following it can catch up. In cases where bunches are present, steps will tend to move away from the front of a bunch [step 1 in Fig. 3.10(a)] and be added to the back of a bunch [step 4 in Fig. 3.10(a)]. This is because the step velocity will be dependent on the area that "services" it, that is, the area from which molecules can two-dimensionally diffuse to the step. Because step 1 has a large adjacent terrace it can move away from steps 2 and 3, which have small adjacent terraces. Step 4 has a large terrace but cannot overtake the slower-moving step 3.

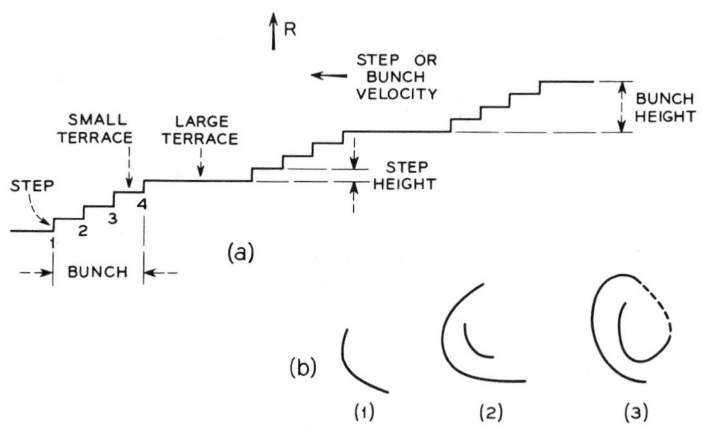

Fig. 3.10 (a) Cross section of growing crystal containing steps and (b) coalescence of bunches to form a spiral.

If for some reason a step is slowed down it will collect a bunch behind it. One mechanism that can slow down a step is the presence of adsorbed impurities on the surface.

It should be pointed out that the presence of a spiral on the surface of a crystal is not proof that it grew from the beginning by means of a screw dislocation. Lang (1957) has pointed out that a spiral step may be generated by the coalescence of two bunches, as shown in Fig. 3.10(b). In Fig. 3.10(b), step 1, a bunch has formed by the coalescence of a number of steps. In step 2, a second bunch is formed. Both bunches exhibit curvature, perhaps caused by the fact that they are not able to advance uniformly across the surface because of local supersaturation variation or the presence of adsorbed impurities. In step 3, the second bunch has coalesced with the first to form a spiral.

The above mechanism explains the presence of spirals on crystals grown at such high supersaturation that island nucleation can occur easily and a repeatable step is not required. Once such a spiral is generated it may persist and act as a repeatable step.

Conventional screw dislocations are thought to be initiated at low supersaturation by dust and impurities. Jackson (1962) has pointed out that the supersaturation of vacancies occurring during ordinary crystal growth is not sufficient to nucleate dislocation loops, partial-dislocation loops, or vacancy discs. However, by quenching from near the melting point, sufficient vacancy supersaturation is produced to nucleate dislocations. Thus if dislocations are operative during growth and seeds that have not been quenched are used, the dislocations cannot have been formed by vacancy aggregation. Soluble impurities are possible nucleators of dislocations. However, the dislocation density in carefully grown pure crystals is similar to that in carefully grown crystals where soluble impurities are present in concentrations high enough to cause constitutional supercooling and even cellular substructure (see Sec. 3.13 for a discussion of cellular substructure). Thermal stresses may nucleate dislocations but crystals are observed to grow by a dislocation mechanism even in extremely low thermal gradients, so thermal stresses are probably not the only nucleating agents. By elimination, Jackson concludes that, in those systems where no other nucleating agent is possible, dislocations are probably nucleated by foreign particles, such as dust. Indeed, because it is virtually impossible to reduce the dust concentration sufficiently it is probable that homogeneous nucleation in crystal growth is extremely rare and that most nucleation is heterogeneous on dust.

3.9 Nucleation

The formation of a crystal in the absence of a seed involves nucleation. If the nucleation occurs on a surface, such as a wall of the system, or on a foreign body, such as a dust particle, it is heterogeneous nucleation. If the structure and interatomic spacing of the surface on which nucleation takes place approximate those of the crystal, growth on the surface can resemble growth on a normal seed. This is called *epitaxial* growth.† If the nucleation occurs in the absence of a surface, that is, in the bulk of a liquid, it is homogeneous.

It is well known that the vapor pressure of a small droplet and the solubility of a small particle exceed the equilibrium values for macroscopic quantities of the liquid or solid.

If the radius, r, of a spherical drop is increased an infinitesimal amount, dr, by the addition of dn moles of liquid from a planar surface by distillation, the increase in surface area is

---

†Present common usage also includes epitaxy to mean the growth of a thin layer of a material on a seed of the same material. In particular, epitaxial growth is often used to mean growth by vapor-phase reaction of a thin layer on a seed.

$$4\pi(r + dr)^2 - 4\pi r^2 = 8\pi r dr \tag{3.32}$$

The increase in surface free energy is

$$\Delta G_S = 8\pi r dr \sigma \tag{3.33}$$

where σ is the surface tension. The free energy of transfer, ΔG_T, of dn molecules in the droplet is

$$\Delta G_T = RT \ln \frac{P_d}{P_p} dn \tag{3.34}$$

where P_d is the vapor pressure of the drop and P_p is the vapor pressure at a planar interface. At equilibrium $\Delta G_S = \Delta G_T$,

$$dn = \frac{4\pi r^2}{v} dr \tag{3.35}$$

where v is the molar volume.

Equating Eqs. (3.32) and (3.34) and substituting for dn, we obtain the Kelvin equation (Van Hook, 1961), which relates the vapor pressure, P_d, of a droplet to the vapor pressure, P_p, of a planar surface:

$$RT \ln \frac{P_d}{P_p} = \frac{2\sigma v}{r} \tag{3.36}$$

The analogous equation for the dependence of solubility on particle size is

$$RT \ln \frac{s_d}{s_p} = \frac{2\sigma v}{r} \tag{3.37}$$

where s_d is the particle solubility, s_p is the solubility of a planar surface, σ is the interfacial tension, v is the molar volume, and r is the particle radius. Because the interfacial tension will be directionally dependent, the rigorous expression for Eq. (3.37) should take this into account together with corrections for edges and corners in a nonspherical particle. Particle size must be quite small before these effects become appreciable. For instance, for water [from Eq. (3.36)], when $r = 10^{-4}$ cm, $P_d/P_p \simeq 1.001$, and when $r = 10^{-6}$ cm, $P_d/P_p \simeq 1.11$. The result of Eqs. (3.36) and (3.37) is that small particles are metastable and if large particles are present in a system they will grow at the expense of small particles.

In some cases nucleation can be looked on as the aggregation of particles, and the precursors of nuclei that are large enough to grow will be formed by association of particles in the system. Thus if A is the material nucleating in some medium there is a family of reactions represented by

$$m\text{A} \rightleftharpoons \text{A}_m \tag{3.38}$$

for different values of m. If m and the equilibrium constant for the reaction are large enough, the nucleus will tend to grow instead of redissolving. Aggregates too small to grow in the average supersaturation in the medium are sometimes called *embryos*. Most embryos never survive to grow but because of statistical fluctuations of supersaturation or because energy is added locally to assist their growth (scratching the beaker with the stirring rod, using ultrasound, etc.) a few survive to grow. The higher the supersaturation, the larger the average embryo size and the greater the chance that statistical fluctuations will allow nucleation. Given enough time there is a statistical chance that nucleation will occur in any supersaturated medium. Of course, in the presence of dust, nucleation is heterogeneous and comparatively easy. Indeed, the scratch of the stirring rod may provide glass particles and aid nucleation by this means. Some form of particulate contamination appears to be invitable in most actual nucleation.

3.10 Effect of Rate on Distribution Constant

When a material is crystallizing in the presence of a dopant or impurity (or from any polycomponent system), as was discussed in Sec. 2.5, segregation will occur and the equation

$$k_0 \equiv \frac{a_{s(\text{eq})}}{a_{l(\text{eq})}} = \frac{c_{s(\text{eq})}}{c_{l(\text{eq})}} \tag{3.39}$$

is obeyed, where k_0 is the equilibrium distribution constant (or coefficient), $a_{l(\text{eq})}$ is the equilibrium activity of the impurity in the liquid, and $a_{s(\text{eq})}$ is the equilibrium activity of the impurity in the solid. In most cases (particularly at low concentrations) activities can be replaced by concentrations. Even at moderate rates of growth the concentration of impurity close to the growing crystal is usually different from that in the bulk of the system so that it is convenient to define the effective distribution constant (or coefficient), k_{eff} (see Thurmond, 1959; Pfann, 1966, and references therein), as

$$k_{\text{eff}} = \frac{c_{s(\text{act})}}{c_{l(\text{act})}} \tag{3.40}$$

where $c_{s(\text{act})}$ and $c_{l(\text{act})}$ are the actual impurity concentrations in the solid and liquid. Burton et al. (Thurmond, 1959) have shown that

$$k_{\text{eff}} = \frac{k_0}{k_0(1 + k_0)e^{-(\mathscr{R}t/D)}} \tag{3.41}$$

where $\mathscr{R}$ is the growth rate on the face under consideration, t is the thickness of the diffusion layer, and D is the diffusion constant at any given time. Equation (3.39) will be obeyed even when growth is rapid if we substitute for $a_{l(\text{eq})}$ the activity or, as an approximation, the concentration in the liquid close to the growing interface. Thus if impurity is rejected from the growing crystal ($k_0 < 1$), $c_{l(\text{act})}$, close to the growing interface, will be greater than the concentration in the bulk. Consequently $k_{\text{eff}} > k_0$ and, as the rate increases, the concentration of impurity near the interface will increase and k_{eff} may approach 1. Similar reasoning for the case of $k_0 > 1$ leads to $k_0 > k_{eff} > 1$ and $k_{eff} \rightarrow 1$ as the rate increases. In general, the higher the viscosity, the lower the diffusion constant; if the solution is violently stirred, if the impurity concentration is low, or if the diffusion constants for impurity and crystal species are large, the width of the diffusion layer and the concentration gradients across it will be smaller.

It should be pointed out that if an adsorbed layer of dopant or impurity exists on the growing interface, $c_{l(act)}$, close to the growing interface, will be greater than the concentration in the bulk and $k_{eff} > k_0$ (where k_0 is defined neglecting any adsorbed layer), regardless of whether $k_0 < 1$ or $k_0 > 1$.† As the rate is increased, concentration gradients caused by diffusion will be present and can affect the concentration in the adsorbed layer. The behavior of k_{eff} as a function of rate can be quite complex under such conditions.

3.11 Constitutional Supercooling

When growth is from a polycomponent system there will be a buildup of the components that tend to be rejected from the solid close to the growing crystal interface. That is, if k, the distribution constant for a given component, is less than 1, the concentration close to the growing interface will be greater than the bulk concentration. Conversely, if k for a given component is greater than 1, there will be a depletion in that component close to the growing interface. It will be necessary for the component to diffuse through the layer to the bulk if $k < 1$ or to diffuse from the bulk if $k > 1$. Usually, in a polycomponent system, $k > 1$ for the species forming the crystal and $k < 1$ for the solvent. Other impurities may have any value of k. If crystal growth is so slow as to be almost at equilibrium, the effect of diffusion is negligible. In many cases, however, the diffusion problem is severe. If we plot concentration of an impurity vs. distance in a system where diffusion is important, as in Fig. 3.11, we see that, because of the buildup of impurity, c_i, the concentration at the growing interface, is greater than the concentration in the bulk of the liquid, c_l. Figure 3.12 shows the phase diagram of the crystal-impurity system. Using the melting-point data of Fig. 3.12 and the composition data of Fig. 3.11, Fig. 3.13, which shows the melting points of the compositions existing near the growing crystal as a function of distance, can be constructed. As can be seen, the melting point of the solution decreases as the interface is approached. The dashed lines AB and $A'B$ are two different possible temperature gradients in the solution. For the larger gradient, AB, at all positions in front of the interface, there is no supercooling; for the smaller gradient, $A'B$, the region CB is supercooled and crystalli-

†For further discussion of adsorption see Sec. 2.5.

zation will tend to take place in front of the solid interface. The region *CB* is referred to as *constitutionally supercooled* (Rutter and Chalmers, 1953) and, as can be seen, large temperature gradients, such as can be obtained easily in crystal pulling, will lessen the tendency toward constitutional supercooling.

If the growing interface is not bounded by a surface (or surfaces) close in orientation to a surface that would bound the crystal if it grew close to equilibrium, the problems associated with constitutional supercooling will be even more severe. Where the growing interface is not an "equilibrium form" of the crystal, it will usually change to a surface composed of facets (see below) whose faces are equilibrium faces (or at least faces of lower

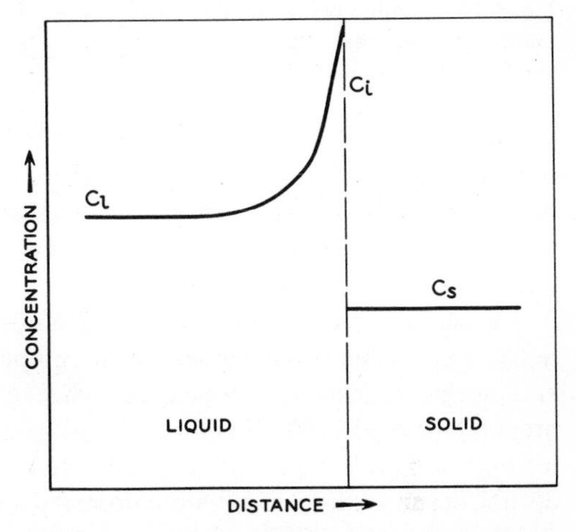

Fig. 3.11 Concentration vs. distance—constitutional supercooling.

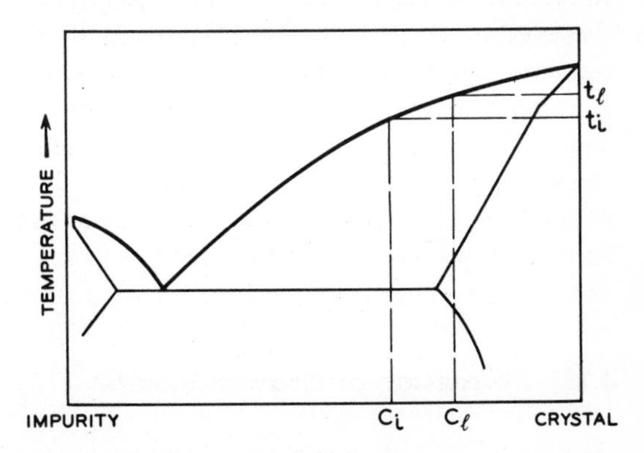

Fig. 3.12 Crystal-impurity phase diagram—constitutional supercooling.

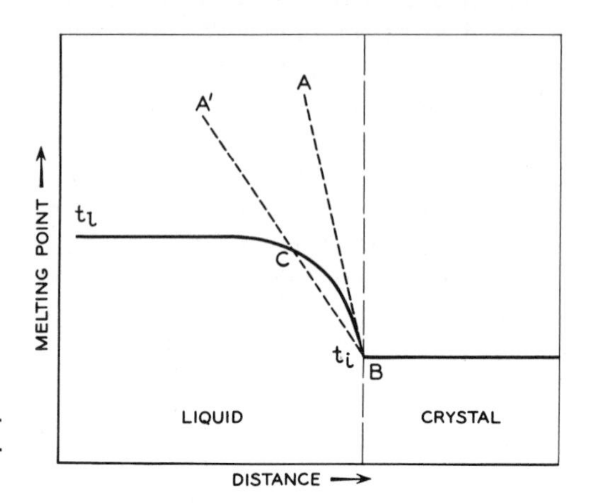

Fig. 3.13 Melting point vs. distance—constitutional supercooling.

specific interfacial surface free energy). Under these conditions, the peaks of some of the facets will be in regions where the supercooling, or supersaturation, is greater than the valleys. The peaks will grow more rapidly with a tendency toward runaway dendritic growth. This is the usual cause of unintentional dendritic growth.

Constitutional supercooling is ordinarily not important in pure or nearly pure melts (for instance, in the preparation of most semiconductors and inorganic oxides). However, at high levels of doping [for instance, in the preparation of Esaki (1958) diodes] and in the preparation of incongruently melting inorganic oxides and compounds whose maximum melting points do not occur at the desired stoichiometry and whose liquid and solid compositions are quite different, diffusion problems and associated constitutional supercooling are likely to be troublesome. If a melt is stirred, the width of the diffusion layer is narrowed and constitutional supercooling is less likely. However, as Hurle (1961, and references therein) has shown, constitutional supercooling can occur even in stirred melts. When $k_0 < 1$, the concentration of impurity or dopant close to the growing interface builds up to a steady-state value, c_0, as growth progresses. Extension and modification of the treatment of Burton et al. (Thurmond, 1959) for the dependence of k_{eff} on stirring speed led Hurle (1961) to equations relating the onset of constitutional supercooling to stirring speed.

3.12 Formation of Growth Facets

The genesis of facets appears in some cases to depend on selective interfacial adsorption on a particular crystallographic face (Tiller, 1963). When adsorption occurs, the specific interfacial surface free energy, σ, decreases, because the process would not take place unless ΔG were negative. If σ is reduced at a particular orientation, this will correspond to a cusp in the Wulff plot in that direction and thus, according to the arguments of Sec. 3.3, there will exist a flat surface more stable than the hill-and-valley surface that would have bounded that direction. In growth from the melt, particularly by crystal pulling, the solid–liquid interface is often a curved surface, such as Fig. 3.14(a), because the isotherm corresponding to the melting point is curved. If $\langle a \rangle$ is a direction where adsorption can take place, then a flat "facet" *a–b* will form in that direction.† The size of the flat is such that $\sigma \cos \theta = \sigma_a$, where σ is the surface free energy without adsorption and σ_a is the surface free energy with adsorption. The area of the facet decreases as shown in Fig. 3.14(b) as the solid–liquid interface becomes more convex. The area of the facet will also decrease when the angle between the reference axis and $\langle a \rangle$ decreases. If the temperature isotherm is concave, facets are difficult to form because they would be superheated.

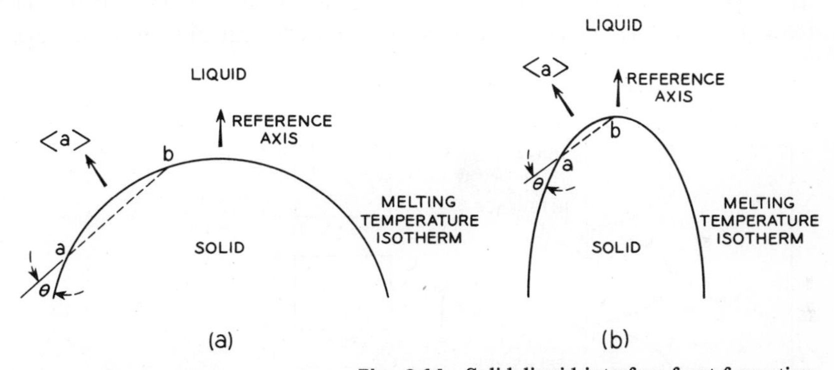

Fig. 3.14 Solid–liquid interface-facet formation.

The adsorption mechanism for facet formation predicts a higher impurity uptake when growth occurs on facets. This is usually observed. However, the connection between impurities and facets may not always be this straightforward. Dikhoff (1960, and references therein) has studied the {111} facets in germanium and the increase in impurity uptake associated with the formation of these facets in Czochralski growth. He agrees that k_{eff} (when $k_{eff} < 1$) will increase if growth is at a lower temperature, as would be required if a higher supercooling were needed to produce growth on a facet face. [The increase in k_{eff} should be apparent from a consideration of typical phase diagrams, such as Fig. 2.3(a) or (b).] Although this mechanism may

†A facet may also form at *a–b*, as described in Sec. 3.3-1. (p. 85), without impurities. Once formed, it may have a larger distribution constant for impurities than neighboring directions.

be operative in some cases, the adsorption mechanism is believed (Dikhoff, 1960) to be more general. Facet formation may also interact with constitutional supercooling and further investigations of their interconnection are clearly needed.

3.13 Formation of Cells (Tiller, 1963)

When constitutional supercooling takes place, the stable configuration may be either cellular or dendritic. If a curved projection forms on a flat interface when constitutional supercooling is present, such as shown in Fig. 3.15, then the tips of the projections, *D*, will tend to project into the solution to the point *D*, where the temperature of the solution is equal to the melting point. The temperature of the solution is indicated by *AB* and the melting point by *CB*; that is, the interface will try to grow into the solution to remove the constitutional supercooling (the region *DB* is constitutionally supercooled). Now, because the surface is no longer planar, lateral diffusion can furnish

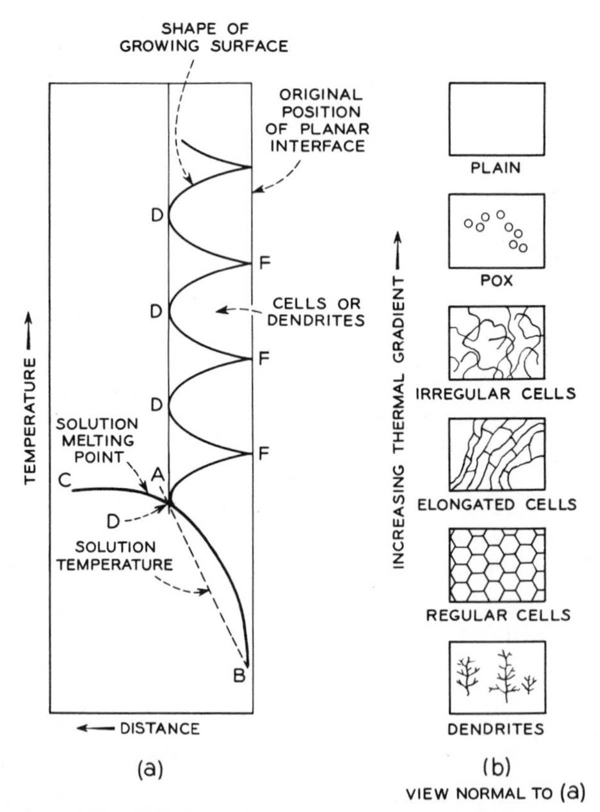

Fig. 3.15 Cellular growth.

solute, to remove constitutional supercooling in the regions *DF*. The shape of the crystal in the *DF* regions will adjust itself so that lateral diffusion will be sufficient to remove constitutional supercooling. Thus the composition will adjust itself along *DF* so that it will just equal the melting point dictated by the temperature profile in that region. The cell shape will depend on the thermal gradient, the diffusion field (concentration difference and diffusion constant), and the surface free energies of the various orientations of the solid–liquid interface. According to Tiller (1963), the surface will often exhibit the morphologies shown in Fig. 3.15(b) as the thermal gradient is increased. Cellular growth will tend to become fernlike or dendritic with large projections into the melt when the supercooling is very large in front of the interface. Frost patterns on windowpanes are excellent examples of dendrites.

The formation of dendrites and cells can occur in growth from the melt when impurities are present in aqueous and other solution growth, and in growth from the vapor where a diffusion field exists. The older literature sometimes refers to nonplanar faces as cobbled or containing growth hillocks. On occasion the growth hillocks are bounded by planar faces.

3.14 Correlations with Entropy

Jackson (1967) has shown that the shape of growth surfaces in macroscopic crystals may be correlated with their entropy of fusion, ΔS ($\Delta S = L_f/kT_m$, where L_f is the latent heat of fusion, k is the Boltzmann constant, and T_m is the melting point). Four classes of materials can be distinguished on the basis of ΔS:

1. $\Delta S < 1$, growth isotropic, cells and dendrites occur only with impurity addition
2. $2 < \Delta S \lesssim 4$, facet growth, faceted "pseudo dendrites" with sufficient impurity addition
3. $\Delta S \gtrsim 4$, facets always observed
4. $\Delta S \gtrsim 10$, spherulitic growth usual

Jackson's correlation is plausible because, in essence, it states that the smaller the ΔS, the smoother a growth surface will be. High ΔS suggests high randomness or roughness in growth surfaces with a resultant tendency toward dendrites, facet effects, and spherulitic growth. *Spherulitic growth* (or growth of many crystals radially outward from a point) is most commonly observed in polymers, facet growth in semiconductors, dendrites in metals, and isotropic growth in inorganic and short-chain organic materials. If ΔS^*, the entropy for the actual growth process, were known, it might be better used for correlations than ΔS, the entropy at equilibrium.

REFERENCES

Becker, Von R., and W. Döring, *Ann. Physik* **24**, 719 (1935).

Burton, W. K., and N. Cabrera, *Discussions Faraday Soc.* **5**, 33 (1949).

Cabrera, N., and R. V. Coleman, *The Art and Science of Growing Crystals*, Ed. by J. J. Gilman, Wiley, New York, 1963, p. 3.

Cabrera, N., and D. A. Vermilyea, *Growth and Perfection of Crystals*, Ed. by R. H. Doremus, B. W. Roberts, and David Turnbull, Wiley, New York, 1958, p. 393.

Crank, J., *Mathematics of Diffusion*, Oxford Univ. Press, New York, 1956.

Dikhoff, J. A. M., *Solid State Electron.* **1**, 202 (1960).

Donnay, J. D. H., and D. Harker, *Am. Mineralogist* **22**, 446 (1937).

Esaki, L., *Phys. Rev.* **109**, 603 (1958).

Frank, F. C., *Discussions Faraday Soc.* **5**, 48 (1949).

Frank, F. C., *Advan. Phys.* **1**, 91 (1952).

Frank, F. C., *Growth and Perfection of Crystals*, Ed. by R. H. Doremus, B. W. Roberts, and David Turnbull, Wiley, New York, 1958, p. 411.

Frost, Arthur A., and Ralph G. Pearson, *Kinetics and Mechanism*, Wiley, New York, 1953.

Gibbs, J. W., "The Collected Works of J. Willard Gibbs", Vol. 1, Longmans, London, 1928, p. 55.

Glasstone, G., K. Laidler, and H. Eyring, *The Theory of Rate Processes*, McGraw-Hill, New York, 1941.

Hartman, P., and W. G. Perdok, *Acta. Cryst.* **8**, 49, 521, 525 (1955).

Herring, Conyers, *Phys. Rev.* **82**, 87 (1951).

Hurle, D. T. J., *Solid State Electron.* **3**, 37 (1961).

Jackson, K. A., *Phil. Mag.* **7**, 1117, 1615 (1962).

Jackson, K. A., in *Crystal Growth*, Ed. by H. S. Peiser, Pergamon, New York, 1967, pp. 17ff.

Jost, W., *Diffusion in Solids, Liquids, Gases*, Academic Press, Inc., New York, rev. ed., 1960.

Kossel, W., *Nachr. Ges. Wiss. Göttingen, Math. Physik. Kl.* 135 (1927); *Ann. Physik* **21**, 457 (1934); **33**, 651 (1938).

Lang, A. R., *J. Appl. Phys.* **28** (4), 497 (1957).

Lighthill, M. J., and G. B. Whitham, *Proc. Roy. Soc.* (*London*) **A229**, 281, 317 (1955).

Pfann, W. G., *Zone Melting*, Wiley, New York, 2nd ed., 1966.

Rutter, J. W., and B. Chalmers, *Can. J. Phys.* **31**, 15 (1953).

Slater, Noel B., *Theory of Unimolecular Reactions*, Cornell Univ. Press, Ithaca, N.Y., 1959.

Thurmond, C. D., *Semiconductors*, Ed. by N. B. Hannay, Reinhold, New York, 1959, p. 145.

Tiller, W. A., *The Art and Science of Growing Crystals*, Ed. by J. J. Gilman, Wiley, New York, 1963, pp. 277ff.

Van Hook, Andrew, *Crystallization—Theory and Practice*, Reinhold, New York, 1961, p. 73, pp. 92 ff.

Volmer, M., and W. Schultze, *Z. Physik. Chem.* (*Leipzig*) **A156**, 1 (1931).

Wulff, G., *Z. Krist.* **34**, 449 (1901).

4

CRYSTAL GROWTH BY SOLID–SOLID EQUILIBRIA

4.1 Solid–Solid Growth Methods

Chapters 1–3 laid the groundwork for a discussion of the specific methods of growth, which comprises the rest of this book. The remaining chapters are arranged according to our classification of growth methods. Each chapter will give the general considerations and details of theory not covered in the introductory chapters for the methods discussed in that chapter. Following this, there will be a description of the application of the technique (including equipment necessary where equipment is unusual) to specific representative crystals and a brief discussion of other systems where growth has been achieved by the method.

This chapter will discuss solid–solid growth methods often called *recrystallization* growth methods by metallurgists and will include both mono- and polycomponent methods because the techniques and problems in solid–solid growth are similar regardless of the number of components and because polycomponent solid–solid growth has so far seen such limited applicability as to make an individual chapter on it unwarranted.

There are five principal methods of solid–solid growth:

1. Recrystallization by annealing out strain (so-called "strain-anneal" techniques)

2. Recrystallization by sintering
3. Recrystallization by polymorphic transition
4. Crystallization by devitrification
5. Recrystallization by solid-state precipitation (sometimes called growth by exsolution)

General considerations applicable to these techniques are discussed in Sec. 2.3.

The main advantages of solid–solid growth methods are as follows: They permit growth at low temperatures without the presence of additional components (see Sec. 2.1 for a discussion of the advantages of low-temperature growth); the shape of the grown crystal is fixed beforehand so that wires, foils, etc., are easily grown; and often orientation can be easily controlled. A desired orientation can be obtained by bending a specimen so that its single-crystal region bears the desired spatial relation to the axes of the specimen. A single crystal can then be formed by continuing growth of the single-crystal region into the rest of the specimen. An additional advantage is that, except in growth by exsolution, the distribution of impurities and other additional components is fixed before growth and not altered by the growth process (except slightly by the relatively slow process of solid-state diffusion). The main disadvantage is that, because crystal growth takes place in the solid, the density of sites for nucleation is high and it is difficult to control nucleation to produce large single crystals.

4.2 Recrystallization by Annealing Out Strain

4.2.1 GENERAL CONSIDERATIONS

When a force of any magnitude is applied to a solid body the body becomes distorted; that is, some part of the body moves with respect to some neighboring portion. As a result of this displacement the atomic attractive forces in the body set up restoring forces that resist the alteration and tend to restore the body to its original shape. The restoring force in a deformed body is termed *stress* and has the units of force/area. The dimensional change produced by an applied force is called *strain*. For a force that produces elongation (stretching force) or shortening (compressing force) the strain is longitudinal. Strain is sometimes expressed as change in dimension/original dimension. Thus a given stretching force per unit area produces a given *percent elongation*.

If hydrostatic pressure is applied to a body there will be a volume change and the change in volume/original volume is the *bulk strain*. If a shearing force is applied to a body the amount of slide or shear between two layers unit distance apart is the *shear strain.* Figure 4.1 shows a shearing force applied to a body. *Hooke's law* states that stress is proportional to strain within the *elastic limit,* that is, within the range of forces where the body will recover its original shape when the forces are removed.

The constant of proportionality between stress and strain for longitudinal strain is Young's modulus, Y, where

$$Y = \frac{F/A}{e/l} \tag{4.1}$$

and F is the force causing the strain ($-F$ is the stress), A is the area, e is the elongation, and l is the length of the specimen. Young's modulus is known for most materials of commerce. In high-precision calculations it must be remembered that Y is dependent on structure, perfection, surface condition, purity, temperature, direction (in single crystals), etc., but, for ordinary calculations, handbook values of Y (Lyman, 1961, 1964; Shaffer, 1964; Samsonov, 1964) are satisfactory.

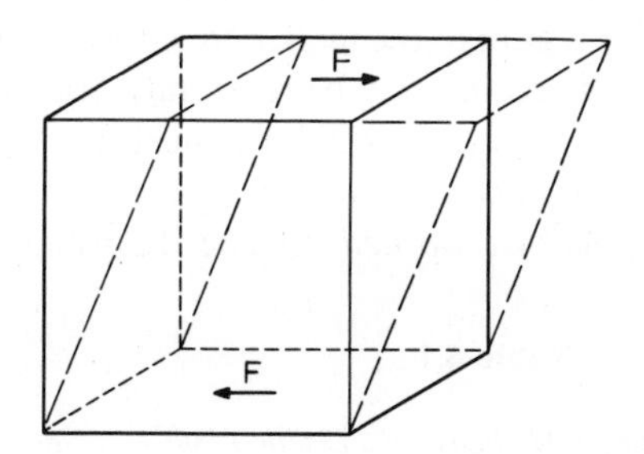

Fig. 4.1 Shearing forces on a body.

If one plots F vs. e for a material, Hooke's law will break down when the elastic limit is exceeded, as shown in Fig. 4.2 at point a. Beyond this point the material does not regain its original shape when the tension is removed, so that it is *plastically deformed.* Beyond the point b, the *yield point,* the sample begins to lengthen considerably and the tension must be reduced in order to maintain any given value of strain. After a certain stretch has occurred, at the point c (a, b, and c are usually closer together than shown on Fig. 4.2) the material is able to withstand larger tensions until d is reached, after which the material parts at e. The strain for strain-annealing crystal growth is usually in the region a–b and for strain annealing to be practical a–b must be large enough in extent and far enough from e to be experimentally attainable. The stress in the region a–b (a function of Y) must also be large enough to provide sufficient driving force for recrystallization. Table 4.1 lists some approximate room-temperature values for Young's modulus, the elastic limit, and the breaking point for common materials. When other data are unavailable, the tensile strength is a fair measure of the elastic limit. As is common with nonmetallic materials (notable exceptions are silver chloride

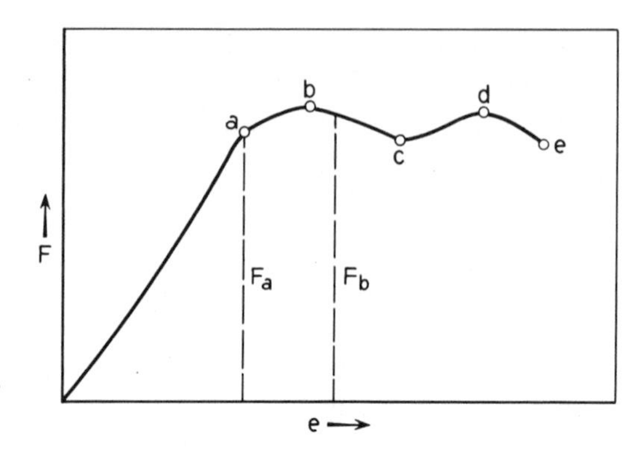

Fig. 4.2 Elongation vs. force for a tensile stress.

and thallium chloride), the breaking point is so close to the elastic limit that straining the crystal permanently is not practical.

Most crystal growth by strain annealing has been of metals. Typically metals are formed by casting the molten metal into a mold. The resultant solid is a polycrystalline mass of grains. Nucleation occurs either randomly at the walls or in some particular region of the melt, depending on the thermal gradients during cooling. Thus the grains may be randomly oriented or there may be considerable preferential orientation. If bar, sheet, plate, wire, etc., stock is made from the ingot, the next step is to deform the metal. When a material is mechanically deformed or *worked*, plastic deformation occurs,

Table 4.1

Material	*Young's Modulus* (lb/in.^{2})	*Elastic Limit* (lb/in.^{2})	*Breaking Point* (lb/in.^{2})
Aluminum	10×10^6	19×10^3	22×10^3
Copper	18×10^6	10×10^3	50×10^3
Iron	30×10^6	23×10^3	50×10^3
Mild steel	27×10^6	27×10^3	55×10^3
Al_2O_3 (sapphire)	$60^+ \times 10^6$	38.7×10^3†	—‡
MgO (periclase)	30×10^6	14×10^3†	—‡

†Tensile strength = maximum load reached/original cross-section area. In most brittle materials, such as sapphire and periclase, the tensile strength is probably quite close to the breaking point but somewhat greater than the elastic limit.

‡So close to the elastic limit as to be indistinguishable.

the grain shape is changed, dislocations and, in some cases, twinning and slip are introduced. Very often substantial changes in the strength and hardness of the material occur, particularly if the work occurs at a temperature below that where recrystallization takes place at an appreciable velocity. *Cold working* thus causes *work hardening*. Among the common techniques used for working metals are rolling, drawing, forging, and extruding. Figure 4.3 shows the structure of a test piece after it has been drawn to a wire. Many studies of recrystallization and grain growth have been made on aluminum because its low melting point (660°C) leads to a low-recrystallization-temperature range and permits experimentation at convenient temperatures. Grains are generally revealed in metal surfaces by etching. Hydrofluoric acid is a convenient etch for revealing the grain structure of Al. Figure 4.3 shows that the grains in the wire have been oriented by the drawing process so that their long axis is in the drawn direction. If a material is strained in a particular direction the crystallites will usually be elongated in the same direction.† In addition, the strain will usually tend to partially orient the distorted crystallites. The glide planes and glide directions in crystallites tend to become parallel to the deformation direction. Such preferential orientation is called *texture*. Texture will often persist through subsequent recrystallization.

Fig. 4.3 Effect of drawing 4% Mo-79% Ni-17% Fe alloy into wire—x-ray transmission pinhole pattern. If there were no preferential orientation (texture), each Debye ring would be of equal intensity about its circumference. (Courtesy of G. Y. Chin.)

One way of achieving a high degree of orientation in a polycrystalline mass is by working a monocrystal. The work will cause the monocrystal to become polycrystalline but the polycrystals will have orientations close to one another because their orientation is influenced by the orientation of the original monocrystal. If one works a monocrystal so that stresses are applied essentially in only one or a few directions the resulting polycrystals will be highly oriented. Highly oriented polycrystalline samples are said to have *sharp texture*.

After plastically deforming a specimen, a material has a large amount

†Etching studies reveal this elongation in the grains. However, such elongation is not proof of orientation. X-ray studies are required to establish texture.

of strain and consequently a large amount of stored "strain energy." Equation (2.9) indicates that the change in free energy in straining a crystal is approximately equal to the work done minus the energy released as heat. This free energy is usually the principal driving force for recrystallization in strain annealing.

A large part of the strain energy resides in dislocation arrays associated with grain boundaries. Grain boundaries also contribute excess free energy because of their surface free energy. In the same way that the solubility of small particles is high and the vapor pressure of small droplets is high, the surface free energy of small crystallites is high. This effect will, however, be of importance as a contributor to the driving force for recrystallization only if the crystallite size is quite small. In addition, grain-boundary energy will depend on the orientation of the two crystallites that form the boundary with respect to each other. Crystallites with low energy because of their orientation with respect to their neighbors will tend to grow at the expense of those unfavorably oriented. Thus ΔG, the driving force for recrystallization by strain annealing, is given by

$$\Delta G = w - q + G_s + \Delta G_0 \tag{4.2}$$

where w is the work done either in deliberate straining or fabrication (most of w resides at grain boundaries), q is the energy released as heat, G_s is the surface free energy of the grains, and ΔG_0 is the difference in free energy between the grain orientation existing in the specimen and the free energy of some other orientation. By decreasing grain-boundary area a material decreases its excess free energy. A strained specimen is thermodynamically unstable with respect to an unstrained specimen. At room temperature the rate at which materials relieve strain is usually very slow. However, if the temperature is raised to increase the atomic mobility and the amplitude of lattice vibrations, the rate of strain relief is markedly increased. The purpose of annealing is to accelerate strain relieving. Thus during annealing, grain size increases, *primary recrystallization* takes place, and this process will be accelerated by an increase in temperature.

Additional factors that are important in the ease of grain growth are the coherence of atoms across boundaries that are growing and the nature of the impurities present in the lattice and in the boundaries. The effect of these factors is hard to estimate in any quantitative way. However, it is clear that, because atoms must move for grain growth to take place, growth will be easier (all other things being equal) across a boundary where the "register" or coherence of atoms is good so that only small movement is required. Similarly, impurities that inhibit motion either in the lattice or across the boundary will hinder growth.

In some cases sufficient strain will have been introduced during the commercial fabrication of a material so that annealing will result in considerable grain growth. More often than not, however, if really large crystals are desired it will be necessary to strain the material in the laboratory and then anneal it.† In fact several strain-anneal cycles are usually required. Usually if one studies grain growth as a function of annealing time (Fig. 4.4), grain growth will fall off rapidly with time because the driving force decreases rapidly with increasing grain size. Behavior of this sort is called *primary recrystallization, normal or continuous grain growth, normal coarsening*, or *coalescence*.

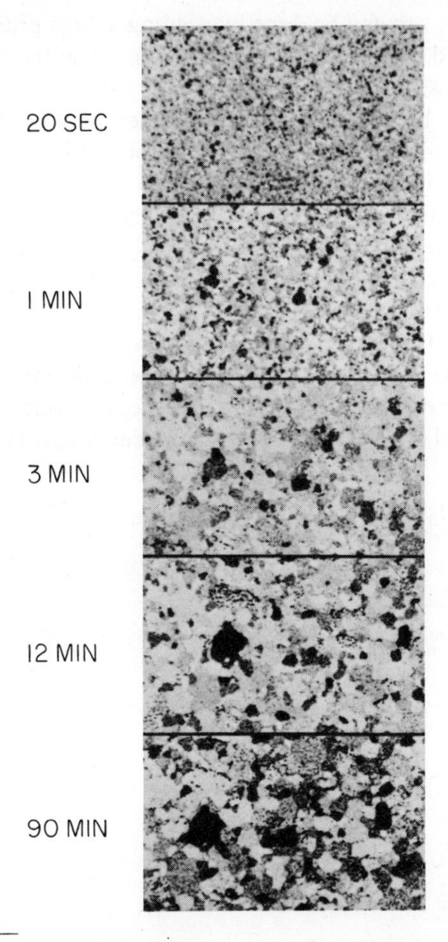

Fig. 4.4 Grain growth as a function of annealing time in aluminum (after Burgers, 1963). (Courtesy of John Wiley & Sons, Inc.)

†If no deliberate straining is carried out, then the driving force for recrystallization includes only components caused by fabrication-introduced strains and components caused by the surface free energy, small particle size, and orientation effects. Growth during the annealing of such a specimen probably is properly considered growth by sintering, although the term is usually reserved for nonmetals where fabrication-introduced strains are small.

Sometimes, however, a few grains will continue to grow at the expense of others during prolonged heating. Under such conditions the whole specimen may become composed of only two or three large grains. This behavior is called *secondary recrystallization*. The fine-grained matrix from which such growth starts is obtained by primary recrystallization. Growth of a few large crystals is sometimes called *exaggerated, discontinuous*, or *abnormal* grain growth and is discussed further below. Grain growth may occur by the growth of existing grains during annealing or by the nucleation of new grains that grow during annealing.

Seeding of solid–solid growth can be done by welding a large grain onto a polycrystalline specimen and causing the large grain to grow at the expense of the adjacent smaller grains.

Grain growth takes place by grain-boundary movement, not by the capture of very labile atoms or molecules, as in liquid–solid or gas–solid growth. The driving force is the reduction of the extra free energy stored in the grain boundary. Thus grain-boundary movement acts to shorten boundaries. The boundary energy may be thought of as a sort of interfacial tension between the crystallites, and coarsening of the grains reduces this tension. It should be apparent from this that the growth of a grain from many very small grains will be rapid, but that the growth rate of a grain from a few only slightly smaller grains will be negligible. In growth of a large grain at the expense of small grains, as shown in Fig. 4.5, if $\sigma_{S\text{-}S}$ is the interfacial tension between the small grains and $\sigma_{S\text{-}L}$ between the small grains and the large grain, then for growth,

$$\Delta A_{S\text{-}L}\sigma_{S\text{-}L} < \Delta A_{S\text{-}S}\sigma_{S\text{-}S} \tag{4.3}$$

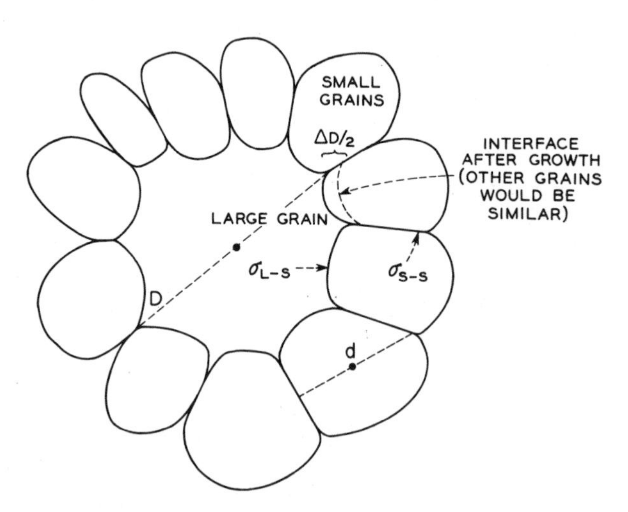

Fig. 4.5 Schematic of grain growth.

where $\Delta A_{S\text{-}S}$ is the change in area of the grain boundaries between small grains and $\Delta A_{S\text{-}L}$ is the change in area of the grain boundaries between the small grains and the large grain. If we assume that the grains are roughly circular and that the diameter of the large grain is D, then

$$\Delta A_{S\text{-}S} = \frac{\Delta D}{2} n \tag{4.4}$$

$$\Delta A_{S\text{-}L} = \pi \Delta D \tag{4.5}$$

where n is the number of small grains in contact with the large grain. Now, if $\bar{d}$ is the small-grain average diameter,

$$n \simeq \frac{\pi(D + \bar{d}/2)}{\bar{d}} \simeq \frac{D}{\bar{d}} \tag{4.6}$$

because the numerator is the circumference of the circle that is the locus of the centers of the small grains, and because $D \gg \bar{d}$. Substituting the appropriate quantities in Eq. (4.3),

$$D > \frac{2\sigma_{S\text{-}L}\bar{d}}{\sigma_{S\text{-}S}} \tag{4.7}$$

for growth to take place. Smith (1951) and Burgers (1963) have also derived Eq. (4.7).†

The foregoing assumes no directional dependence of the interfacial energies. Grain boundaries will have σ's depending in detail on the orientation of the grains that they delineate and on the orientation of the boundary relative to the grains. Boundaries may be large-angle or small-angle and they may involve a twist or a tilt between the grains. In the growth of large crystals one is interested in boundary mobility whose rate, $\mathscr{R}$, is given by the relation (Burgers, 1963):

$$\mathscr{R} \propto (\sigma/R)M \tag{4.8}$$

where R is the radius of curvature of the boundary, σ is the interfacial energy, and M is the mobility. The boundary area is decreased when movement is toward the direction of the radius of curvature, as shown in Fig. 4.6. Depending on the boundary and grain geometry, boundary motion may involve slip or glide or may require dislocation movement. If individual atoms must move, the process will be slow unless the temperature is an appreciable fraction of the melting point.

Secondary recrystallization is favored under conditions where there is a fine-grained strong texture containing a few larger crystals with slightly different orientation. If a material has pronounced texture, most of the crystals are preferentially oriented. Thus the driving force for the recrystallization is furnished by the removal of strain, by the size difference, and by

---

†The remainder of this section is based to a considerable extent on the reviews of Burgers (1963) and Aust (1963).

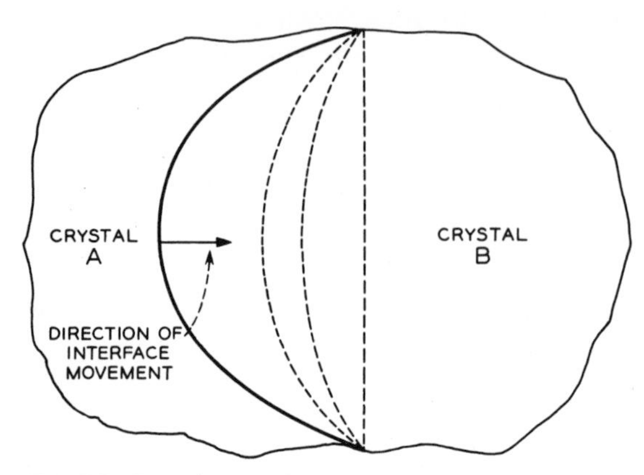

Fig. 4.6 Boundary motion related to curvature of boundary (after Burgers, 1963). (Courtesy of John Wiley & Sons, Inc.)

the orientation difference of the crystals that grow, because, in Eq. (4.2), w, G_s, and ΔG_0 are all large. In particular, even after primary recrystallization has taken place, G_s and ΔG_0 will still be large enough to provide a substantial driving force. Pronounced texture will assure that only a few crystals will have the orientational driving force.

It is probable that grain growth can occur in many cases without need of nucleation. The nuclei for growth are the already present grains. Indeed one of the problems in controlling growth by strain annealing is preventing growth at the many potential growth centers. However, under some conditions, it is observed that nucleation of new grains occurs during annealing and it is these grains that subsequently grow at the expense of their neighbors. One way of looking at the situation is to consider the lattice regions, which might eventually serve as nuclei, as the solid-state analogue of embryos. Making a particular region (embryo) grow to a size where it will act as a nucleus is required. The driving force for this growth in ordinary-sized grains is caused by orientation and size differences. In the deformed matrix near or in a grain boundary where nucleation usually occurs, the additional driving force caused by internal-energy differences brought about by dislocation density differences is important. Thus regions relatively free of dislocation networks will grow at the expense of regions with high-dislocation concentrations. In a polygonized† specimen there are regions different in orientation from their neighbors and free of dislocations that may serve as fast-growing embryos. Indeed the incubation time required for nucleation in some systems is thought to be the time required for dislocation nucleation in a strained region to produce polygonization. Figure 4.7 shows a schematic of grain formation by nucleation in a grain boundary. Figure 4.8 shows polygoniza-

†See Section 1.3.3 for a discussion of polygonization.

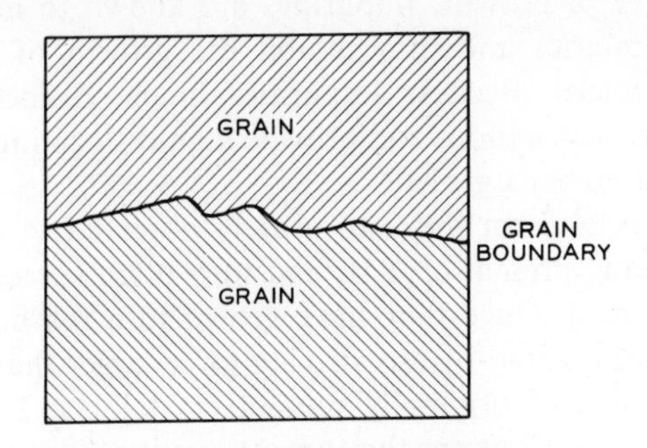

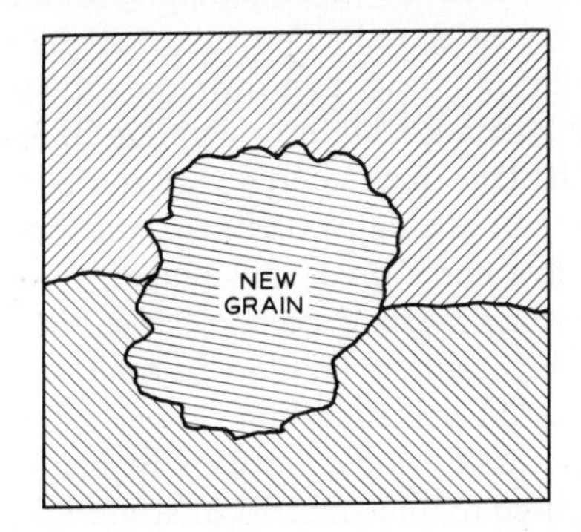

Fig. 4.7 Schematic of nucleation at grain boundary.

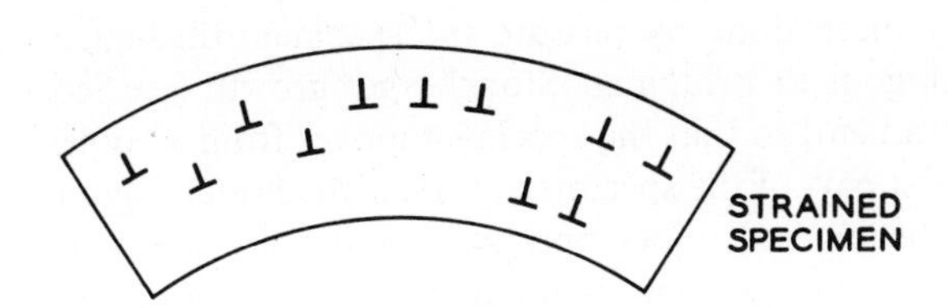

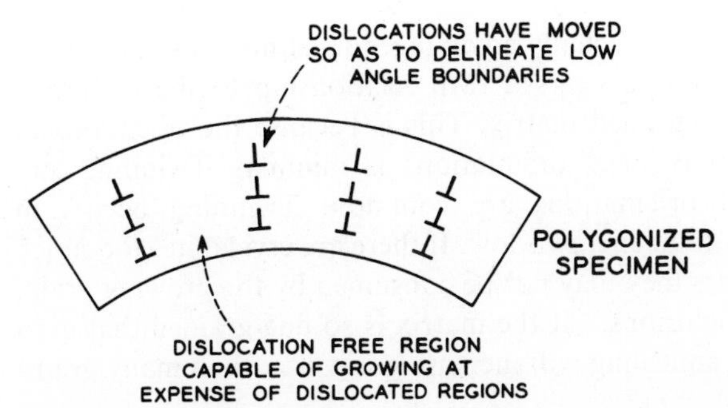

Fig. 4.8 Polygonization.

tion-producing regions capable of growth. Impurities are known to inhibit the movement of grain boundaries and thus inhibit the growth of once formed or already present nuclei. Because impurities hinder dislocation movement they help to localize dislocations and provide high-dislocation and low-dislocation regions to aid embryo growth.

In systems where new nuclei form they are usually observed to grow rapidly at the expense of already present crystals and they often succeed in dominating most of the specimen. Once they have grown to a reasonable size, their further growth is not favored by any driving force larger than the driving force causing the growth of originally present crystallites. This is because their size is then similar to the grains they are attempting to consume. Their growth does not result in any greater reduction in strain than does the growth of already present crystallites. In order to grow more rapidly than their neighbors their principal advantage must lie in orientation differences from the older grains. This is especially true in a material with sharp texture. In such a material nearly all the old grains are highly oriented. A new nucleus that forms with a new orientation thus has a great advantage.

In practice, in the strain-anneal method one usually varies the strain in a series of test pieces to discover the optimum or *critical strain* necessary to cause one or at most a few grains during annealing. Usually a 1–10% strain is sufficient. Critical strain is usually not controlled to better than $\pm 0.25\%$. Tapered specimens can be used to find the critical strain for a particular material because, when they are subjected to an elongating force, a strain gradient is automatically produced. Following annealing, the region of best growth can be observed and the strain in that region calculated, because the cross-section area and elongating force are known. As shown in Fig. 4.9, annealing is very often done by passing the specimen through a temperature gradient [analogous to Bridgman–Stockbarger growth (see Sec. 5.3), but with a reversed gradient] so that the specimen moves from a cooler to a hotter region. In the first part of the specimen to enter the hotter region, exaggerated grain growth begins and under optimum conditions one grain will grow to include the whole cross section. Sometimes the region *A* of Fig. 4.9 is severely deformed before annealing (for instance, by pinching it) to promote initial nucleation.

As stated above, texture is usually advantageous, although sometimes if the growing crystal orientation bears a twin relationship to the texture it will not consume the fine-grained matrix. This is because the orientational driving force for certain twinned orientations is minimal. Twinning and "inclusion" of different orientations are common. Twinning occurs in materials where stacking-fault energy is low. If there are crystallites too much larger than their neighbors they may not be consumed by the growing grain, with the result being "inclusions." If the matrix is so fine-grained that even in the undeformed state annealing will increase grain size, then many grains will grow during the growth period when the driving force is large and the driving force will be too small during the later growth period to cause any one grain to consume its neighbors. A grain size of 0.1 mm, according to Burgers (1963), is often optimal.

Twinning is sometimes controlled by increasing the stacking-fault energy by impurity addition or by starting with a matrix having a sharp texture and consuming it with a crystallite whose lattice orientation is neither so different from the matrix that twinning is favored† nor so close to it that the driving force for matrix consumption is too low.

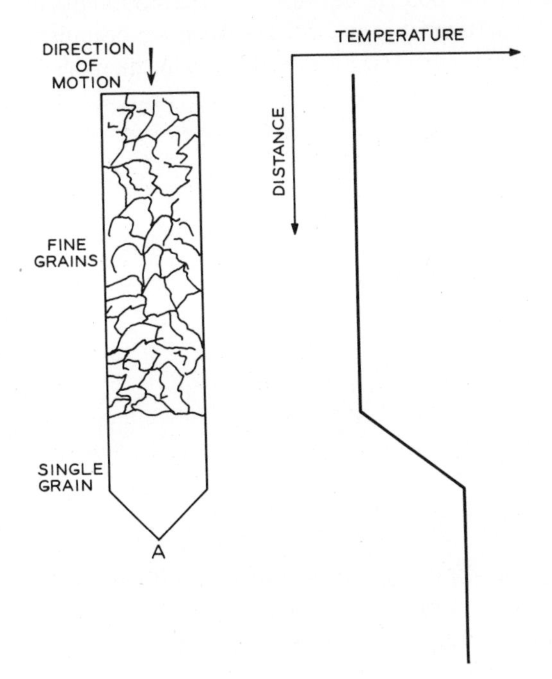

Fig. 4.9 Annealing in a gradient.

Nonmetallic crystals are much more difficult to grow by strain annealing than are metallic crystals, principally because it is not easy to deform them plastically. Deformation usually results in fracture, so that one is usually limited to the increase in grain size that can be obtained by annealing with particle-size differences the main driving force (essentially sintering). High purity tends to increase grain size and to make the preparation of proper structure for use in secondary recrystallization difficult. In some cases, if a material is very pure, large crystals are formed by primary recrystallization. In the case of nonmetals this is often the only technique possible because deformation is so difficult.

†Twinning will be favored when the twin orientation is closer to the texture orientation than the texture orientation is to the growing crystal orientation.

It has been observed (Aust, 1963) that maximum grain size is a linear function of specimen thickness. This is caused by the fact that *thermal grooves* (Mullins, 1958) occur between the grains where they meet the surface. Figure 4.10(a) shows a thermal groove at a grain boundary *ABCD*. Thermal grooves form at grain boundaries at temperatures where the surface mobility is high. The driving force for thermal-groove formation is provided by the grain-boundary area reduction that occurs when the groove forms. Because the interfacial free energy of grain boundaries is high, thermal-groove formation is usually quite easy at temperatures approaching the melting point. Thermal grooves ordinarily anchor grain boundaries because, for the boundary to move away from the groove its area must grow, at least when the boundary $C'A$ does not make an angle larger than θ [see Fig. 4.10(b)] with the normal

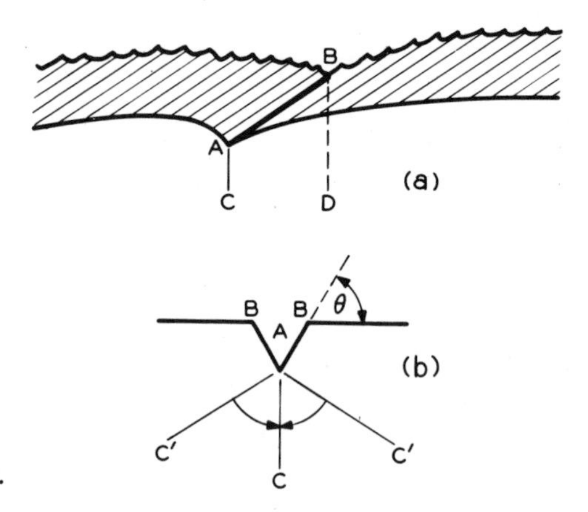

Fig. 4.10 Thermal grooves.

CA of the groove. Other conditions for boundary motion are discussed by Mullins (1958), who also derives a formula that predicts the experimental observation that grain growth falls off when the grain size approaches the specimen thickness. Qualitatively this is plausible on the simple assumption that, once grain size approaches specimen thickness, most of the grains will intercept the surface and their growth will tend to be inhibited by their own thermal grooves. However, one grain can grow at the expense of others even under these conditions if its specific surface energy is a few percent less than the grains it consumes. It is usual for the specific surface energy to exhibit a directional dependence of several percent so that, even in thin samples containing thermal grooves, notably oriented grains will often consume their neighbors.

Low-angle grain-boundary migration is a particularly powerful technique for growth by recrystallization. It has been applied almost entirely to metals but could be applicable in some circumstances to nonmetals. Most crystals prepared by liquid–solid and vapor–solid techniques usually have considerable lineage or low-angle grain boundaries. The free energy of a lineage-free specimen is lower than that of a lineage-containing specimen, so there is a driving force for the reduction of lineage. If, in a high-purity crystal containing lineage, lineage-free grains are introduced during annealing, the lineage-free grains will often consume the rest of the material. Lineage-free grains can be produced by deformation of a small region followed by a recrystallization anneal. A lineage-free crystal can be welded onto the lineaged crystal and in this manner the orientation of the grown lineage-free crystal can be controlled.

4.2.2 EQUIPMENT USED IN STRAIN ANNEALING

Strain annealing does not require highly specialized equipment beyond that available in the typical metallurgical processing laboratory. Furnaces for annealing capable of producing gradients and various cooling and heating rates in different atmospheres are required. Facilities for rolling specimens, for straining specimens in compression, tension, and torsion, and for drawing wire are usually required.

Metal working is adequately discussed in the literature (Sachs and Van Horn, 1940; Lyman, 1961, 1964), so only a few remarks are needed here. Similarly, metal-working equipment suitable for the laboratory is available from a variety of suppliers (Fenn Manufacturing Co., Newington, Conn.; H. J. Ruesch Machine Co., Springfield, N.J.; Stanat Manufacturing Co., Long Island City, N.Y.; and others), so no elaborate equipment descriptions are necessary.

In the case of metals the starting material often will have been subjected to some treatment by the processor that will affect its properties relative to crystal growth. Common forms for starting materials are

1. *Castings*. Castings are made by pouring molten metal into a mold and allowing it to freeze. The mold may fill by gravity, the metal may be forced into the mold by a die, or the mold may be filled by centrifugal force. The grain size and orientation will depend on the purity, the shape of the casting, the cooling rate, the heat-flow pattern during cooling, etc. Cast materials have no strain caused by work hardening but may be strained because of temperature gradients and differential contraction during cooling. In the case of metals this strain is usually negligible but for nonmetals it may be large and because strain through plastic deformation is difficult to achieve with nonmetals this strain is often the principal driving force for later recrystallization. It is unfortunate that in much strain annealing the details of the fabrication treatment and the initial condition of the specimens used are not adequately reported. It would seem that a photomicrograph of a suitably etched specimen in its original condition showing grain size and orientation should be a minimum reporting requirement in strain-anneal work. Better

yet, detailed data on cooling rate, texture, shape, etc., of the casting should be given. Usually rectangular or cylindrical pieces that result from casting are called ingots.

2. *Forgings*. Forgings result from hot- and cold-working ingots with drop hammers, presses, and forging machines. The primary forces applied are compressive and the pounding of hot metal on an anvil is the simplest forging operation. Forging will introduce strain and sometimes work hardening. The strain will not, in general, be uniform. In pounding, all areas of the pounded surface are often not worked evenly and, even when they are, there will be a gradient in compression away from the pounded surface. Not only are forgings sometimes the starting material for strain annealing, but sometimes material is forged by the crystal grower to strain it. In view of the nonuniformity of forging processes forging is not appropriate for producing homogeneous strain. However, if one wishes to strain a local region severely to induce nucleation there, pounding and heating the region (which is essentially local forging) is often employed.

3. *Rollings*. Figure 4.11 illustrates the rolling process. Most metallic-sheet stock is produced by rolling. Rolling may be carried out hot or cold. Metal is deformed rather uniformly by rolling, compared to other working processes. Only stretching in tension provides more homogeneity. Several passes may be made to achieve the desired decrease in thickness (*reduction*). In addition to using rolled stock as a starting material the crystal grower

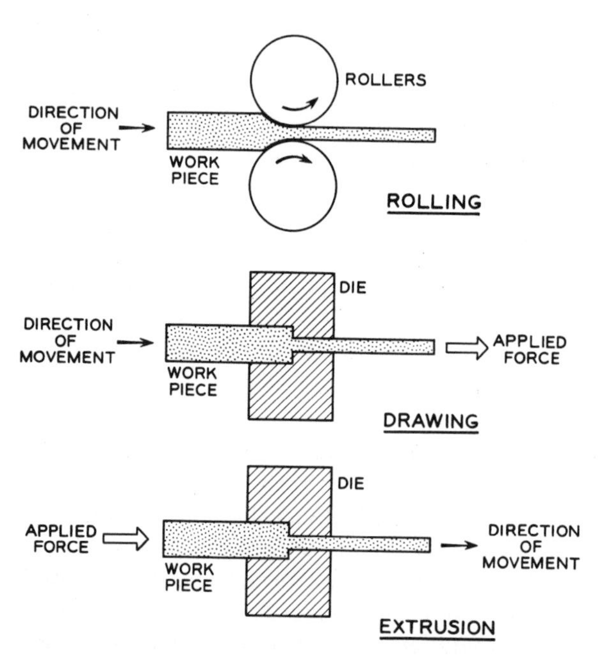

Fig. 4.11 Rolling, extrusion and drawing.

often rolls materials to produce strain and texture. Texture is increased if all the passes are made in the same direction. Commercial laboratory rolling machines are available from several suppliers, including Fenn Manufacturing Co. (Newington, Conn.) and H. J. Ruesch Machine Co. (Springfield, N.J.). Figure 4.12 shows a commercially available rolling mill suitable for the laboratory. (Ingots are sometimes rolled into square or rectangular cross-section *billets* or *blooms* in the commercial fabrication of metals.)

Fig. 4.12 Laboratory apparatus for rolling.

4. *Extrusions*. The extrusion process is illustrated in Fig. 4.11. Extrusion may be hot or cold. Rods and tubing are often made by extrusion. The strains are nonuniform and the process is not usually used by the crystal grower.

5. *Drawings*. Figure 4.11 shows drawing. It is generally used for making wire and is most often done cold. The resultant material is subject to fairly homogeneous tensile strain and the process is sometimes used by the crystal grower to induce strain, if commercial wire cannot be obtained.

The techniques and equipment used in deliberately inducing strain by tension and compression are usually the same as those used in mechanical testing. Mechanical-testing techniques are described by Lyman (1961, 1964) and Muhlenbruch (1944, 1955) and equipment is available from Instron Engineering Corporation (Canton, Mass.), Tinius Olsen Testing Machine Co. (Willow Grove, Pa.), Wiedeman Machine Co. (King of Prussia, Pa.), and others, so only a few brief statements are required here. Only in tension and compression can one achieve homogeneous strain throughout a substantial body of the material. Tension machines consist of (1) a device for straining the specimen, (2) a device for measuring the load applied to the specimen, and (3) a device for measuring the elongation of the specimen. A schematic of a tension or compression machine is shown in Fig. 4.13, and Fig. 4.14 shows a typical commercially available machine. For compression, the holder could consist of simple flat plates that press against the ends of the specimen. Even though the specimen is made larger at the ends to minimize strain inhomogeneities induced by the holder, homogeneous strain will exist only near the middle. The strain may be applied by a simple device such as the screw-thread arrangement shown or it may involve a lever or a hydraulic system.

If a hydraulic system is used, the stress is usually calculated from the measurement of the pressure in the system by means of a Bourdon gauge. This pressure allows a calculation of the tensile or compressive force that permits a calculation of stress because the cross-section area of the homogeneous-strain region is known. If a thread or lever system is used, the tensile or compressive force is determined by calibration with an appropriate gauge, or by a consideration of the mechanical arrangement of the system and the applied force. The strain (elongation) is usually measured by observing the movement of reference marks on the specimen or by the use of strain gauges. To assure that the strain is homogeneous the forces should be applied parallel to the specimen axis and it is useful to measure strain on all sides of a square or rectangular cross-section specimen or at several points on the circumference of a circular cross-section specimen to insure uniformity of strain. If one desires higher strain in a given region the specimen may be suitably tapered. If no quantitative knowledge of stress or strain is desired a simple vise arrangement may be employed. If no knowledge of strain is desired a press may be used for compressive straining. Suitable arrangements for heating and atmosphere control will depend on the problem at hand.

The principal other equipment needed for growth by strain annealing is a high-temperature furnace, associated controls, and temperature-measuring and -recording equipment. Facilities for heating during rolling and other working processes are quite specialized and are usually part of the commercially-purchased metal-working equipment mentioned above, so they will not be described here. Furnaces for heating during deliberate straining are usually quite simple "jury-rigged" affairs. Furnace construction is described in Paschkis and Persson (1960) and Trinks and Mawhinney (1951) and is also discussed in Chapters 5, 6, and 7. Radio-frequency heating is discussed in Langton (1949) and in May (1950). Most texts on furnaces and heating are from the industrial viewpoint, so the laboratory worker will often find it necessary to make extensive alterations in the designs suggested. A simple

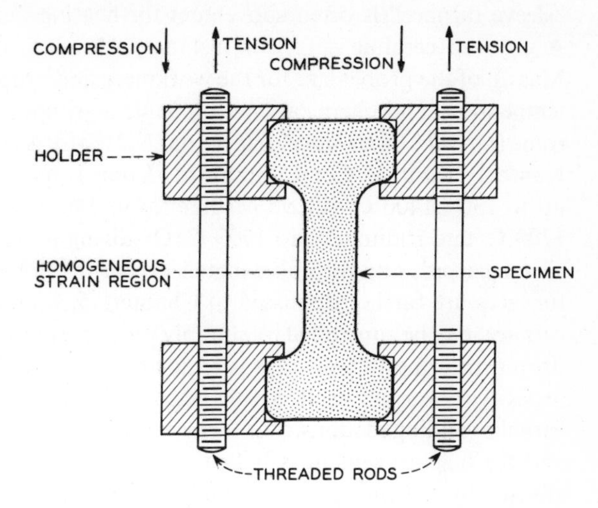

Fig. 4.13 Schematic of tension or compression straining machine.

Fig. 4.14 Laboratory tensile test apparatus. (Courtesy of Instron Co.)

"sleeve furnace" is often convenient for heating during deliberate straining. A grooved ceramic core, for instance, Alundum (Norton Co., Worcester, Mass.), of the proper size for the workpiece and proper in composition for the temperature and chemical environment, is wound with resistive wire. Nichrome (Driver Harris Corp., Harrison, N.J.) wire is suitable up to 900–1000°C, Kanthal (Kanthal Corp., Stanford, Conn.) up to 1100–1200°C, platinum up to 1300–1400°C, silicon carbide up to 1400°C, Pt–20% Rh up to 1600–1700°C, and iridium up to 1900°C. Oxidizing atmospheres are advisable for silicon carbide and the noble metals except Ir. Noble metals and noble-metal furnaces are further discussed in Chapters 5, 6, and 7. The gradient in the furnace may be controlled by suitably spaced taps on the winding. The voltage drop across the various taps is regulated by connecting dropping resistors in series with various sections of the furnace or by separately controlling the furnace winding sections. The furnace is coated with a suitable furnace cement and the furnace cement is baked on. Temperature control and recording is effected by cementing thermocouples close to the winding. Chromel–Alumel thermocouples are suitable for temperatures up to about 1100–1200°C and Pt vs. Pt–10% Rh couples for temperatures up to about 1600–1700°C. Thermocouples are discussed in Herzfeld (1962). For crude control, constant power at a desired power level (neglecting the voltage fluctuations) can be obtained by the use of a variable autotransformer, such as Variac (General Radio Corp., Concord, Mass.) or Powerstat (Superior Electric Corp., Bristol, Conn.). More sophisticated controllers such as on–off controllers and proportional controllers are discussed in Paschkis and Persson (1960), as are recorders.

Annealing may take place in a conventional-hinge furnace (furnace manufacturers include Hevi Duty Electric Co., Milwaukee, Wis.; Burrel Corp., Pittsburgh, Pa.; and Harper Electric Co., Buffalo, N.Y.), in a tube furnace, or in a homemade furnace such as described above. Silicon carbide-heater furnaces are conveniently made by sawing, drilling, etc. bricks to make a "brick-pile" furnace of the desired geometry into which conventional SiC heaters, for instance Globar (Norton Co., Worcester, Mass.), are inserted. Very high temperatures may be obtained with flame-sprayed, coated, noble-metal, resistance-wire furnaces (Engelhard Industries, Newark, N.J., produces furnaces rated up to 1900°C). Temperatures above 2000°C are obtained by, for example, gas-fired and plasma furnaces or by radio-frequency heating.

4.2.3 GROWTH OF SPECIFIC CRYSTALS BY STRAIN ANNEALING

4.2.3-1 Aluminum

It was stated earlier that because of the convenient annealing temperature, the growth of aluminum has been perhaps the most studied strain-anneal process. In Al the grain sizes should be about 0.1 mm before the critical-strain and growth-anneal processes. Buckley (1951) credits A. Sauveur in 1912 with the first growth of any material by strain annealing. The material grown was aluminum and Carpenter and Elam (1921) gave the first complete report of the conditions necessary for growth. They annealed 99.6% pure Al at 550°C to remove the effects of previous strain and to provide the desired grain size. The strain-free fine-grained Al was given a deformation to produce 1–2% strain and then annealed by raising the temperature from 450°C to 550°C at a rate of 25°/day. A final anneal at 600°C for 1 hr was used in some cases (Buckley, 1951).

Nucleation sometimes occurs preferentially at the surface during annealing. Graham and Maddin (1955) felt that nuclei in Al originate at dislocation pileups against surface-oxide film. Etching away a surface layer of $\sim 100\ \mu$ after critical straining helps to prevent surface nucleation.

After initial annealing, a so-called *recovery anneal* at a lower temperature will reduce the number of grains and aid more rapid grain growth during later annealing. Thus annealing at 320°C for 4 hr to obtain recovery followed by heating the specimen to 450°C with a 2-hr soak at that temperature will grow crystals 15 cm long in 1-mm-diameter wire (Bagaryatskii and Kolonstova, 1959).

The best treatment, however, is thermal-gradient annealing. Longitudinal gradients of about 20°/cm are satisfactory for small rods (Honeycombe, 1959). For sheets and other geometries it is also important to control the radial gradient at several points across the sheet to ensure that nucleation does not occur.

If the Al purity is greater than 99.99%, strain is apparently relieved mostly by polygonization and low-angle grain-boundary formation rather than by recrystallization. Gradient annealing thus results in negligible crystal growth. Surprisingly, isothermal annealing promotes growth of suitably strained high-purity Al (Montuelle, 1959). Traces of Li^+ ($\sim$0.04%) and Fe ($\sim$0.035%) will permit easy recrystallization of Al, probably because they pin (hold in place) dislocations to prevent polygonization (Montuelle and Chaudron, 1955).

Strong texture also aids in the growth of Al. Lommel (1960; see also Aust, 1963) prepared Al for recrystallization by cold rolling near liquid-N_2 temperatures followed by a 10-sec anneal at 640°C and a water quench. A specimen prepared in this manner had 2-mm grains and a strong texture. The strip was then initially strained and annealed by passage through a temperature gradient. It was then heated to 640°C and the resultant crystals were found to be $\sim$1 m long.

High-purity single-crystal Al strips under 2.5 cm in width were made easily by Leighly and Perkins (1960) by alternate application of straining and annealing. The strain was not severe enough to nucleate new grains and the annealing was at 640°C.

If there is little texture before annealing, then the resulting Al monocrystal orientation will not be predictable. However, if there is a texture, then the successful crystal that grows will have to have an orientational relationship to the texture that favors growth. Good growth seems to occur only with certain texture orientations because many texture orientations are such that no possible single-crystal orientation provides a sufficient driving force to cause the single crystal to consume the matrix. Texture orientations lying within about 40° of $\langle 111 \rangle$ in Al are such that the rate of growth of a single crystal into such texture is rapid (Beck et al., 1950).

Crystals of Al have also been prepared by strain annealing by Fujiwara (1939), Fujiwara and Ichiki (1955), Tiedema (1949), and Hägg and Karlsson (1952). Table 4.2 gives a summary of the conditions used by various researchers.

Often small crystals not consumed by the single-crystal growth are found on the surfaces of specimens. These may be twins or small regions misoriented by low-angle grain boundaries. Such crystals will form if the initial grain size is too large (too-long or too-high temperatures in preannealing) or if the final-growth anneal was too short or too low (so that it did not give time to consume surface crystals).

The perfection of well-made strain-annealed crystals is surprisingly good. Because Al has high-stacking fault and twin-boundary energies, strain annealing tends to produce untwinned crystals. Guinier and Tennevin (1950), in their X-ray studies, examined carefully prepared strain-annealed Al crystals and could find no detectable misorientation. Al crystals with misori-

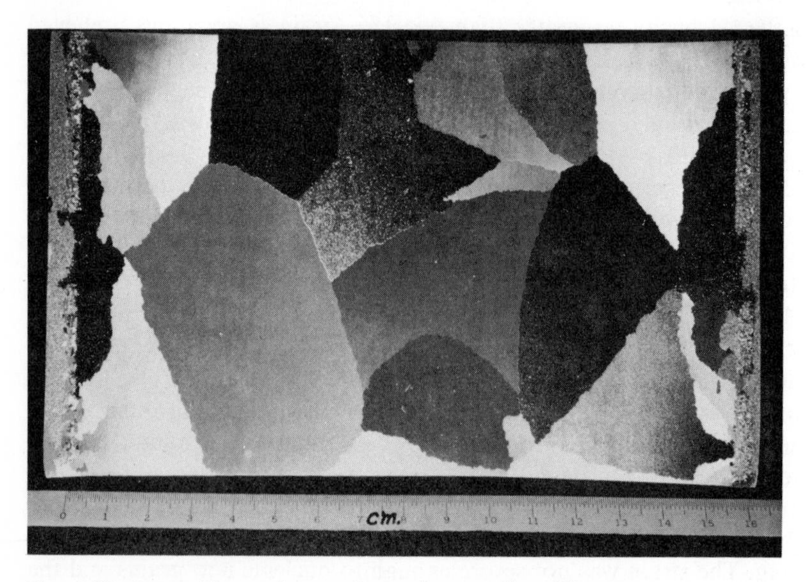

Fig. 4.15 Large Al crystal grown by strain annealing. (Courtesy of W. G. Burgers.)

entations of < 1 min of arc are commonly prepared by strain annealing. If the purity is over 99.997%, substructure is more difficult to avoid. Misorientations of the order of a few degrees are common in such crystals. The absence of pinning impurities apparently makes polygonization easy. Figure 4.15 shows an Al crystal grown by strain annealing.

4.2.3-2 Copper

Strain annealing tends to produce twinned orientations in copper, apparently because of the low stacking-fault energies (Carpenter and Tamura, 1927).

However, secondary recrystallization will produce good copper crystals. It will be recalled that secondary recrystallization is the growth of a few grains from the matrix formed during primary recrystallization. Secondary recrystallization is brought about by annealing the strained specimen at a temperature higher than the temperature where primary recrystallization took place. Secondary recrystallization in copper is typically brought about as follows:

1. Annealed Cu strip is straight-rolled to a reduction of ~90% at room temperature.
2. Specimens are slowly heated in vacuo to 1000–1040°C and held at temperature for 2–3 hours.

Step 1 results in strong texture, which is consumed by one or a few crystals in step 2. Too rapid heating in step 2 results in twins. Sometimes before step 2 the specimen is examined and deviating-orientation grains are removed (if on the surface) by etching or (if near an edge of a sheet) by cutting out. No subboundaries can be detected with the Schulz technique (see Sec. 1.4.1) in carefully grown, stain-annealed Cu crystals (Aust, 1963). Table 4.2 gives further details of the growth methods for Cu.

4.2.3-3 Iron

Edwards and Pfeil (1924; see also Buckley, 1951) were probably the first to grow iron by the strain-anneal method. The strain-anneal method is capable of producing good crystals of iron but the conditions required are strongly dependent on the purity. Iron containing more than about 0.05% carbon (mild steel) cannot be recrystallized. Decarbonizing in a reducing atmosphere so that C is reduced to about 0.01% or starting with Armco or similar iron or vacuum-melted iron of 99.99% purity is required. The grain size prior to critical straining should be about 0.1 mm. A reduction by rolling of about 50% is required followed by a critical strain of about 3% in tension (Aust, 1963). The critically strained region may be limited to a small volume of the sample for better control of nucleation (Holden and Holloman, 1949).

Table 4.2

CRYSTALS GROWN BY STRAIN ANNEALING

Material	Prestraining and Other Pretreatment	Percent Critical Strain	Growth-Anneal Conditions	Comments	References
Aluminum (99.6%)	Anneal heavily worked Al at 550°C for several hours (principal impurities Fe and Si) (sheets 3-mm thick × 2.5-cm wide or cylinders 2.5-cm diameter)	1–3% by extension (strain controlled ±0.25%) (remove 75-μ surface layer by electropolishing)	Heat slowly to 450–550°C at 15–20°/day		Carpenter and Elam, 1921; Graham and Maddin, 1955
Aluminum (99.5%)	Wire	4%, 320°C for 4 hr to "recover"	1 hr to 450°C; soak at 450°C for 2 hr (in gradient)	15-cm × 1-mm single crystals	Bargaryatskii and Kolonstova, 1959
	Strip 1.2 cm × 1 mm × 5 cm		10 cm/min through gradient furnace (linear and radial gradient)		Williamson and Smallman, 1953
	Rod 3 mm × 5 cm		Longitudinal gradient 20°/cm		Honeycombe, 1959
Aluminum (99.99%)	Deliberately add 0.04% Li or 0.035% Fe to inhibit polygonization			Tends to polygonize rather than grow if treated like 99.6% Al	Montuelle, 1959; Montuelle and Chaudron, 1955
Aluminum (99.993%)	Heavily cold rolled at −196°C; annealed for 10 sec at 640°C; water quenched—yields strong texture, which helps growth anneal; grain size a few mm	Critically strained	4 cm/hr through temperature gradient of 100°/cm; heated to 640°C	Crystals ~1 m long	Lommel, 1960

Aluminum (99.99%)	Alternate strain and anneal to produce 5-mm-diameter grains	Not enough to nucleate new grains	640°C growth by induced grain-boundary migration	Sheets narrower than 2.5 cm	Leighly and Perkins, 1960; Beck and Sperry, 1950; and Aust and Dunn, 1957 discuss strain-induced grain-boundary migration
Aluminum 0.5% Ag	15 min at 500°C	2%	6.5 cm/hr—90°/cm		Aust, 1963
Aluminum 4–5% Cu 1.3% Si 4.2% Ge 4.5% Mg 8.6% Ag	Flash-anneal 10° below melting point of composition	1.25%	10–30° below melting temperature of composition, 0.5 cm/hr steep temperature gradient		Gane, 1958
Copper				Strain anneal often yields twins because of low-fault and twin-boundary energies (similar problems with Pb, Ag, and other face-centered cubic metals)	Carpenter and Tamura, 1927
Copper	Annealed copper strip straight rolled to a reduction of 90% at room temperature	No deliberate critical straining; growth essentially by secondary recrystallization	Heated slowly to 1000–1040°C and soaked several hours		Aust, 1963
Copper electrolytic purity	Cold-rolled reduction of 95%, annealed for 10 hr at 600°C forming strong texture; deviating orientations etched off	No deliberate critical straining, growth essentially by secondary recrystallization	Anneal near melting point, e.g., 72 hr at 1030°C		Parthasarathi and Beck, 1958

Table 4.2—*Cont.*

Material	*Prestraining and Other Pretreatment*	*Percent Critical Strain*	*Growth-Anneal Conditions*	*Comments*	*References*
Au Cu–up to 1% Zn Cu–0.2% Al Cu–0.1% Cd Cu–0.1% O Ni–1% Mn Fe–30–100% Ni Some ternary Fe–Ni–Cu compositions; probably also useful for Pt	Cold rolled to a reduction of 80–95% to get strong texture	No deliberate critical straining; growth essentially by secondary recrystallization	Heat following primary recrystallization	Conventional strain anneal yields twins in these and other face-centered cubic materials; difficult to use in very concentrated alloys	Barrett and Massalski, 1966
Iron < 99.99%	Must reduce C to < 0.1% by decarbonizing anneal in wet H_2; require grain size ~0.1 mm; optimum reduction by rolling of about 50% in "Ferrovac E", cold-worked rods heated to 700°C, and water quenched	3% in tension, advantage if critical strain in small volume of material; slow strain rate ~0.01%/hr prevents Lüders' bands and inhomogeneous deformation; hard to get uniform strain if yield point exceeded; remove surface layers by electropolishing or etching	Anneal below 910°C, then 880°C in H_2 for 72 hr; temperature-gradient anneal helpful		Stein and Low, 1961; Holden and Holloman, 1949; Allen et al., 1956; gradient-furnace design used by Dunn and Nonken, 1953, for Fe–Si growth

Iron 99.99+%	Add 0.035% C, use 0.01% C if iron less pure, also can add 0.03% Si but cannot decarbonize out in growth anneal; carbonize with acetylene 4 hr at 920°C	2%–0.06% C 5%–0.02% C 3%–0.035% C	Must keep carbon in long enough to prevent polygonization but must eventually get it out of final product	Usual strain anneal leads to polygonization	Talbot, 1956, 1960
Iron 99.9% Fe	Mild steel–hot-rolled 1/8″ plate, 1/2″ bars—decarbonize at 950°C + 48 hr in H_2; slow-cool or at 750°C for 2 weeks in wet H_2	3%	880°C for 72 hr		Edwards and Pfeil, 1924; Gensamer and Mehl, 1938
~99.9% Iron (Armco Iron)	Anneal less than 24 hr in H_2 at 950°C	3%	850°C for 72 hr in H_2		Talbot, 1956; Stone, 1948
Fe–3% Si plate	20 hr in H_2 at 870°C	2.5%	1150°C—1 cm/hr, gradient—1000°/cm		Dunn and Nonken, 1953
Fe–0.15% P	Annealed to produce ~700 grains/mm²	3% followed by chemical machining	800°C—1 cm/hr in a steep temperature gradient		Honeycombe, 1959
Fe–up to 6% Al rod	Decarbonize at 900–970°C	2.4%	800–1000°C for 1–2 days		Yamamoto and Miyasawa, 1954
Fe–18% Cr–8% Ni	Cold reduction to 70%, recrystallize at 1000°C for 1 hr	2.5%	Heat rapidly to 1350°C, soak 100 hr (1200°C for Fe–20% Cr)		Legett et al., 1959
Molybdenum—wires	Mechanically polished	No critical strain; method is modification of Andrade's technique	Heated to 1000°C in furnace and raised to 1700°C by passing current; hot zone moved down wire by moving furnace	Crystals of several cm	Andrade, 1937; Tsien and Chow, 1937
Molybdenum—small rods			Stationary temperature gradient in furnace; furnace temperature held 7 hr at 2000°C		Chen et al., 1951

Table 4.2—*Cont.*

Material	*Prestraining and Other Pretreatment*	*Percent Critical Strain*	*Growth-Anneal Conditions*	*Comments*	*References*
Molybdenum—large rods	Preanneal	Strain 1% during growth anneal	Anneal as above	Rod crystals up to 1.2 cm	Chen et al., 1951
Niobium 3-mm diameter × 18-cm rods			Anneal at 2000°C for 2–4 hr under axial tension less than 2%	Crystals ~1 cm (similar procedure applicable to Mo)	Maddin and Chen, 1953
Tantalum 5–40-mm wire 99.9%	Anneal at 1800°C several min	2–3% strain	2200°C, lowered at 0.3–1 cm/hr, through steep temperature gradient	Crystals with length/diameter ~100 obtained	Nichols, 1954; Seraphim et al., 1960
Thorium (prepared by I_2 process)	Arc-melted bar 2.5 × 10 × 0.75 cm cold-compressed 5%, heated 50° above phase-transition temperature		Lowered through temperature gradient at 225°/cm at 1.5 cm/hr	2.5 × 1.3 × 0.75 cm in grains	Armstrong et al., 1959
Titanium (~99.9%) (Van Arkel process)	Specimens cut, so preferred orientation with plane of strip	1.5–2.0% tensile or 0.25–1.0% transverse compressive strains	Annealed in Ar 200 hr at 860°C	Twins easily, important to exclude O_2 and N_2 throughout	Churchman, 1954
Titanium–13% Mo	0.1-cm-diameter rods		Heated to 1600°C, soaked $\frac{1}{2}$ hr, quenched on cold plate	Form β phase with slight loss of Mo; transform traces of α by heating to 1000°C and water-quenching, crystals ≈ rod diameter	Spachner and Rostoker, 1959
Tungsten–2% ThO_2	Filamentary wire made from W powder made by reducing WO_3 containing 2% ThO_2		Wire drawn through 2000–2200°C zone in dry hydrogen, steep temperature gradient	Best for wires, single crystals–meters in length; ThO_2 inhibits growth of too many grains; contaminated with ThO_2; technique is modification of Pintsch process	Smithells, 1952; Pintsch, 1918

Tungsten—99.99%	Add a few % of K_2SiO_3 + $AlCl_3$ to WO_3; reduce and make 0.90-in. wire		Wire drawn through gradient as above or heat to 2300°C in dry H_2 and then use moving gradient at speed of 0.4–4 cm/hr followed by heating to 2700°C to absorb small crystals	1 cm long, impurities removed by processing	Rieck, 1958; Swalin and Geisler, 1957–58; Millner, 1957; Andrade's method described for molybdenum by Nichols, 1940 and Hughes et al., 1959
Tungsten–99.99%		3–10 mil wire, small amount of plastic twist	Nichols' moving gradient melted or grain-growth method of Robinson		Gifford and Coomes, 1960; Nichols, 1940; Robinson, 1942
Uranium	Sheet heavily rolled at room temperature to produce strong primary recrystallization texture containing about 35 ppm C, anneal in α region, rapid heating to 650°C, heat 100°/min, hold ~10 hr	No deliberate strain, essentially secondary recrystallization	Anneal 48 hr at 650°C	α crystals	Lacombe et al., 1959
Uranium	Preferred orientation in matrix required	6%	650°C	α crystals not possible in very high-purity U; Fisher has described a technique that controls excessive grain growth so that a few large grains coarsen	Lacombe et al., 1959, 1961; Cahn, 1953; Fisher, 1957

After critical straining, surface layers are etched or electropolished away. Then the sample is annealed at temperatures in the 880–900°C range for about 72 hr. Annealing in a temperature gradient probably improves the product (Stein and Low, 1961; Dunn and Nonken, 1953). After the final growth anneal it is sometimes necessary to etch away surface polycrystals to reveal the large crystals.

If the purity of iron is much above 99.99% the deformed matrix (as with Al) tends to polygonize and special techniques are required to grow large crystals. In general, carbon is added and the procedures are as above except that the growth anneal is conducted in a decarbonizing atmosphere to eliminate most of the carbon from the grown crystals (Aust, 1963).

There seems to be some tendency for the growth of twin crystals of preferred orientation. Allen et al. (1956) observed that directions close to $\langle 011 \rangle$ were most often obtained. Controlled-orientation crystals have been prepared by the method of Fujiwara (1939; see also Dohi and Yamoshita, 1957; Stein and Low, 1961). In Fujiwara's method the sample is critically strained and one end is annealed in a temperature gradient. Growth is interrupted and the sample is cut so that a grain close to the desired orientation dominates the growing interface. The sample is then bent so that the "seed" grain bears the desired orientation relative to the rest of the sample and annealing is continued. By careful manipulation the whole specimen can be transformed to the desired orientation and even bicrystal samples can be prepared (Dunn, 1949). Table 4.2 summarizes the methods for Fe growth.

4.2.3-4 Other Materials

Single-phase aluminum alloys are very conveniently grown by strain annealing. The growth of such alloys is a notable example of solid–solid growth from a polycomponent system. Some other systems where polycomponent growth has been achieved are mentioned in Table 4.2. Because melting does not take place, segregation is not possible and the single crystals preserve the composition of the starting ingot. Large temperature gradients are required in the growth anneal to obtain good recrystallization. The higher the concentration of alloying elements the larger the gradient and the lower the growth speed permissible. Polygonization is still troublesome (Montuelle, 1959) even in Al–Zn 6–15% alloys unless an impurity such as 0.15% Fe is present. The amount of iron necessary to repress polygonization in Al–Zn high-purity alloys is much greater than in high-purity Al (Aust, 1963).

Tungsten-wire single crystals are conveniently grown by the Pintsch process (1918; see also Buckley, 1951). Tungsten wire is unrolled from one spool through a tube furnace onto another spool. The furnace temperature is about 2500°C. Squirted filaments were found to be more suitable than drawn wire, and no deliberate straining was required. Speeds as high as 3 m/hr produce long single-crystal regions in the annealed wire. Table 4.2 summarizes the conditions for growth by strain annealing and lists pertinent literature references for a variety of materials not discussed in the text.

4.3 Growth by Sintering

Sintering is the heating of a compacted polycrystalline body. During sintering the driving force for grain growth is mainly caused by

1. Residual strains [w in Eq. (4.2)]
2. Orientation effects [G_s in Eq. (4.2)]
3. Grain-size effects [ΔG_0 in Eq. (4.2)]

It is probable that, in inorganic materials, causes 2 and 3 are most important because large strains (without exceeding the breaking strength) are not possible. The term *sintering* is usually reserved for the process of grain growth in *nonmetals.* When grain growth is observed during the heating of a polycrystalline *metallic* body the process is usually considered as a sort of special case of strain annealing in which the strains were not induced deliberately, but occurred during the preparation of the material. Consequently, several metallic systems where grain growth would formally be considered to be by sintering were discussed in Sec. 4.2. In this section we will therefore principally discuss the sintering of nonmetals, although, where appropriate, metals will be mentioned.

Garnet crystals up to 5 mm in size have been formed by sintering polycrystalline $Y_3Fe_5O_{12}$, yttrium–iron garnet (YIG) above 1450°C (Van Uitert et al., 1959). Grain growth of considerable size has been observed by Harrison (1959) in copper manganese ferrite. BeO (Duwez et al., 1949), Al_2O_3 (Wilder and Fitzsimmons, 1955), and Zn (Burke, 1951; Aust, 1963), have all been grown to fairly large grain sizes by sintering. In general, grain growth by sintering has been so far observed to be more effective in nonmetals. This may be because metals have been less studied because strain annealing and other methods can be used to produce crystals. Burke (1959) has found that inorganic ceramics have many more pores than do metals. He feels that pores inhibit all but a few grains from growing so that large grain sizes occur in more porous materials. Additives that can inhibit sintering include MgO in Al_2O_3 (Coble, 1961) and Ag in Au (Burke, 1951). Additives may also accelerate grain growth, for example, ZnO in Zn (Burke, 1951). The mechanism of these effects is not clear. Initial grain size has an effect. Fine grains seem unaffected by sintering in Al_2O_3 and seeding has been tried with some success (Coble and Burke, unpublished, reported in Aust, 1963).

Hot pressing is sintering under compression. Hot pressing has been used mainly for the densification of ceramics. In ceramic processing, large grains are usually undesirable because many small grains fulfill the ceramist's objective of nondirectional dependence of properties. Thus in ordinary hot pressing the pressure is made large to encourage densification but the temperature is made only high enough to provide a reasonable rate of pore removal but not high enough to cause appreciable grain-boundary movement. (Murray et al., 1958 have reviewed hot-pressing techniques.) MgO and Al_2O_3 have been hot-pressed with great success. However, if the temperature were raised during hot pressing, grain growth by sintering might be appreciable and perhaps useful single crystals could result. The value of w in Eq. (4.2) might be increased to the magnitude it reaches in strain annealing. Longitudinal compressive or tensile stresses rather than the essentially hydrostatic stress of hot pressing might be more effective. Laudise (1967) has observed grain growth in $ZnWO_4$ while annealing under compression and Sellers et al. (1967) have grown Al_2O_3 crystals up to 7 cm^3 by the technique.

In growth by strain annealing, w in Eq. (4.2) is principally caused by the strain that remains as a result of the imperfections introduced in connection with the plastic deformation once the deforming force is removed. In Fig. 4.2 the force still remaining on the specimen must be considerably less than F_b once the deforming force is removed. If the elastic limit is not exceeded (exceeding the elastic limit will usually cause rupture in nonmetals) then the force responsible for w can approach F_a. Thus because $F_a \approx F_b$ and all of F_b is not effective in contributing to w once the deforming force is removed, w will be larger if the force is kept on the specimen during annealing. This technique might even be more effective than strain annealing for metals and is the only practical approach for nonmetals. This was the approach used by Laudise (1967) in growing $ZnWO_4$ crystals and Sellers et al. (1967) for Al_2O_3.

In general, useful crystals have not been deliberately grown by sintering. Crystal growth by sintering has been used to study the sintering process and some single crystals have been formed more or less inadvertently during the formation of ceramics.

4.4 Growth by Polymorphic Transition

Growth by polymorphic transition of an undesired polymorph to a desired polymorph implies a technique for the formation of the "undesired" polymorph as a single crystal. Thus polymorphic transition is in a sense a second step after the initial growth. Experience has shown that if possible it is easier to grow the desired polymorph directly. Nevertheless, there are some cases when it is essential to conduct a polymorphic transition and form a single crystal as a result. Some background on polymorphic transitions was given in Sec. 2.3. To effect the desired phase change either pressure or temperature or in some cases both must be varied. The usual case is to vary temperature at ambient pressure.†

In the case of minor symmetry transitions (see Sec. 2.3) it is often possible to bring a single crystal of the undesired polymorph through the transition and form a single crystal of the desired polymorph without the formation of polycrystals, twins, severe strain, or other imperfections. In the case of major symmetry transitions, stacking faults and polytypism often result, and in the case of major structural transformations single crystallinity is probably hardly ever preserved and metastability is common in the absence of solvents. Thus the more different the structures of the two phases, the more difficult a transformation to produce a single crystal becomes. The usual method is to keep the crystal isothermal and to raise or lower the temperature of the entire furnace. The result is often nucleation of the new phase at many sites in the matrix with resultant twinning or polycrystal formation. Although not usually practiced, it is obvious that it would be more fruitful to cause nucleation of the new phase at one site and to cause one nucleus to dominate the solid–solid growth interface. Thus polymorphic transitions using the geometry employed in Bridgman–Stockbarger growth (see Sec. 5.3) would be appropriate. A crystal pointed on the end might be lowered through a temperature gradient or the temperature of a furnace having a gradient might be lowered. The usual case is to grow a high-temperature polymorph and then cautiously lower the temperature to room temperature to form a room-temperature polymorph. However, low-temperature polymorphs are sometimes transformed to high-temperature stable polymorphs, which are "frozen" to stability by quenching. Provided the kinetics of transformation of the high-temperature polymorph is slow it may be preserved indefinitely at room temperature. Diamond is an example.

4.4.1 IRON

The phase transitions for Fe at atmospheric pressure are

$$\alpha \xrightleftharpoons{910^\circ C} \gamma \xrightleftharpoons{1400^\circ C} \Delta \xrightleftharpoons{1539^\circ C} \text{liquid} \qquad \text{(Lyman, 1961)}$$

It is therefore not possible to form α-Fe by processes above 910°C and difficult to cool high-temperature modifications through the transition and preserve single crystallinity. McKeehan (1927) and Andrade and Greenland (Andrade, 1937) passed a temperature gradient along a wire at such a rate that the $\gamma \rightarrow \alpha$ transformation produced single crystals of α-Fe.

†The most notable practical exception is graphite–diamond. Here, however, growth of usable crystals is probably most often from solution.

4.4.2 URANIUM

The phase transitions of U at atmospheric pressure are

$$\alpha \underset{}{\overset{\sim 670°C}{\rightleftharpoons}} \beta \overset{1130°C}{\rightleftharpoons} \text{liquid} \qquad \text{(Lyman, 1961)}$$

Mercier et al. (1958) made poor-quality α crystals by passing a β sample through a temperature-gradient furnace. Crystals of high perfection were made by critically straining the imperfect α crystals and annealing below the transition.

4.4.3 QUARTZ

The phase transitions for SiO_2 at atmospheric pressure are

$$\alpha \text{ (low quartz)} \overset{573°C}{\rightleftharpoons} \beta \text{ (high quartz)} \overset{867°C}{\rightleftharpoons} \text{tridymite}$$

$$\overset{1470°C}{\rightleftharpoons} \text{cristobalite} \overset{1723°C}{\rightleftharpoons} \text{liquid} \qquad \text{(Levin et al. 1964)}$$

There are several forms of tridymite and cristobalite that tend to exhibit metastability (Levin et al., 1964). Quartz itself exhibits several metastable polymorphs (Frondel, 1962). In addition, there are at least two high-pressure polymorphs of SiO_2, coesite and stishovite (see Frondel, 1962). Coesite and stishovite exhibit enough metastability at ambient conditions that they can be "brought back alive"; that is, they can be formed at elevated temperature and pressure and quenched back to ambient conditions and will preserve their structure. In most cases and especially if precautions in crossing transition curves are not exercised the result will not be good single crystals but instead will be highly twinned and often even polycrystalline.

Considerable effort has been devoted to the detwinning of α-quartz by solid-solid transformations. M. V. Klassen-Neklyudova (1964) discusses twin modes in quartz and reviews detwinning procedures. Tsinzerling (Schubnikov and Tsinzerling, 1932, and other papers by Tsinzerling cited in Klassen-Neklyudova, 1964), Frondel (1946), Wooster and Wooster (1946), Thomas, et al. (1946), and Thomas and Wooster (1951) have also studied the problem of detwinning.

One method (Thomas and Wooster, 1951) is to heat a twinned plate above the α–β transition. The β-quartz is then cooled slowly through the transition in hopes that α nucleation will start at a single site and form a large untwinned crystal. A better method is to anneal a plate below the α–β transition while subjecting it to a torsional stress. It is then slowly cooled to room temperature. This method can make 100% of the area free of twins. Stresses of $\sim$200 kg/cm^2 are needed and the capacity of quartz plates to detwin is directionally dependent (Thomas and Wooster, 1951). The plate length is perpendicular to c and the torque is applied about the long axis of the plate.

The effectiveness of torque in detwinning depends on the crystal orientation but the sense of the applied torque (clockwise vs. counterclockwise) has no effect. Typical plate sizes used by Thomas and Wooster (1951) were $45 \times 30 \times 4$ mm and typical times were 3 hr. The torque was maintained while the crystals were cooled. The effect was the same whether the temperature was above or below the α–β transition. A quartz plate can be specified by three orientation angles as shown in Fig. 4.16. X, Y, and Z are alternative designations (Frondel, 1962) for the axes that are particularly useful in piezoelectric applications, while a and c are the conventional crystallographic axes. As shown, $X(a)$ is the electric axis. Some authors have made Y and a coincident (Frondel, 1962). φ and θ define the plate normal and ψ is the angle of rotation of the plate from a given initial position about its normal. In Fig. 4.16, P is the direction of the normal to the plate and L is the direction of the long axis of the plate. Table 4.3 shows the results of applied torque on various orientations and Fig. 4.17 shows the ease of detwinning for plates where $\psi = 0$ (Thomas and Wooster, 1951). The length of the radius vector is inversely proportional to the untwinning torque at a temperature of 450°C.

Because the rate and direction of movement of the twin interface do not depend on the sign of the applied stress, Thomas and Wooster (1951) concluded that the rate of movement of the twin interface must depend on the second power of the stress component. The elastic energy is a function of the second power of stress (or strain) and thus the twin boundary moves so that the crystal yields as much as possible under the stress; that is, it maximizes its stored elastic energy. Thomas and Wooster (1951) have shown that a detailed consideration of this energy maximization predicts the observed orientation dependence for the ease of detwinning by means of a torque.

Torque detwinning and other stress-induced solid–solid polymorphic transitions should be possible whenever the forms are closely related struc-

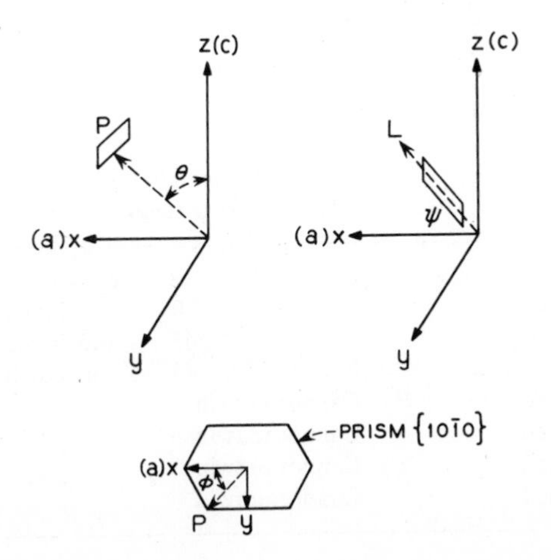

Fig. 4.16 Orientation of quartz.

Fig. 4.17 Ease of detwinning of quartz as a function of orientation (after Thomas, Rycroft & Fielding, 1946). (Courtesy of Macmillan Journals Ltd.)

turally, no bond breaking or formation is required in the transformation, and there exists a sufficient difference in elastic energy in the two forms of the structure.

Fay and Brandle (1967) have shown that Czochralski-grown $LaAlO_3$ is twinned by passage through a cubic–rhombohedral transition at ~435°C. The dominant twin planes are those of the {100} system. Detwinning was found to be possible by compressing a hexagonal-shaped crystal in a ⟨111⟩ direction. $Ba_2NaNb_5O_{15}$, which is of great interest as a nonlinear optic material, twins easily when passing through a tetragonal–orthorhombic transition at 260°C. Twins are easily removed when a pressure of about 1000 psi is applied along ⟨100⟩ of the orthorhombic cell at ~250°C (Van Uitert et al., 1968).

Table 4.3

DETWINNING QUARTZ

Initial Orientation				*Orientation After Torque*			
φ (deg)	θ (deg)	ψ (deg)		φ (deg)	θ (deg)	ψ (deg)	
0	128	0	twinned	0	128	0	untwinned
0	52	0	twinned	0	128	0	untwinned
0	128	45	twinned	0	52	45	untwinned
0	52	45	twinned	0	52	45	untwinned
0	0	0	twinned	Cannot untwin			
0	90	0	twinned	Cannot untwin			
30	0	0	twinned	Cannot untwin			
30	90	0	twinned	Cannot untwin			

4.4.4 FERROELECTRICS

In ordinary dielectrics the polarization and electric field show a linear relationship. In a ferroelectric, the polarization and electric field exhibit the relationship shown in Fig. 4.18. In general, the direction of spontaneous polarization throughout the crystal is not the same. The crystal consists of a number of *domains* within which the direction of the electric dipoles is the same, but this direction varies from domain to domain. Increasing the applied field tends to orient all the domains similarly so that at A in Fig. 4.18 the crystal is a single domain. Extrapolating P to zero field gives the *spontaneous polarization* or the polarization existing in an unaligned domain in zero field. If the field is reduced the polarization falls to the *remnant polarization* at zero field. To remove the remnant polarization the polarization of about half the crystal must be reversed and this occurs when a field in the opposite direction (the *coercive field*) is applied.

To show ferroelectricity (1) a crystal must lack a center of symmetry (thus leading to an electric dipole) and (2) there must be two equilibrium arrangements for the dipole separated by a potential-energy barrier, that is, two positions for an atom in the crystal, and the atom must be "switchable" between these positions by an electric field.

Twenty-one of the crystal classes lack a center of symmetry. Of these, 20 are piezoelectric; that is, they become polarized under stress. Of the piezoelectric classes ten are *pyroelectric*; that is, they are spontaneously polarized. Polarization is usually marked by surface charges that are changed by heating. These changes are easily detectable and such crystals are termed pyroelectric. Ferroelectrics are part of the class of spontaneously polarized pyroelectrics. Thus criterion (1) can be predicted from structure and criterion (2) must be determined experimentally.

The movement of ferroelectric domains is thus a particular solid–solid

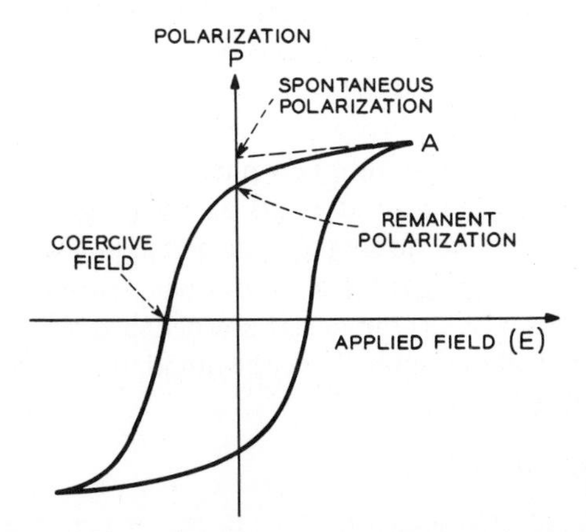

Fig. 4.18 Polarization vs. applied field for a ferroelectric material.

transition appropriately considered under polymorphism, with some analogues to the removal of electric twinning. At the ferroelectric Curie temperature, ferroelectric properties disappear and a true polymorphic change takes place.

The alignment of ferroelectric domains requires a lower field at high temperatures and often can be observed by viewing the crystal between crossed polarizers.

Triglycine sulfate is monoclinic and ferroelectric below 49°C. Twins are usually present in crystals grown from aqueous solutions. The twins are not observable in polarized light, so the optical axes must be parallel across twin boundaries. The twins are ferroelectric domains and can be revealed by etching. A field of 1200–2400 V/cm applied to the faces normal to the *b* axes produces untwinned crystals. Triglycine sulfate differs from Rochelle salt, $NaKC_4H_4O_6 \cdot 4\,H_2O$, in that the domains will reappear after several days (Klassen-Neklyudova, 1964).

Aligning domains by an electric field is sometimes called poling and can be applied to most ferroelectrics. In some cases the field necessary for alignment nearly exceeds the breakdown voltage (e.g., $LiNbO_3$) so that poling by conventional techniques is not possible except at temperatures close to the Curie temperature. However, special procedures, such as poling with liquid electrodes well below the Curie temperature (Wemple et al., 1968) and poling during Czochralski growth (Nassau et al., 1966), can be used to make single-domain material.

It is probable that in some nonferroelectric crystals electric twinning could be prevented by growth in an electric field, while, in a ferroelectric, domains might be aligned by the proper application of stresses. The alignment of ferromagnetic domains by a magnetic field is in some respects analogous and will not be further discussed here.

4.4.5 HIGH-PRESSURE POLYMORPHIC TRANSITIONS

For most high-pressure polymorphic transitions the phase transition occurs rapidly and in an uncontrolled manner. This is because the apparatus for reaching high pressures is unwieldy and not easily capable of being modified to produce controlled nucleation. In early experiments the object usually is to map the phase boundaries and to produce new phases. For such work, powder specimens are usually used and no attempts at crystal growth are made. It is, therefore, not surprising that no single crystals of appreciable size have been made by high-pressure polymorphic transitions. Nevertheless, it is worthwhile to review systems where ceramic and powder specimens have been formed, because this work will undoubtedly be the foundation on which future single-crystal syntheses will rest.

4.4.5-1 Diamond (*see also Sec. 7.5*)

The graphite–diamond-phase diagram was given in Sec. 2.3. It is interesting to point out that diamond is stable at pressures as low as ~10,000 atm if the temperature is near room temperature. However, at low temperatures, the rate of transformation is impractically slow. To accelerate the kinetics the temperature is increased and then the pressure must be increased to stay within the diamond-stability field. Apparatus for achieving superpressure has been described in the literature (Bundy, 1962; Hall, 1960) and will not be reviewed here. Until recently, no apparatus was available that was capable of reaching a high enough pressure to be within the diamond field at a temperature where the graphite → diamond conversion rate was observable. The first diamond synthesis, achieved by Bundy et al. (1955), was achieved in the so-called belt apparatus (Hall, 1960), and employed a "catalyst" that accelerated the conversion rate at a low-enough temperature that the diamond field was accessible at "moderate" pressures. Small diamonds were made by the high-pressure decomposition of lithium carbonate but in general the family of "catalysts" that were discovered were all solvents for carbon so that practical diamond growth is actually from solution. Chromium, manganese, cobalt, nickel, and palladium are among the successful catalyst–solvents used. Graphite is the usual source of carbon but other sources including peanut butter (!) have been used. The catalyst is molten and acts as a thin-film transport medium between the phases. It may also help to nucleate the diamond. Because successful diamond growth is from solution it will be further discussed in Chapter 7. Although apparatus capable of forming a diamond in the absence of solvents is now available, larger diamonds are made by the catalyst process. It is probable, however, that, by employing thermal gradients, suitably oriented "ceramic" starting material, seeding, etc., further increases in size could be achieved even without the catalyst. The principal difficulty is experimental.

4.4.5-2 Boron Nitride (borazon)

Boron nitride is isoelectronic with carbon and the compound has two layerlike modifications (Pease, 1952; Hérold et al., 1958). Wentorf (1957, 1961) showed that BN transforms to a hard, dense *cubic* (zinc-blende structure) phase at 45,000 atm at 1700°C. This phase has been dubbed "borazon" by the General Electric Company and is of interest as an abrasive and cutting material that may be superior to diamond for some applications. The methods and difficulties of single-crystal growth are similar to those in diamond growth.

4.4.5-3 Silica

The high-pressure form of SiO_2, known as coesite, was first produced by L. Coes, Jr. (1953) at 750°C and 35,000 atm. Coes' starting materials were sodium or potassium metasilicate, and various "mineralizers" such as $(NH_4)_2HPO_4$, NH_4Cl, and KBF_4 were employed, so the growth may have been from a thin solution layer on the surface of his starting materials. Coesite has since been found in various meteorites and the p–T curve shown in Fig.

4.19 was determined by Boyd and England (1960). Coesite can be formed from pure α-quartz above 1200°C at the appropriate pressure but below this temperature a mineralizer such as H_2O or those mentioned above must be employed.

Two other high-pressure silica polymorphs are known, keatite (Keat, 1954) produced at 5000–18,000 psi at 380–585°C in the presence of alkali and stishovite (Stishov and Popova, 1961, quoted in Chao et al., 1962) at 160,000 atm at 1200–1400°C. Stishovite occurs in nature in certain meteorite craters. Several of the compounds whose structures are analogous to SiO_2, such as BPO_4, $MnPO_4$, $GaPO_4$, $AlPO_4$, and BeF_2 exhibit polymorphism similar to SiO_2 (Dachille and Roy, 1959).

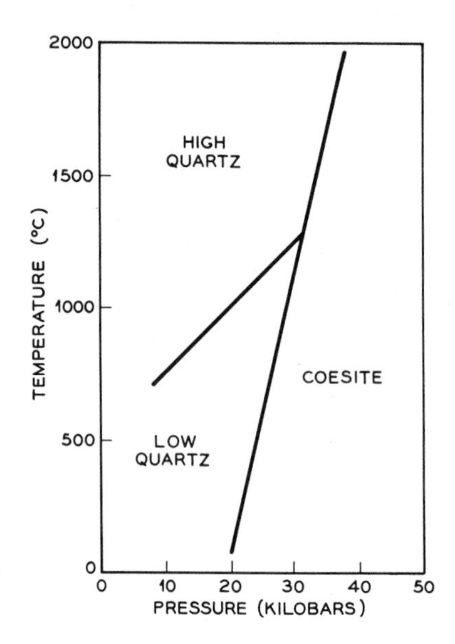

Fig. 4.19 Transition curve for coesite.

4.4.5-4 Other Materials

Ice, NH_3, CBr_4, CS_2, P, alkali metals, Bi (Bridgman, 1949; Hamann, 1957; Bradley, 1963), etc., undergo polymorphic modification at high pressure. Anomalies in resistivity at a phase transformation are often used as pressure-calibration points in high-pressure work. Single-crystal growth is subject to the difficulties discussed above.

4.5 Devitrification

When many glasses are heated, local recrystallization takes place and this process is called *devitrification*. It is usually undesired and glass compositions are chosen so that they will not devitrify easily. However, some glass compositions are chosen to accelerate devitrification deliberately. Such compositions contain nucleating agents and the devitrification is controlled to provide crystalline regions in a glassy matrix. Even when devitrification is nearly complete the crystallite size is small. The most important commercial example of such controlled devitrification is Pyroceram (Corning Glass Co.). So far no deliberate crystal growth by devitrification has been reported; however, with suitable control the process is not an impossible one. Very few attempts to grow single crystals on a seed from glass-forming melts have been made. Glass-forming melts do not easily form crystals by homogeneous nucleation, because at supercoolings large enough for embryo growth the viscosity is so high that the structure cannot order itself. However, R. L. Barns (private communication) has pointed out that in the presence of a seed, growth could perhaps be made to take place at temperatures high enough so that the viscosity was reasonably low. Monocomponent systems would be more likely to produce crystals because diffusion problems would be absent. Crystal pulling on a seed from monocomponent melts ordinarily thought to be glass formers is one attractive approach. Se is one example of a high-viscosity melt where glass formation prevents crystallization. However, when the melting point is raised (by the application of pressure) (Harrison, 1965), or when chain-breaking additives (Keezer, et al., 1967) are added to reduce the viscosity, crystals can be formed.

4.6 Polycomponent Solid–Solid Growth

Growth by solid-state precipitation or exsolution is the primary example of polycomponent solid–solid growth. Although extremely important in determining the properties and structure of alloys, the problem of nucleation control is so severe that it has not so far been applied to single-crystal growth.

A few materials have been grown from polycomponent systems by strain annealing. In particular, single-crystal Al–M solid solutions have been prepared, where M is Ag, Cu, Zn, Si, Ge, and Mg; Cu–M solid solutions, where M is Zn, Al, Cd, Ni, Mn, and Fe; and Fe–M solid solutions, where M is Si, Al, Cr, and Ni, have been grown as single crystals by strain annealing. Strain annealing presents the advantage that the distribution of the second component throughout the specimen is not altered by the crystal-growth process. The details of the growth conditions and pertinent references are given in Sec. 4.2.3 and Table 4.2.

REFERENCES

Allen, N. P., B. E. Hopkins, and J. E. McLennan, *Proc. Roy. Soc.* (*London*) **A234,** 231 (1956).

Andrade, E. N. da C., *Proc. Roy. Soc.* (*London*) **A163,** 16 (1937).

Armstrong, P. E., O. N. Carlson, and J. F. Smith, *J. Appl. Phys.* **30**, 36 (1959).

Aust, K. T., in *The Art and Science of Growing Crystals*, Ed. by J. J. Gilman, Wiley, New York, 1963, pp. 452ff.

Aust, K. T., and C. G. Dunn, *Trans. AIME* **209**, 472 (1957).

Bagaryatskii, Y. A., and E. V. Kolonstova, *Soviet Phys.-Cryst. English Transl.* **4**, 892 (1959).

Barrett, C. S., and T. B. Massalski, *Structure of Metals*, McGraw-Hill, New York, 3rd ed., 1966.

Beck, P. A., and P. R. Sperry, *J. Appl. Phys.* **21**, 150 (1950).

Beck, P. A., P. R. Sperry, and H. Hu, *J. Appl. Phys.* **21**, 420 (1950).

Boyd, F. R., and J. L. England, *J. Geophys. Res.* **65**, 749 (1960).

Bradley, R. S., *High Pressure Physics and Chemistry*, Vols. 1 and 2., Academic Press, New York, 1963.

Bridgman, P. W., *The Physics of High Pressure*, G. Bell, London, 1949.

Buckley, H. E., *Crystal Growth*, Wiley, New York, 1951, pp. 92ff.

Bundy, F. P., in *Modern Very High Pressure Techniques*, Ed. by R. H. Wentorf, Jr., Butterworth, London, 1962.

Bundy, F. P., H. T. Hall, H. M. Strong, and R. H. Wentorf, *Nature* **176**, 51 (1955).

Burgers, W. G., in *The Art and Science of Growing Crystals*, Ed. by J. J. Gilman, Wiley, New York, 1963, pp. 417ff.

Burke, J. E., *Atom Movements*, A.S.M., Cleveland, Ohio, 1951, p. 209.

Burke, J. E., *Conference on Kinetics of High Temperature Processes*, M.I.T Technical Press, Cambridge, Mass., and Wiley, New York, 1959, p. 109.

Cahn, R. W., *Acta Met.* **1**, 176 (1953).

Carpenter, H. C. H., and C. F. Elam, *Proc. Roy. Soc.* (*London*) **100**, 329 (1921).

Carpenter, H. C. H., and S. Tamura, *Proc. Roy. Soc.* (*London*) **113**, 28 (1927).

Chao, E. C. T., J. J. Fahey, J. Littler, and D. J. Milton, *J. Geophys. Res.* **67**, 419 (1962).

Chen, N. K., R. Maddin, and R. Pond, *Trans. AIME* **191**, 461 (1951).

Churchman, A. T., *Proc. Roy. Soc.* (*London*) **226**, 216 (1954).

Coble, R. L., *J. Appl. Phys.* **32**, 793 (1961).

Coes, L., Jr., *Science* **118**, 131 (1953).

Dachille, F., and R. Roy, *Z. Krist.* **111**, 459 (1959); *J. Am. Ceram. Soc.* **42**, 78 (1959).

Dohi, S., and T. Yamoshita, *Mem. Defense Acad. Yokosuka, Japan* **2**(2) (1957).

Dunn, C. G., *Trans. AIME* **185**, 72 (1949); *Cold Working of Metals*, A.S.M Cleveland, Ohio, 1949, p. 121.

Dunn, C. G., and G. C. Nonken, *Metals Progr.* **64**, 71 (1953).

Duwez, P., F. Odell, and J. L. Taylor, *J. Amr. Ceram. Soc.* **32**, 1 (1949).

Edwards, C. A., and L. H. Pfeil, *J. Iron Steel Inst. London* **1**, 129 (1924).

Fay, H., and C. D. Brandle, in *Crystal Growth*, Ed. by H. S. Peiser, Pergamon Press, New York, 1967, pp. 51ff.

Fisher, E. S., *Trans. AIME* **209**, 882 (1957).

Frondel, Clifford, *Am. Mineralogist* **31**, 58 (1946).

Frondel, Clifford, *Dana's System of Mineralogy*, Vol. III, Wiley, New York, 7th ed., 1962.

Fujiwara, T., *J. Sci. Hiroshima Univ.* **A9**, 227 (1939).

Fujiwara, T., and T. Ichiki, *J. Phys. Soc. Japan* **10**, 468 (1955).

Gane, N., *Bull. Inst. Met.* **4**, 94 (1958).

Gensamer, M., and R. F. Mehl, *Trans. AIME* **131**, 372 (1938).

Gifford, F. E., and E. A. Coomes, *J. Appl. Phys.* **31**, 235 (1960).

Graham, C. D., and R. Maddin, *J. Inst. Metals* **83**, 169 (1955).

Guinier, A., and J. Tennevin, *Progr. Metal Phys.* **2**, 177 (1950).

Hägg, G., and N. Karlsson, *Acta Cryst.* **5**, 728 (1952).

Hall, H. T., *Rev. Sci. Instr.* **31**, 125 (1960).

Hamann, S. D., *Physico-Chemical Effects of Pressure*, Academic Press, New York, 1957.

Harrison, D. E., *J. Appl. Phys.* **36**, 1680 (1965).

Harrison, F. W., *Research* **12**, 395 (1959).

Hérold, A., B. Marzluf, and P. Pério, *Compt. Rend. Acad. Sci. Paris* **246**, 1866 (1958).

Herzfeld, Charles M., Ed.-in-Chief, *Temperature—Its Measurement and Control in Science and Industry*, Vols. 1, 2, 1962, Vol. 3, 1963, Reinhold, New York.

Holden, A. N., and J. H. Holloman, *Trans. AIME* **185**, 179 (1949).

Honeycombe, R. W. K., *Met. Rev.* **4**, 1 (1959).

Hughes, F. L., H. Levinstein, and R. Kaplan, *Phys. Rev.* **113**, 1023 (1959).

Keat, P. P., *Science* **120**, 328 (1954).

Keezer, R. C., C. Wood, and J. W. Moody, *Crystal Growth*, Ed. by H. S. Peiser, Pergamon Press, New York, 1967, pp. 119ff.

Klassen-Neklyudova, M. V., *Mechanical Twinning of Crystals*, Translated from the Russian and published by Consultants Bureau, New York, 1964, pp. 93ff.

Lacombe, P., N. Ambrosis de Libanati, and D. Calais, *Compt. Rend.* **249**, 2769 (1959).

Lacombe, P., et. al., *J. Inst. Met.* **89**, 358 (1961).

Langton, L. L., *Radio Frequency Heating Equipment*, Pitman, London, 1949.

Laudise, R. A., in *Crystal Growth*, Ed. by H. S. Peiser, Pergamon, New York, 1967, pp. 5ff.

Legett, R. D., R. E. Read, and H. W. Paxton, *Trans. AIME* **215**, 679 (1959).

Leighly, H. P., Jr., and F. C. Perkins, *Trans. AIME* **218**, 379 (1960).

Levin, Ernest M., Carl R. Robbins, and Howard F. McMurdie, *Phase Diagrams for Ceramists*, American Ceramic Society, Columbus, Ohio, 1964, pp. 84ff.

Lommel, J. M., *Trans. AIME* **218**, 374 (1960).

Lyman, Taylor, Ed., *Metals Handbook*, 8th ed., Vol. 1, 1961, Vol. 2, 1964, American Society for Metals, Metals Park, Novelty, Ohio.

McKeehan, L. W., *Nature* **119**, 705 (1927).

Maddin, R., and N. K. Chen, *Trans. AIME* **197**, 1131 (1953).

May, E., *Industrial High Frequency Electric Power*, Wiley, New York, 1950.

Mercier, J., D. Calais, and P. Lacombe, *Compt. Rend.* **246**, 110 (1958).

Millner, T., *Acta Tech. Acad. Sci. Hung.* **17**, 67 (1957).

Montuelle, J., *Compt. Rend.* **248**, 1174 (1959).

Montuelle, J., and G. Chaudron, *Compt. Rend.* **240**, 1167 (1955).

Muhlenbruch, Carl W., *Testing of Engineering Materials*, Van Nostrand, Princeton, N.J., 1944.

Muhlenbruch, Carl W., *Experimental Mechanics and Properties of Materials*, Van Nostrand, Princeton, N.J., 1955.

Mullins, W. W., *Acta Met.* **6**, 414 (1958).

Murray, P., D. T. Livey, and J. Williams, in *Ceramic Fabrication Processes*, Ed. by W. D. Kingery, Technology Press of M.I.T, Cambridge, Mass., and Wiley, New York, 1958, pp. 147ff.

Nassau, K., H. J. Levinstein, and G. M. Loiacono, *J. Phys. Chem. Solids* **27**, 983, 989 (1966).

Nichols, M. H., *Phys. Rev.* **57**, 297 (1940).

Nichols, M. H., *Phys. Rev.* **94**, 309 (1954).

Parthasarathi, M. N., and P. A. Beck, *Trans. AIME* **212**, 821 (1958).

Paschkis, V., and J. Persson, *Industrial Electric Furnaces and Applications*, Wiley-Interscience, New York, 2nd ed., 1960.

Pease, R. S., *Acta Cryst.* **5**, 356 (1952).

Pintsch, J., *Gross Jahrb. Radiotech. Elektrotechnik.* **15**, 270 (1918).

Rieck, G. D., *Acta Met.* **6**, 360 (1958).

Robinson, J., *J. Appl. Phys.* **13**, 647 (1942).

Sachs, G., and K. R. Van Horn, *Practical Metallurgy*, American Society for Metals,

Cleveland, 1940.

Samsonov, G. V., *High Temperature Materials—Properties Index*, Plenum Press, New York, 1964.

Schubnikov, A., and E. V. Tsinzerling, *Z. Krist.* **83**, 243 (1932).

Sellers, D. J., A. H. Heuer, W. H. Rhodes, and T. Vasilos, *J. Am. Ceram. Soc.* **50**, 217 (1967).

Seraphim, D. P., J. I. Budnick, and N. B. Ittner, III, *Trans. AIME* **218**, 527 (1960).

Shaffer, Peter T. B., *High Temperature Materials—Materials Index*, Plenum Press, New York, 1964.

Smith, C. S., *Trans. AIME* **175**, 15 (1951).

Smithells, C. J., *Tungsten*, Chapman & Hall, London, 1952.

Spachner, S. A., and W. Rostoker, *Trans. AIME* **215**, 463 (1959).

Stein, D. F., and J. R. Low, Jr., *Trans. AIME* **221**, 744 (1961).

Stone, F. G., *Trans. AIME* **175**, 908 (1948).

Swalin, R. A., and A. H. Geisler, *J. Inst. Metals* **86**, 129 (1958).

Talbot, J., *Inst. de Recherche de la Siderurgie Rept. Series A*, No. 137 (1956).

Talbot, J., *Compt. Rend.* **251**, 243 (1960).

Thomas, L. A., J. L. Rycroft, and Edward A. Fielding, *Nature* **157**, 405–406 (1946).

Thomas L. A., and W. A. Wooster, *Proc. Roy. Soc.* (*London*) **A208**, 43 (1951).

Tiedema, T. J., *Acta Cryst.* **2**, 261 (1949).

Trinks, W., and M. H. Mawhinney, *Industrial Furnaces*, Vol. I, 4th ed., 1951, Vol. II, 2nd Ed., 1942, Wiley, New York.

Tsien, L. C., and Y. S. Chow, *Proc. Roy. Soc.* (*London*) **A163**, 19 (1937).

Van Uitert, L. G., F. W. Swanekamp, and S. E. Haszko, *J. Appl. Phys.* **30**, 363 (1959).

Van Uitert, L. G., J. J. Rubin, and W. A. Bonner, *IEEE J. Quantum Electronics* QE4, 622 (1968).

Wemple, S. H., M. DiDomenico, and I. Camlibel, *Appl. Phys. Lett.* **12**, 209 (1968).

Wentorf, R. H., Jr., *J. Chem. Phys.* **26**, 956 (1957).

Wentorf, R. H., Jr., *J. Chem. Phys.* **34**, 809 (1961).

Wilder, D. R., and F. S. Fitzsimmons, *J. Am. Ceram. Soc.* **38**, 66 (1955).

Williamson, G. K., and R. E. Smallman, *Acta Met.* **1**, 487 (1953).

Wooster, W. A., and Nora Wooster, *Nature* **157**, 405 (1946).

Yamamoto, M., and R. Miyasawa, *Sci. Rept. Res. Inst. Tohoku Univ.* **A6**, 333 (1954).

5

CRYSTAL GROWTH BY MONOCOMPONENT LIQUID–SOLID EQUILIBRIA

5.1 Introduction

In many ways monocomponent growth by a liquid–solid equilibrium is the preferred method of crystal growth. It is essentially growth by controlled freezing and consequently in comparison to other growth techniques is an uncomplicated, readily controllable process. Crystal growth by liquid–solid equilibrium is probably the most widely practiced commercial process for single-crystal growth. Except for growth from aqueous solution it was the earliest studied technique and it is likely that it has been the most extensively investigated. It would perhaps appear that liquid–solid growth is the only weapon the crystal grower might need in his armory. This statement, of course, is not true because many materials cannot be grown from their own pure melts. Among the reasons why liquid–solid growth may not be applicable to a given material are:

1. The material decomposes before it melts or melts incongruently.
2. The material sublimes before it melts or its vapor pressure is too high at the melting point.

3. The polymorphic modification desired is not the structure in equilibrium with the melt and solid–solid transformations do not yield good crystals of the desired polymorph.

4. The melting point is so high that it makes growth experimentally impractical.

5. The growth conditions are not compatible with some dopant that must be included in the crystal.

However, when reasons 1–5 are not applicable liquid–solid growth is usually the first technique the crystal grower will utilize. Liquid–solid growth principles have been discussed in Secs. 2.4, 2.5, 2.7, 2.8, 3.5, and 3.11. We have previously discussed the fact that there is some arbitrariness in distinguishing between mono- and polycomponent liquid–solid growth in certain borderline cases. Nevertheless, the classification is useful. Strictly speaking, monocomponent growth should not include systems where *any* second components—either accidental impurities or dopants—are present. Because all materials contain impurities, this viewpoint would eliminate monocomponent growth as a real process. It should be clear that when a second component is deliberately added to lower the melting point of the crystallizable material, the growth (ordinarily designated as growth from solution) should be considered polycomponent. A practical definition of monocomponent growth would allow the presence of impurities and dopants, provided their concentration is low enough that diffusion processes are not of overriding importance.

Ideally then, in monocomponent liquid–solid growth the rates can be quite rapid (diffusion is not severely growth-rate limiting) and the purity of the resultant crystals can be quite high (additional components are not present to cause contamination). In the simplest cases, all that one need know for growth is the melting point of the material. Thus liquid–solid techniques are a mainstay of the crystal grower. The remainder of this chapter will discuss the principal techniques.

5.2 Uncontrolled Freezing

The simplest method of forming crystals by liquid–solid equilibrium is by uncontrolled freezing of a melt. The difficulty is that the resultant product is generally a fine-grained polycrystalline matrix with few single crystals of any size. Metal castings are a typical example. Drinkers of lemonade and stronger beverages may be interested in the example of Fig. 5.1. Sometimes it is possible to identify single-crystal grains of appropriate size and remove them from the matrix for study. Standard metallographic techniques, including polishing and appropriate etching, are generally used. However, because

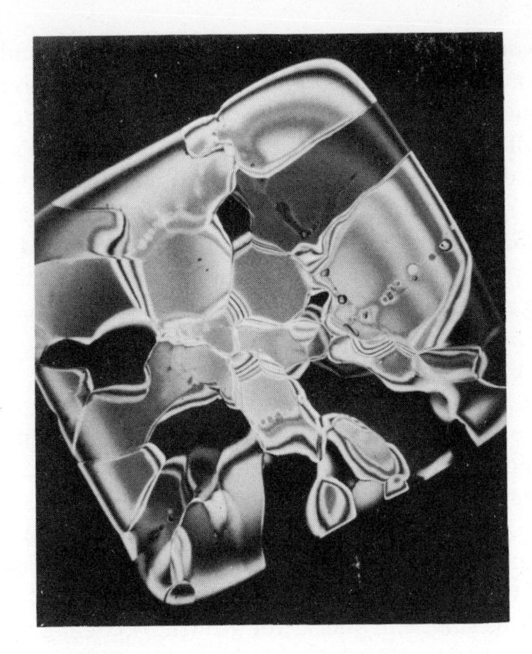

Fig. 5.1 Single-crystal grains in ordinary ice cube viewed between crossed polarizers. (Courtesy of R. L. Barns.)

the initial nucleation was random, it is rare that crystallites of very large size are formed.

Indeed, when large crystallites are observed, it is usually an indication that the material will form single crystals with great ease by the more sophisticated liquid–solid techniques. All of the remaining techniques are thus directed toward controlling nucleation so that one nucleus (or at worst a few) will be the seed on which all growth will take place. Hence, these techniques all make use of controlled thermal gradients such that the largest supercooling will be confined to a local region of the melt near the nucleus and most of the techniques make use of deliberate seeding.

5.3 Bridgman–Stockbarger and Related Techniques

A particularly useful method of controlling the supercooling so that a single crystal results is that first used by Bridgman (1925), and especially exploited by Stockbarger (1936, 1949). This technique is commonly known as the Bridgman–Stockbarger method, although sometimes the names of Tammann (1925) and Obreimov and Schubnikov (1924)† are associated with the tech-

†It is probable that Tammann or Obreimov and Schubnikov first practiced the technique and in justice it probably should receive their names (as it does in some European publications). We have adopted Bridgman–Stockbarger as the most generally used designation.

nique. Buckley (1951) discusses the historical aspects of the technique and distinguishes among various modifications introduced by successive researchers. He assigns the names of the original researchers to a number of these subtechniques. We shall discuss the technique and its modifications as a whole without reference to developments of only historical interest. Bridgman–Stockbarger growth is reviewed in most of the standard treatises on crystal growth (Smakula, 1962; Buckley, 1951; Gilman, 1963) and considerable discussion of the technique can be found in monographs devoted to the growth of particular classes of materials (Tanenbaum, 1959; Lawson and Nielsen, 1958; Rhodes, 1964).

In essence, the Bridgman–Stockbarger technique produces nucleation on a single solid–liquid interface by carrying on the crystallization in a temperature gradient. The material to be crystallized is usually contained in a cylindrical crucible [Fig. 5.2(a) and (b)], which is lowered through a temperature gradient [Fig. 5.2(h)] or a heater is raised along the crucible. In some cases the crucible is held stationary in a furnace designed to produce a temperature profile approximating a linear gradient and the furnace is then cooled [Fig. 5.2(i)]. Indeed, a portion of the gradient normally found in a furnace [the region *a–b* in Fig. 5.2(j)] may be sufficiently linear for growth by cooling. In either case an isotherm normal to the axis of the crucible is caused to move through the crucible slowly enough that the melt interface follows it. Usually the whole crucible is initially molten and the first nucleation will be several crystallites. Among the various expedients employed to cause one of these crystallites to dominate the solid–liquid interface are the following:

1. The tip of the crucible is conical [Fig. 5.2(b)] so that initially only a small volume of the melt is supercooled. Thus only one nucleus (or at worst several nuclei) will be formed. If one nucleus is suitably oriented it may dominate the growing interface.
2. The tip of the crucible is a capillary [Fig. 5.2(c)]. Only a very small volume of the melt is initially supercooled. If several crystallites are formed, there is an increased chance that, as the growth interface moves through the capillary, one will dominate the interface.
3. The tip of the crucible is conical [Fig. 5.2(d)] and the conical region is connected to the main volume of the crucible by a capillary. This gives the advantages of both expedients 1 and 2.
4. The tip of the crucible is conical [Fig. 5.2(e)]. The crucible flares to a reasonable volume, which is connected by a capillary to the main volume of the crucible or to another flared region, which is in turn connected by a capillary to the main volume of the crucible. Obviously as many bulb–capillary pairs as the researcher finds useful may be connected in series ahead of the main crucible volume. This arrangement encourages initial nucleation in a very small volume (the conical tip) and tends to select a single-crystal region from the bulb as the seed for growth in the capillary.

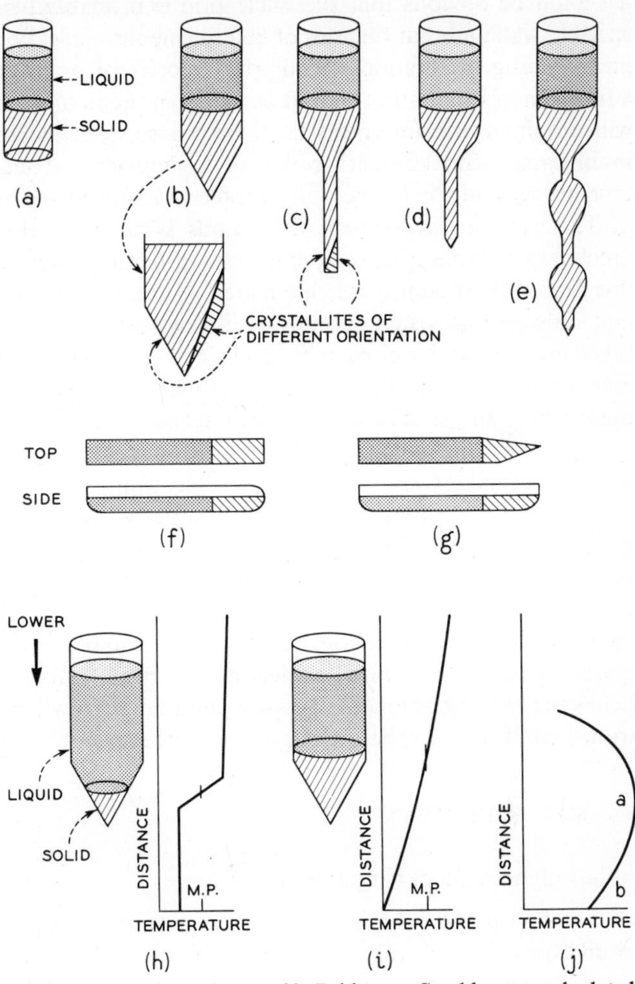

Fig. 5.2 Configuration used in Bridgman–Stockbarger and related techniques.

It should be apparent that the geometries of Fig. 5.2(a)–(e) can be adapted to growth in open boats (sometimes called Chalmers' method) [Fig. 5.2(f) and (g)]. In this case the furnace axis is horizontal rather than vertical. Whether or not a single crystal dominates the interface will be determined by the initial orientation of the first nucleated crystallites and the orientation of the grain boundaries between these crystallites. These factors have not been studied in any detail in practical Bridgman–Stockbarger growth principally because it is almost always possible to find empirically a crucible geometry, gradient, and lowering rate (or cooling rate) that will produce single crystals or at least large single-crystal regions in the volume of the crucible. However, it should be obvious that the nucleation is probably heterogeneous on the crucible wall and that theories of heterogeneous nucleation should be helpful in predicting orientations. Similarly, theories of grain-boundary energy as a function of orientation should be useful in predicting whether single grains will dominate the interface. It is, of course, possible to seed deliberately in Bridgman–Stockbarger growth, if a single-crystal seed is placed in the crucible tip and the temperature profile is arranged so that it does not melt it. This is often experimentally tedious because, in the usual Bridgman–Stockbarger apparatus, temperatures are neither known nor controlled to this degree and because crucible materials are usually opaque, as are furnace materials, so that visual observation is not possible. However, vitreous-silica tubes grooved and wound with resistance wire together with vitreous-silica crucibles such as are used in vapor growth (see Sec. 6.4.2-2) would permit observation and should prove advantageous in preliminary experiments.

A common difficulty in Bridgman–Stockbarger growth is to have too small a thermal gradient along the crucible. Many melts will supercool appreciably before nucleating. If the melt can supercool sufficiently and the thermal gradient is relatively small, often the whole sample will be below the melting point before the first solid nucleates. When nucleation occurs under these conditions, growth through the rest of the melt is very rapid and small crystals or polycrystals of poor perfection are the inevitable product. Large thermal gradients ensure that initial nucleation will begin before the entire sample is below the melting point. Under such conditions growth proceeds under controlled conditions as the melting-point isotherm moves through the sample.

5.3.1 EQUIPMENT

Bridgman–Stockbarger growth requires

1. A crucible (or boat) of suitable geometry compatible with the compound being grown, the growth atmosphere, and the temperature
2. A suitable furnace capable of producing the desired gradient
3. Temperature-measuring and control equipment and in some cases equipment for programming temperature or equipment for lowering the crucible

Because the Bridgman–Stockbarger technique requires a crucible, the melt must be unreactive with the crucible material at the growth temperature. It is also helpful if the solid does not stick to the crucible because this helps to minimize strains and makes the removal of the grown crystal after the growth possible without destroying the crucible. The crucibles of Fig. 5.2(a)–(e) are easily closed so that one advantage of Bridgman–Stockbarger growth using those geometries is that moderately volatile compounds can sometimes be grown and that atmosphere control can be achieved without additional difficulty. It is more difficult to control volatility or regulate the atmosphere in open boats. However, open boats present the advantage that the solid–liquid interface can usually be observed.

Crucibles, sometimes called molds in Bridgman–Stockbarger work, have been made of Pyrex, Vycor, vitreous silica, alumina, noble metals, or graphite and other materials. Pyrex (softening point ~600°C), Vycor (softening point ~1000°C) and vitreous silica (softening point ~1200°C) are limited to low melters, and have the advantage that in a suitably designed furnace they permit visual observation of the growth process. Split molds can be made from these materials for easy removal of the grown crystals, but breaking the mold is often a satisfactory expedient. Alumina fired with various binders has been used for the growth of aluminum. Graphite has been used as a crucible material for the growth of metals that do not easily form carbides and for the growth of some nonmetals. Graphite is useful up to at least 2500°C in nonoxidizing atmospheres. It is usually necessary to sweep an inert gas through the furnace when a graphite crucible is used. Noble metals are discussed in Secs. 4.2.2, 5.4.1, 7.3.1 and 7.4.1. For unreactive materials base metal and various ceramic crucibles have sometimes been used. In a few cases, crucible materials such as carbides and even single-crystal fluorides have been used.

To prevent strains in materials that wet the crucible, sometimes so-called "soft molds" are used. One version of a soft mold is a very thin, easily deformable platinum container. Even though such a mold is wet by the melt, it deforms as the material contracts during freezing resulting in no introduction of strain in the grown crystal. If such a mold is so thin that it is incapable of supporting the melt, it may be supported by a stronger external crucible. Sometimes it is convenient to pack loosely a material such as bubbled alumina between the two containers.

Two sorts of furnace gradients have been used in Bridgman–Stockbarger growth. When the gradient is caused to move through the crucible by cooling the whole furnace a gradient like that of Fig. 5.2(i) is required. In a resistance-wire-wound furnace where the windings are evenly spaced, the hottest region is in the center of the furnace and the gradient is like that shown in Fig. 5.2(j). In such a furnace the region *a–b* approximates a linear gradient and can usually be used where one is required. Where the crucible is lowered through the furnace (or the furnace is raised about the crucible) it is usually advantageous to have two isothermal regions with a gradient between them. This permits annealing the crystal after growth without introducing large thermal strains caused by excessive temperature gradients. Such a furnace set up ideally should consist of two furnace regions with a minimum of thermal coupling between them. The regions should be capable of independent control and a thermal barrier should separate them. To ensure isothermal conditions in the separate regions it is helpful if the inner furnace walls are made of a material of high thermal conductivity. It is also useful if the gradients in the crucible are controlled essentially by the thermal conditions in the furnace so that thermally nonconducting crucibles would be advantageous (in practice, crucible materials compatible with other conditions are almost always of rather high thermal conductivity and for practical reasons this criterion is not fulfilled). A reflecting baffle (often of Pt sheet) whose size just permits the passage of the crucible is often used to separate the two zones. Where temperatures are low enough it is sometimes possible to use glass or vitreous silica for the core and to wind heating tape or resistance wire so that with a transparent crucible or an open boat visual observation is easy. In horizontal Bridgman–Stockbarger the same criteria are applicable to furnace design. In some cases radio-frequency heating may be used. This again aids in visual observation. It is usual to melt the whole boat contents (except perhaps for a seed) but on some occasions a narrow zone is passed through the boat. This sort of growth is considered under crystal growth by zone melting in Sec. 5.5. Furnace design, radio-frequency heating, temperature measurement, and temperature control are further discussed in Secs. 4.2.2, 5.4.1, 7.3.1, and 7.4.1.

In some cases it is advantageous to cool the conical tip or the end of the capillary of the crucible. Heat leaks are provided by placing one end of a conductive material in contact with the region that is to be cooled and maintaining the other end in a cooler region of the furnace or letting it project outside the furnace. "Spot cooling" is also possible by directing cool gases (in the simplest case an air blast) against the desired region. In extreme cases one region of the crucible may be in contact with a water-cooled coil or block.

The usual method of lowering crucibles is to attach them to a wire or chain and to pay out the chain by means of a suitable-size sprocket (Fig. 5.3) attached to a clock motor (available from Cramer Division of Giannini Control Corporation, Old Saybrook, Conn., and Automatic Timing and Controls, Inc., King of Prussia, Pa.). Clock motors with speeds from a fraction of a revolution up to many revolutions per hour are available and by choosing suitable-size gears desired lowering rates are easily obtained. Constancy of rate is probably no better than $\pm 0.1\%$ over periods of several hours.† Improved control can be obtained by shock-mounting the apparatus so that vibrations are not transmitted to the crucible. Transmitting the crucible

---

†Because a clock motor is a synchronous motor, its accuracy is determined by the accuracy of the average line frequency of the ac source. The line frequency is usually maintained to much better than $\pm 0.1\%$ over periods of the order of 24 hr but short-term variations are common. Nevertheless, the greatest control difficulties in Bridgman–Stockbarger growth are undoubtedly caused by variations in temperature control, vibrations carried to the crucible from the environment, and sloppy transmission of the relatively precise movement of the clock motor through the mechanical system employed to the crucible.

motion through a rigid support, such as a rod attached to the crucible, also tends to eliminate accidental crucible motion. Where a heat leak is desired the rod may be attached to the crucible bottom. More precise control over lowering rate (with a commensurate increase in expense and complexity of equipment) can be obtained by using a conventional crystal puller to provide the motion. Rotation of the crucible to even out thermal asymmetry in the furnace is sometimes helpful. Lowering apparatus where the drive is by means of rotating lead screws (as in a conventional lathe) is often used in crystal pullers and can be adapted for Bridgman–Stockbarger work. Similarly, the apparatus used to obtain travel in zone refiners can be adapted for horizontal Bridgman–Stockbarger-type growth. Crystal pullers are further discussed in Sec. 5.4.1 and zone-melting apparatus in Sec. 5.5.

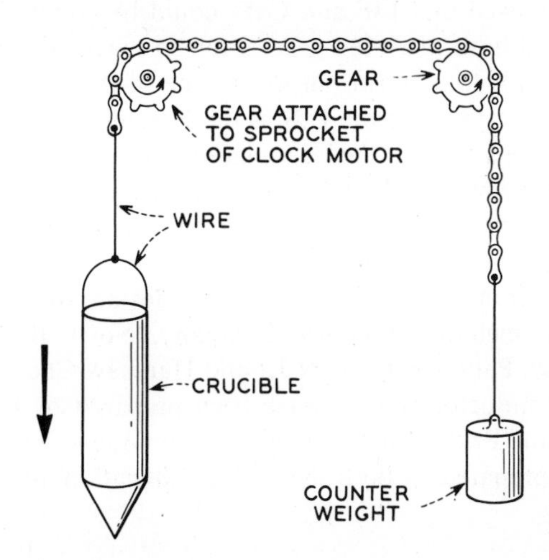

Fig. 5.3 Bridgman–Stockbarger growth—method of lowering crucible.

Aust and Chalmers (1958) have shown that lineage in boat-grown crystals can be reduced when the solid–liquid interface is *not* normal to the boat axis. Because low-angle grain boundaries (lineage) can form by the coalescence of dislocations reducing the dislocation density will reduce lineage. Dislocations tend to propagate normal to the growth interface. Consequently, if the interface is not normal to the boat axis (Aust and Chalmers call this a "tilted interface"), dislocations will grow out of the crystal before they reach a high enough density to interact and form low-angle grain boundaries. Similarly, a convex (with respect to the solid) interface is better than a concave interface†; and these arguments will, of course, hold for liquid–solid growth by other means. Crystals with a deliberate grain boundary (bicrystals) have been

†A convex interface also helps to minimize wall nucleation, because the farthest point of solid advance is farthest from the wall. This helps to prevent nucleation of new crystallites ahead of the main crystal.

grown by Chalmers (1940) and by Fleischer and Davis (1959) by using two seeds of different orientation in boat growth. A change in orientation of the grown crystal from that of the seed may be obtained by appropriate bend in the channel that connects the seed and the main growth region.

5.3.2 GROWTH OF SPECIFIC CRYSTALS

The three classes of materials to which Bridgman–Stockbarger growth has been most frequently applied are metals, semiconductors, and alkali and alkaline earth halides. By far the largest commercial application has been to the alkali and alkaline earth halides. Bridgman's first work (1925) was on bismuth, and during the early years of the technique interest centered on metals. Stockbarger (1936) showed that LiF and CaF_2 could be grown, and the methods that he pioneered have provided a basis for an extensive commercial growth of halide crystals for optical purposes. One of the principal and earliest suppliers of optical crystals in the U.S. is the Harshaw Chemical Company, Cleveland, Ohio. The advent of the laser placed more severe requisites on the optical quality of halide crystals and generated a demand for rare-earth-doped halides with very low light scattering. Guggenheim (1961, 1963) developed techniques for reducing scattering in fluorides and for controlled doping with rare earths in definite oxidation states. Fluorides are available from a number of suppliers, including Optavac Company, North Brookfield, Mass.; Isomet Company, Palisades Park, N.J.; and Harshaw Chemical Co., Cleveland, Ohio. Semiconductor crystals have been prepared by horizontal boat modifications of the Bridgman–Stockbarger technique and many organic materials have been prepared by both the boat modification and the normal technique.

5.3.2-1 Metals and Semiconductors

The first metal grown to any extent was bismuth (Bridgman, 1925). Its melting point is 271°C, so Pyrex crucibles can be used. Bridgman succeeded in growing crystals up to 2.2 cm in diameter at a rate of 4 mm/hr. Speeds up to 60 cm/hr were possible at very narrow diameters. Much of the literature of Bridgman–Stockbarger growth suffers from inadequate reporting of experimental conditions. As a minimum the following are necessary to specify adequately the growth conditions:

1. Temperature profile in the crucible (or at least what the gradient was in the furnace)
2. Rate of growing interface movement (related to lowering rate or cooling rate)
3. Orientation of the grown crystal (orientation of seed if one was used)
4. Purity of the starting materials

5. Information on the stoichiometry, impurity content, and perfection of the grown crystal

6. Obvious experimental details, such as crucible material, precision of temperature control, special problems, etc.

It is quite rare that all of these factors are adequately discussed in the published literature. In the case of bismuth, to the author's knowledge the most perfect single crystals have been produced in horizontal boats (Wernick and Buehler, 1960), using techniques previously reported (Wernick et al., 1957). Because of its low melting point and weak bonding normal to the *c* axis, Bi is very easily damaged. Chilling grown crystals in cold water introduces many dislocations (Wernick and Buehler, 1960), which can be revealed by an etching (Lovell and Wernick, 1959). Differential contraction (and perhaps wetting) during cooling in growth in vertical crucibles causes damage. Other low melters that have been prepared by Bridgman–Stockbarger growth are mentioned in Table 5.1.

Copper is an example of a relatively high-melting metallic crystal (mp 1083°C) that has been grown by the Bridgman–Stockbarger technique. The apparatus employed for the growth of high-melting materials has usually been of rather complicated construction and where oxide formation is a problem, as in copper, growth in vacuo has usually introduced rather severe experimental difficulties (Nix, 1938; Siegel, 1940). The lowering mechanism has often been placed partly within the lowering chamber and the resistance heating elements require large currents (graphite or silicon carbide) or have poor oxidation resistance (tungsten or molybdenum). Wernick and Davis (1956) devised a simple apparatus for the growth of high-temperature metals in vacuo. The crucible is a split graphite mold (Fig. 5.4) that is contained within a porcelain tube. The tube can be evacuated and the tube and mold are arranged for lowering through a gradient. The gradient in the furnace was reported as 30.5°/in. when the porcelain tube was not present, the lowering rate was 2 in./hr and the vacuum $< 25\ \mu$. Single crystals as large as $\sim \frac{3}{8} \times \frac{1}{8} \times 3\frac{1}{2}$ in. were grown without difficulty. It was even possible to lower the crucible intermittently by hand in 1-in. intervals every 10–20 min and still obtain crystals. The copper rod starting material was 99.999 + %, and spectroscopic examination of the grown crystals did not reveal any contamination. The graphite used was $> 99.75\%$ pure. Laue back-reflection pictures were used to establish qualitatively that the whole sample was a single crystal. Electron micrography was used for further study of perfection. The presence or absence of Kikuchi lines on electron diffraction patterns was used as a criterion of perfection, and it was shown that the poorest material was the first to freeze. This effect is probably general in unseeded Bridgman–Stockbarger growth because the supercooling is highest before initial nucleation. Other metals grown by the Bridgman–Stockbarger method are mentioned in Table 5.1.

Table 5.1

SOME REPRESENTATIVE CRYSTALS GROWN BY BRIDGMAN–STOCKBARGER METHOD†

Compound	*Formula*	*Melting Point (°C)*	*Crucible Material*	*Gradient*	*Cooling or Lowering Rate*	*Atmosphere*	*Comments*	*Reference*
Silver bromide	AgBr	434	Pyrex, quartz, Pt	10°/cm or higher	1–5 mm/hr	Cl_2 (HCl or HBr pre-evacuate)	Size doping, etc. given; quartz probably best	Berry et al., 1963
Argon	Ar	−189.4	Glass	5°/mm	1 mm/min	Ar	Polycrystals 4 mm diameter	Stansfield, 1956
Gold	Au	960.5	Graphite	~5°/cm	"Slow"	N_2	Single crystals up to 8 in. long	Elam, 1926
Copper	Cu	1083.2	Graphite	~12°/cm	5–~20 cm/hr	Vacuum	Kikuchi lines prove high perfection	Wernick and Davis, 1956
Nickel	Ni	1455	Recrystallized Al_2O_3		0.1–0.2 mm/hr	Vacuum	No seeding (see Schadler, 1963 for other variations; see	Pearson, 1953
Lithium	Li	179	Steel, stainless steel		2°–30°/hr	Ar	Goss, 1963 for other low melters)	Bowles, 1951; Champier, 1956; Nash and Smith, 1959
Fluorite	CaF_2	1392	Ta, Fe, or Ni	Not given	10 mm/hr	Vacuum	Special techniques to remove CaO‡	Stepanov and Feofilov, 1958
Fluorite	CaF_2	1392	C	"Large"	~1 mm/hr	Vacuum		Stockbarger, 1949
Lithium fluoride	LiF	870	Pt	Not given	Not given	Air or vacuum better		Stockbarger, 1949

†Smakula (1962) gives literature references (but no summary of conditions) to over 100 materials including organic materials grown by the Bridgman–Stockbarger method.

‡For improved techniques see Sec. 5.3-2.

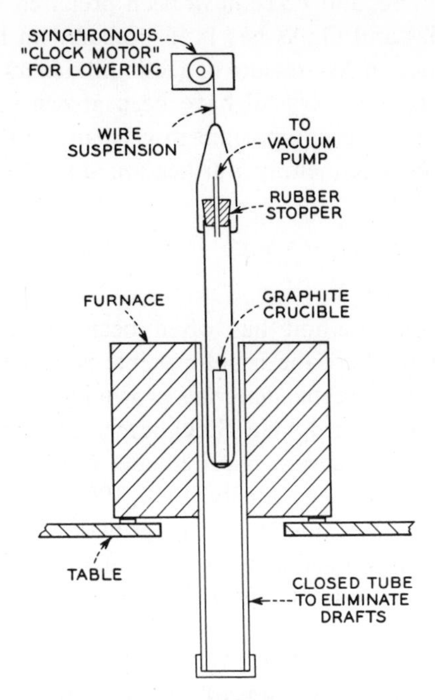

Fig. 5.4 Bridgman–Stockbarger growth of copper (after Wernick and Davis, 1956).

Vertical Bridgman–Stockbarger growth is seldom used for the preparation of the common semiconductors.† This is because materials such as silicon and germanium expand several percent on freezing. The cylindrical crucible used in vertical Bridgman–Stockbarger growth acts as a constraint on the growing crystal and the expansion usually leads to large stresses. The electrical properties of most semiconductors are especially sensitive to perfection, and the strains introduced in vertical Bridgman–Stockbarger growth are severe enough to induce dislocations and in some cases low-angle grain boundaries, which severely degrade the properties. However, where the semiconductor is volatile or a volatile dopant is used, the ease of establishing a closed system might make semiconductor crystal growth by the vertical Bridgman–Stockbarger method attractive. The fact that the shape of the grown crystal is fixed by the crucible geometry might present economic advantages in dicing crystals because a fixed uniform shape and size could easily be grown repeatedly in a commercial process. Powder molds or "soft molds" have been used (Rowland, 1951) to eliminate strain in the growth of some materials. Many of the difficulties associated with strain in the vertical Bridgman–Stockbarger method are eliminated if growth is from an open boat in a horizontal furnace (Kapitza, 1928; Andrade and Roscoe, 1937).

†Commercial suppliers of semiconductor crystals are too numerous to be mentioned here.

Single crystals of PbS, PbSe, and PbTe have been prepared in vertical crucibles (Lawson, 1951, 1952) and GaAs has been prepared in horizontal boats sealed in quartz held under an As pressure of a few millimeters (Edmund et al., 1957). Special shapes of a Ge crystal have been grown in graphite molds. Boat growth is, of course, very important in germanium technology because it is compatible with zone-melting purification and zone-leveling doping processes.

5.3.2-2 Nonmetals

The Bridgman–Stockbarger method has often been applied to the growth of low-melting organics where the usual crucible is a glass tube. It would be superfluous to list all the organic materials grown by this technique.

In many ways the triumph of Bridgman–Stockbarger growth has been the growth of fluorides. The most severe requirements of perfection and optical quality are for laser crystals. Guggenheim (1960, 1961, 1963) has been mainly responsible for developing techniques for the growth of laser quality fluorides. The principal difficulties associated with crystals before his work were control of scattering centers and control of oxidation state of the rare-earth dopants.

Among the crystals grown were Cr, Mn, Co, Ni, Zn, Y, La, Tb, and Ca fluorides. Perhaps the most extensive study was of CaF_2 growth. Guggenheim emphasizes that to prepare first-quality CaF_2 the formation of CaO must be prevented. The starting material must be thoroughly dry and free of surface oxide. This can be accomplished by holding CaF_2 powder for 18 hr at 800°C in an atmosphere of dry HF. If any moisture is present in the starting material or is introduced into the system during growth the reaction

$$CaF_2 + H_2O \longrightarrow 2\,HF_2 + CaO \tag{5.1}$$

results in CaO particles in the grown crystal that severly scatter the laser light and catastrophically raise the threshold for laser action. HF represses this reaction and can react with CaO slowly to remove traces from the starting material. In this connection previously grown crystals of CaF_2 provide a source of CaF_2 that is relatively CaO-free. A better way of ensuring CaO-free starting material is to treat $CaCl_2$ with HF:

$$CaCl_2 + 2\,HF \longrightarrow CaF_2 + 2\,HCl \tag{5.2}$$

CaF_2 prepared in this manner is quite free of oxides that lead to scattering.

During growth the atmosphere should be free of oxidizing agents and thoroughly dry. Argon passed over Linde molecular sieves at liquid-nitrogen temperatures was found generally satisfactory. However, even better results were obtained when helium that had been diffused through fused quartz was used. A glass capillary system for He purification devised by McAfee (1958), which has a large throughput, was especially useful in this regard.

A particularly sensitive test of technique was found to be the growth of $CaF_2: Sm^{2+}$. Even at concentrations less than 0.05 mole % Sm^{2+} gives an intense green color. If oxidizing agents contaminate the growth, Sm^{2+} is changed to Sm^{3+}, which is colorless.

It was found important (Guggenheim, 1963) that the thermal gradient be at least 7°/cm for CaF_2 and 30°/cm for LaF_3 to prevent supercooling. Lowering rates of 1–5 mm/hr were generally used for fluoride growth and in some cases the oxidation state of the rare earth was altered by postgrowth electrolytic reduction (Guggenheim and Kane, 1964; Fong, 1964a). Figure 5.5 shows a schematic of Guggenheim's CaF_2 Bridgman–Stockbarger apparatus.

Fong (1964b) has studied extensively the effect of γ radiation on rare-earth-doped fluorides and has used this technique to produce, for example, Dy^{2+} in CaF_2.

Table 5.1 summarizes a number of representative crystals grown by the Bridgman–Stockbarger technique. Table 5.1 is by no means complete; probably several thousand crystals have been reported as grown by this technique. However, in most cases details are scanty, and Table 5.1 is meant to be indicative of the range of conditions and materials so far exploited.

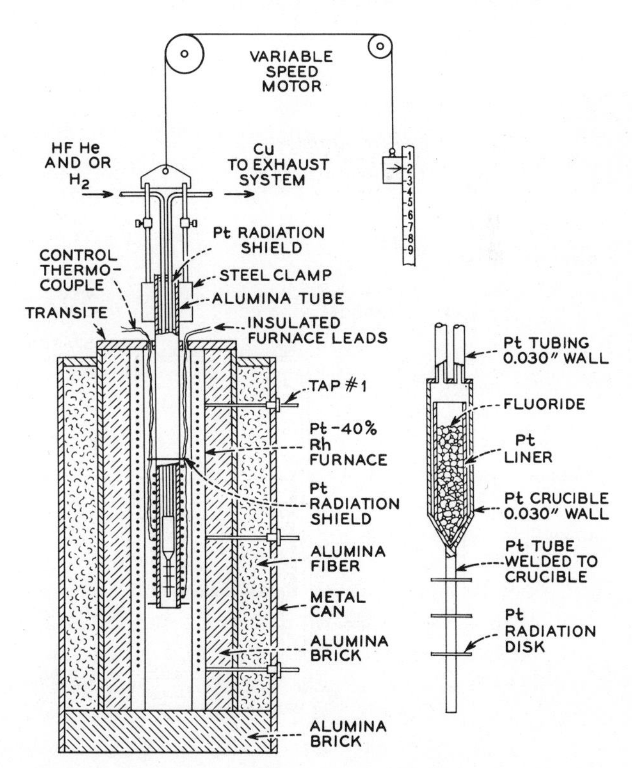

Fig. 5.5 Bridgman–Stockbarger growth of calcium fluoride (after Guggenheim, 1963).

5.4 Czochralski and Related Techniques

The technique of pulling from the melt first practiced by Czochralski (1917) results in a crystal grown free of the physical constraints imposed by the crucible. Figure 5.6(a) shows a schematic of Czochralski growth, and Fig. 5.7 shows a photograph of a typical pulling apparatus. The necessity of having to maintain the melt in a crucible that often acts as a contamination source is one of the main disadvantages of the technique, although some modifications of technique that alleviate this difficulty will be discussed later. The criteria that must be fulfilled for successful pulling are

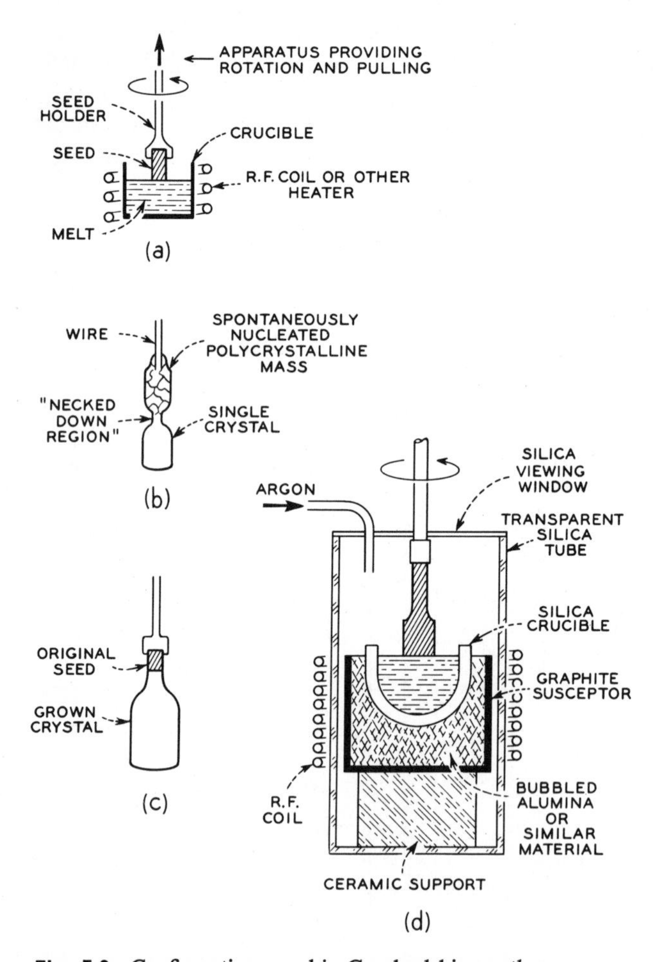

Fig. 5.6 Configurations used in Czochralski growth.

Fig. 5.7 Apparatus used for Czochralski growth of compounds such as $LiNbO_3$, $CaWO_4$, etc. (Courtesy of *Bell Laboratories Record*).

1. The crystal (or crystal plus dopant) should melt congruently without decomposition. If the crystal melts incongruently, it may sometimes be possible to grow from a melt of composition so that the crystal is the stable phase. However, such growth is from a polycomponent system and is subject to the difficulties associated with solution growth. This sort of growth will be discussed in Chapter 7. If the decomposition product is gaseous, it is often possible to use a closed apparatus and establish an equilibrium pressure of the decomposition product to repress decomposition.

2. The crystal should not be reactive with the crucible or the atmosphere present during pulling. In a closed apparatus inert, oxidizing, or reducing atmospheres may be provided.

3. The melting temperature should be attainable with the heaters available and should be below the melting temperature of the crucible.

4. It should be possible to establish a combination of pulling rate and thermal gradients where single-crystal material can be formed.

The principal advantage of pulling is that growth can be achieved on the seed under conditions of very good control. Good control results from the fact that the seed and grown crystal are visible during growth, and the experimentalist can use visual observation of the growth process to guide him in adjusting the process to control perfection. In addition, growth in any given direction is usually easy to obtain when oriented seeds are available. When a seed is not available, spontaneously nucleated crystallization is usually started on a wire [see Fig. 5.6(b)]. The polycrystalline mass formed is grown for a short distance and then its diameter is reduced (it is "necked down"), usually by slightly raising the temperature or by increasing the pulling rate. In practice, increasing the pulling rate is a hard method to use for necking down since often the seed pulls out of the melt. Once the crystal is necked down, a single crystallite will usually dominate the growing interface. The diameter of the crystal is then increased and single-crystal growth results. In recalcitrant substances it is sometimes necessary to select a crystallite from a polycrystalline mass (prepared in preliminary experiments) for a seed for later experiments. Once seeds are available it is preferable to start growth on a seed [Fig. 5.6(c)] of comparatively small diameter and then to increase the diameter of the crystal slowly by lowering temperature until the desired diameter is obtained. Initial nucleation is usually better controlled on small area seeds. If poor growth results, it is easy to melt back (provided the whole seed is not lost) and begin anew.

Czochralski (1917) first applied the pulling technique for low-melting-point metals such as tin, lead, and zinc. Fig. 5.6(d) shows a typical apparatus. Over the years the technique has been used for congruently melting compounds of all classes but perhaps its widest application is to semiconductors. Teal and Little (1950) were the first to produce single crystals of germanium and of silicon and this work provided the best means of preparing well-characterized crystals of these semiconductors for research and applications. Crystal pulling is an important commercial technique in present-day semiconductor technology. Nassau and Van Uitert (1960) applied crystal pulling to inorganic compounds of interest as laser hosts, and Nassau in a series of papers describes the growth and properties of $CaWO_2$: Nd (Nassau and Broyer, 1962; Johnson and Nassau, 1961). Good reviews of some aspects of the technique have been published in Gilman (1963).

An interesting technique in some ways related to Czochralski growth is the Kyropoulos (1926, 1930) technique. A seed is inserted into the melt centered in an appropriate crucible. However, the seed is not withdrawn from the melt; instead, growth is achieved by causing an isotherm corresponding to the melting point of the substance to move from the seed downward into the crucible [Fig. 5.7A]. This is often done by simply cooling the seed by means of a seed holder, which causes a large heat leak from the furnace. Alternatively the crucible is moved through a thermal gradient or the temperature of a furnace with an appropriate gradient near the seed is lowered. The requirements and applicability of Kyropoulos growth are generally similar to Czochralski growth except that

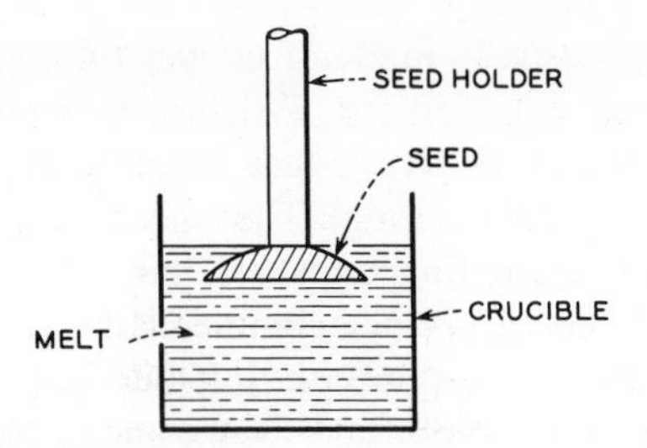

Fig. 5.7A Kyropoulos growth.

1. Kyropoulos growth is best fitted to the growth of crystals of large diameter/height, while Czochralski growth is best fitted for crystals of large height/diameter.

2. Growth direction and size are more subject to control in Czochralski growth.

3. Because pulling is not used, Kyropoulos equipment is simpler.

5.4.1 EQUIPMENT

For Czochralski growth the following equipment is required:

1. A means of heating the melt and controlling melt temperature and gradients.
2. A means of containing the melt.
3. A means of holding, rotating, and withdrawing the seed.
4. A means of controlling the atmosphere if air is not a compatible atmosphere.

The main heating means are radio-frequency heating and resistance heating. Radio-frequency heating requires either that the melt or crucible be conductive enough to couple to the radio frequency field or that a susceptor be used to couple to the field and that the crucible and melt be heated by the susceptor. A common frequency is 450 kc and a useful power to heat a 50-cm^3 quartz crucible contained in a graphite susceptor to the melting point of Ge (937°C) is 5 kW (10-kW generators are more commonly employed because they permit inefficient coupling). A similar frequency is useful for heating inorganics contained in metal crucibles, but a power up to 20 kW would be required to heat a similar-size Ir crucible containing yttrium–aluminum garnet to its melting point ($\sim$1900°C). In some cases auxiliary heating is useful to bring materials to a temperature where their conductivity has increased to the point that they begin to couple to the rf field.

For insulators, high frequencies (the so-called dielectric heating range) such as may be obtained with diathermy units can be used. Even for materials of moderate resistivity such as Si, a frequency of 3–5 Mc makes impedance matching easier and eliminates the need for preheating. In addition, this frequency range decreases mechanical oscillations in zone width in zone refining (see Sec. 5.5).

The work coils, which are usually made of copper tubing, are usually wound into an evenly spaced cylinder for Czochralski work, although shaped coils have sometimes been used to produce special gradients (shaped coils are of more importance for zone melting). The coils are cooled by water circulating through them and circulating units are part of commercial rf generators. Radio frequency generators are commercially available from, for instance, Westinghouse Electric Corporation, Pittsburgh, Pa., and Lepel High Frequency Laboratories, Inc., Woodside, N.Y., and rf heating is discussed in Simpson (1960) and Brown et al. (1947).

It is especially important to "tune" the generator so that its internal impedance matches the combined impedance of the work-coil–susceptor–crucible arrangement. This is generally done by altering the capacitance of a tank circuit or by varying the impedance of an rf transformer associated with the generator. If delicate rf measurements are to be made in the vicinity of the generator, extreme care in shielding will be required to minimize interference. Radio-frequency burns of body tissues are particularly difficult to heal and care should be exercised so that fingers are not accidentally placed close to the work coil.

One means of temperature control in rf heating is to use a thermocouple placed close to or in the melt crucible. The thermocouple emf is used to control the power of the generator. Typical circuits are described in Simpson (1960) and Brown et al. (1947) and such controllers are available from, for instance, Leeds and Northrup Company, Philadelphia, Pa. Shielding against rf pickup is achieved by placing the couple in a grounded sheath. A second means of temperature control is to utilize a circuit that maintains constant power in the rf coil. Such circuits are described in Simpson (1960) and are available from Leeds and Northrup. A constant power circuit compensates for line voltage variations but of course will not maintain constant temperature as the heat loss changes. At temperatures below 1200°C, thermocouples in conjunction with readily available self-balancing potentiometers provide precision of the order of $\pm 2°C$ for periods of many hours. If thermocouple sensing is combined with a feedback system (Simpson, 1960) that maintains constant rf input power at constant thermocouple output (e.g., compensates for line voltage fluctuations), sensitivity of the order of $\pm 1°C$ is attainable.

It should be borne in mind that irregular convection can cause temperature fluctuations of as much as ± 10–$30°$ (see Sec. 5.4.2) if the thermal conditions are appropriate even when external control is near perfect. At temperatures above 1200°C difficulties increase because commonly available thermocouples have shortened lives and radiation pyrometers are troubled by fluctuating emissivities and difficulties associated with the absorption of radiation caused by material intruding into the sight path. Temperature can be determined by an optical pyrometer sighted on the melt surface or the crucible wall, but emissivity corrections may be required if one wishes to know the true temperature (for control no correction is required). However, uncorrected pyrometer temperatures are useful in that they can be approximated at a later time if the experimental arrangement is similar. An interesting pyrometer is the "two-color" pyrometer typical of which is the Rockwell Color-Ratio Pyrometer (Rockwell Manufacturing Company, Chicago, Ill.), which determines temperatures by intercomparing the intensity at two different wavelengths. This reduces (but does not eliminate) the need for emissivity corrections, and the output from such a pyrometer can be used for temperature control. Rubin and Van Uitert (1966) have made use of an rf absorption meter to maintain constant rf power in the growth of high-melting materials.

Resistance heating has been discussed in Secs. 4.2.2 and 5.3.1.

Common crucible materials are vitreous silica, graphite, and the noble metals. Silica does not couple to the rf field, so a graphite susceptor [see Fig. 5.6(d)] must be used. If the melt is conductive once it has been heated by the susceptor, then it will couple to the rf field†; if the melt is not, then the susceptor will heat the melt throughout the experiment. Graphite is the usual susceptor material. To prevent oxidation the crucible and susceptor must be contained in an inert atmosphere. Commonly a vitreous-silica tube serves this purpose and permits observation of growth. The useful range of SiO_2 crucibles ends at about 1000°C. Both silicon and germanium may be grown from a silica crucible. However, at the higher melting point of Si, reaction of the crucible is a problem and oxygen impurities in the grown crystal prevent the attainment of intrinsic resistivity (see below).

Graphite crucibles couple to the rf field and are useful at higher temperatures (up to above 2500°C) than SiO_2 provided the atmosphere is not oxidizing. Noble-metal crucibles couple to the rf field and are quite inert; their useful ranges are

Pt	1500°C	oxidizing and reducing atmospheres
Pt–20%Ir	1700°C	oxidizing and reducing atmospheres
Ir	2100°C	some volatilization—severe volatilization at 1000–1500°C because of volatile oxide formation

Rubin and Van Uitert (1966a) have increased the useful life of Ir crucibles by flame-spraying ZrO_2 on the outside. This greatly reduces Ir volatilization.

Crucibleless techniques that eliminate crucible problems include float zoning were the heat input is by rf coupling, radiant heating and focused image, and Verneuil growth and its modifications. These techniques will be discussed in Secs. 5.5 and 5.6.

---

†The conductive susceptor will partly shield the melt from the rf field. If one desires to couple directly to the melt, preheating without a conductive susceptor (e.g., by flame-heating, etc.) and with a nonconductive crucible is best.

Crystal-pulling devices should provide constant upward uniform motion and stirring without vibration. The crystal should be able to be manually lowered and raised for growth to begin at an appropriate level. Usually, growth is at rates from one-half to several inches per hour and stirring speeds are usually from a few to a few hundred revolutions per minute. Slower pulling rates are sometimes useful for difficult-to-crystallize materials. A few thousands of an inch (mils) per hour may be required for growth from polycomponent systems (see Chapter 7). Very rapid rates are useful in preparing dendrites or webs by the use of a puller (inches per minute; see Sec. 5.8). Pullers are available from a number of suppliers, including National Research Corporation and Arthur D. Little, Cambridge, Mass.; Hansen and Yorke Company, Woodbridge, N.J.; and Lepel High Frequency Laboratories, Inc., Woodside, N.Y.

5.4.2 GROWTH OF CRYSTALS—GENERAL CONDITIONS

One of the objects of pulling is to balance pulling rate and thermal conditions so that sound growth results without withdrawing the seed ("pulling out from the melt"). Once imperfections such as dislocations, lineage, and polycrystallinity are initiated, they are often propagated through the remainder of the growth, so it is especially important to start with seeds of the highest perfection. The thermal gradient across the diameter of the crucible in the region near the seed and the thermal gradient normal to the growth interface are of major importance in establishing the shape and perfection of the grown crystal (and indeed in establishing even whether a crystal can be grown at all). Rhodes (1964) has discussed the effect of various parameters important in crystal pulling, and the following is taken in part from his discussion. The thermal gradient normal to the growing interface is controlled by:

1. *Heater arrangement.* In rf heating, it includes the shape of the work coil, the position of the crucible in the work coil, and the material and dimensions of the crucible and susceptors; in resistance heating, it includes the geometry of the heaters and the position of the crucible in the heaters.

2. *Heat leaks to the environment.* This is affected by the nearness of the crucible to the end of the rf coil or the end of the furnace, the temperature of the room (drafts can cause disastrous irregularities), the size of the crystal, the thermal conductivity of the crystal, the temperature of the chuck that holds the crystal, and the efficiency of thermal contact to the chuck and the rest of the puller (when large gradients are required, water-cooled chucks can be used), the emissivity of the melt surface, the reflectivity of the furnace walls, and the optical path out of the furnace, which determine radiation losses. After-heaters placed around the growing crystal and reflectors (usually

of Pt) that help to prevent radiation losses can be used when small gradients are desired.

3. *Depth of melt in the crucible*. This factor is often neglected. When the crucible is only partly full, the crucible wall above the melt can serve as an after-heater.

4. The rate of pulling and the latent heat of fusion.

To average out radial asymmetries in the thermal environment and to provide mixing when doping is carried out, it is usual to rotate the crystal. For the same reasons the crucible is often rotated. In addition, when reactivity with the crucible is a problem, as in the growth of Si from a SiO_2 crucible, sometimes the crucible and the seed are rotated at the same rate. This tends to average thermal asymmetries without any stirring action that would tend to accelerate crucible attack.

Local regions within the melt called *cells* that mix poorly with one another can be set up under some conditions of seed and crucible rotation (Carruthers and Nassau, 1968). Poor mixing of this sort can account for unexpected distributions of impurities in the grown crystal.

To begin growth the melt temperature is brought slightly above the melting point as the seed is introduced. Thus a small amount of the seed is melted to ensure that growth begins on a clean surface. Growth is begun when the temperature of the melt is lowered by decreasing the power. At exactly the right conditions, as judged by the experience of the grower, pulling is commenced. The diameter is increased to the desired size by careful manipulation of furnace controls. At the conclusion of the run the crystal is usually withdrawn from the melt by raising the melt temperature or increasing the pulling rate. Sometimes when thermal shock results in crystal imperfections the crystal is allowed to cool in contact with the melt or an after-heater arrangement is used as an in situ annealing furnace.

It is important to avoid rapid changes in diameter because they usually are associated with the introduction of imperfections. Furnace control of $\pm 2°$ is often sufficient for preliminary crystal pulling. Furnace control of $\pm 0.5°$ which can be achieved without inordinate complications is usually sufficient for all but crystals requiring extreme homogeneity and it is axiomatic that improved control leads to improved perfection. Slichter and Burton (1958) showed that temperature fluctuations of 1°/sec near the growing interface can produce about 1.8 mil/sec changes in the growth rate.

In addition to controlling temperature, the gas flow, pulling rate, and stirring rate must be under commensurate control. However, little quantitative information on the required precision of such control is available. As the melt level in the crucible falls because of material removal during growth, it is usual to make adjustments in either temperature or pulling rate to compensate for the changed thermal conditions.

Along with a vacuum environment for low-vapor-pressure crystals, the common protective atmospheres used in crystal pulling are He, Ar, H_2, and N_2. At high powers in rf heating (for example, the power levels required for Si pulling) low-pressure discharges are often produced in N_2. These discharges form active N, which is quite reactive with many melts (Si, again, is an example). There is sometimes difficulty in obtaining gases of the required purity and thus growth in vacuo presents some advantages from the viewpoint of contamination. A pressure of 10^{-5} mm Hg of an active gas is equivalent to a gas pressure of 1 atm of an inert gas where the gas has an active gas content of one part in 10^8. However, pressures as low as 10^{-5} mm are not easy to maintain at high temperatures and require high-speed pumps. In vacuum, unwanted impurities with appreciable vapor pressures are removed by evaporation from the melt.

The usual condition for the growth of crystals low in imperfections is the maintenance of a flat liquid–solid interface throughout the growing process. This requires that the isotherms be nearly normal to the growth direction and implies good control of both the axial and longitudinal heat flow. Reed (1967) has analyzed the heat-flow situation in radiating cylinders of Ge, Cr_2O_3, and W in relation to the growth of these materials. Rhodes (1964) has developed equations relating interface shape to the important parameters in growth, and the following is along the lines of his treatment. When the main heat flow is down the growing crystal and the axial heat flow is negligible, the isotherms will be normal to the growth direction. This situation is described by

$$\frac{dT}{dx} = k \tag{5.3}$$

and

$$\frac{dT}{dr} = 0 \tag{5.4}$$

where T is temperature, x is the longitudinal direction, r is the radial direction, and k is a constant. The heat of crystallization must flow from the melt to the crystal and be dissipated by conduction down the crystal to the puller and by radiation to the walls or the ambient. The rate at which this heat is dissipated governs the maximum rate of permitted growth. When growth is under control the growth interface does not move; that is, the pulling rate is made to equal the growth rate. Although in principle the radial thermal gradient, dT/dr, should be negligible, in actual practice, because of heat loss to the surroundings from the grown crystal, this is not the case, Figure 5.8 shows the isotherms as they exist in a typical puller. Figure 5.8(a) shows the situation in a rf-heated conducting crucible without an after-heater. In such a case all the heat input is from rf coupling with the crucible and the melt, and heat loss from the crystal is severe. In the case of growth by rf-heating with an after-heater or growth by resistance-heating when the crucible is placed well within the furnace, radial heat loss from the growing crystal can be minimized to a much greater degree. In Fig. 5.8(a) A is heat loss from the surface of the melt mainly by radiation, B is radial heat loss from the sides of the growing crystal caused mostly by radiation, and C is longitudinal heat loss mostly from conduction.

The isotherms of Fig. 5.8(a) are concave relative to the melt. If the experimental arrangement is such that heat flows to the growing crystal [Fig. 5.8(b)] in the region T_1 because of after-heaters or growth of the crystal deep in a resistance furnace, T_2 is reduced over that in Fig. 5.8(a), and the interface is convex. Consider the magnitude of heat flow at several points across the diameter of the crystal at a short distance from the growing interface [Fig. 5.8(c) and (d)]. Assume that the heat input to the crystal (Q_{in}) across the growing interface is the same in Fig. 5.8(c) and (d) (Q_{in} = heat flow from melt + heat of crystallization). Then if there is radial heat loss to the environment (Q_r) the heat flow down the axis of the crystal at the center (Q_c) will be greater than that at the edge (Q_e), and both the isotherm, T_1, and the interface will be concave [Fig. 5.8(c)]. If, on the other hand, there is radial heat input

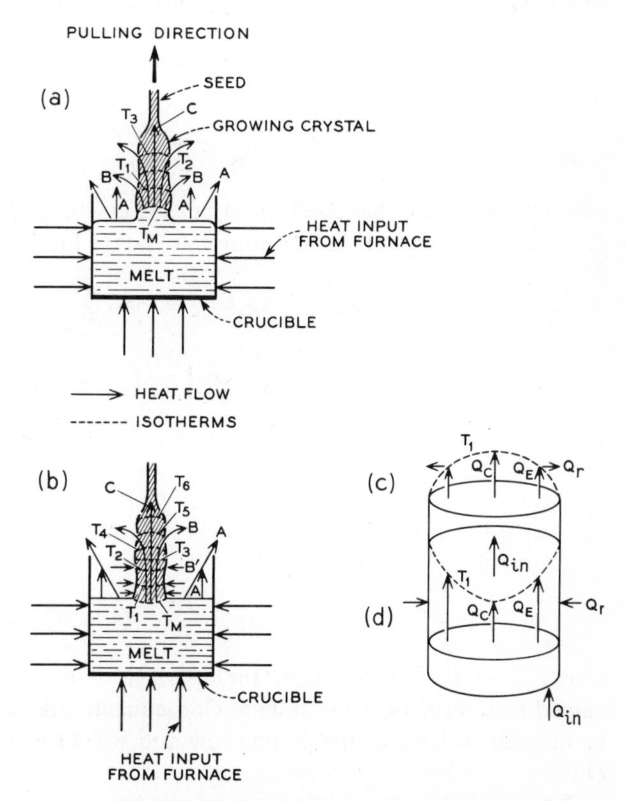

Fig. 5.8 Thermal conditions in Czochralski growth.

from the environment ($Q_e > Q_c$), then the isotherm at T_1 is convex, as is the interface. Thus when $Q_r = 0$, $Q_e = Q_c$ and the interface is flat. The role of after-heaters in interface-shape control then is really to control Q_e.

It is of interest to consider in detail the heat flows composing Q_{in} (in calories per second):

$$Q_{\text{in}} = Q_L + Q_M \tag{5.5}$$

where Q_L is associated with the latent heat of crystallization and Q_M is the heat flow from the melt. Q_L is given by the equation

$$Q_L = A\rho_s L \frac{dx}{dt} \tag{5.6}$$

where A is the external area of the crystal, ρ_s is the density of the solid, L is the heat of crystallization (minus the heat of fusion), and dx/dt is the rate of growth (pulling rate to a first approximation). Q_M is given by the equation

$$Q_M = AK_L \frac{dT}{dx_L} \tag{5.7}$$

where K_L is the thermal conductivity of the liquid (assuming no convection or a correction to K to include convection) and dT/dx_L is the thermal gradient in the liquid. Q_{in} is also the heat flow down the crystal, which is given by the equation

$$Q_{\text{in}} = AK_s \frac{dT}{dx_s} \tag{5.8}$$

where K_s is the thermal conductivity of the crystal and dT/dx_s is the thermal gradient in the crystal. Substituting into Eq. (5.5),

$$AK_s \frac{dT}{dx_s} = A\rho_s L \frac{dx}{dt} + AK_L \frac{dT}{dx_L} \tag{5.9}$$

$$\frac{dx}{dt} = \frac{1}{\rho_s L}\left[K_s\left(\frac{dT}{dx}\right)_s - K_L\left(\frac{dT}{dx}\right)_L\right] \tag{5.10}$$

Equation (5.10) is of use in reasoning regarding the effect of process and material variables on rate. If the pulling rate exceeds the value of dx/dt given by Eq. (5.10) the crystal pulls from the melt. If it is less than that given by Eq. (5.10) the cross-section area tends to grow.

According to Rhodes (1964),

$$Q_{in} = A\rho_s L \frac{dx}{dt} = Q_1 - Q_2 \tag{5.11}$$

where Q_1 is the heat loss from the interface by thermal loss and Q_2 is that gained from heaters in the system. Consequently, the area of the crystal will be inversely related to the pulling rate and will be increased by a decrease in Q_1.

According to Eq. (5.10) the maximum rate possible occurs when dT/dx_L

approaches 0 (if dT/dx_L becomes negative the liquid would be supercooled and the interface would advance rapidly to grow dendritically). Thus

$$\frac{dx}{dt_{\max}} = \frac{1}{\rho_s L} K_s \left(\frac{dT}{dx}\right)_s \tag{5.12}$$

The value of $(dx/dt)_{\max}$ thus depends on the gradient in the solid. The largest gradients in the solid are obtained by deliberately introducing large heat leaks from the growing crystal, such as by water-cooling the seed holder. However, the high rates obtained under such conditions are generally not useful because the perfection of material grown under such conditions usually leaves much to be desired.

Slichter and Burton (1958) analyzed the temperature distribution obtained in Ge growth in some detail. Using a one-dimensional approximation, assuming a large-diameter rotating crystal and negligible temperature gradients they showed that

$$\kappa \frac{d^2T}{dx^2} - V_x \frac{dT}{dx} = 0 \tag{5.13}$$

where the "diffusion coefficient for heat," $\kappa = k/\rho_l C_p$, k is the thermal conductivity, C_p the constant-pressure heat capacity, ρ_l the liquid density, and V_x the sum of normal fluid and growth velocities. Slichter and Burton solved Eq. (5.13), assuming that at $x = 0$ (the solid–liquid interface) $T = T_m$, the melting point, that dT/dx in the crystal is constant, and that the seed end of the crystal is at constant temperature. Slichter and Burton's analysis leads to a temperature profile like that shown in Fig. 5.9. In Fig. 5.9, σ is the width of the diffusion layer. Stirring will reduce σ.

In addition to heat transfer at the growing interface, usually, there is also solute transport, which complicates the overall heat-transport situation (see Secs. 2.5 and 3.10). As the growth temperature increases, radiative heat-transfer processes become of greater importance. Reed (1967), Billig (1955), and Wilcox and Fullmer (1965) have discussed the role of radiation and derived heat-transfer equations for processes where radiation is important.

Beyond σ the transport is dominated by fluid motion. Even when matter transport across the diffusion layer is unimportant (in a highly pure mono-component system), heat transport across the diffusion layer will be more difficult because of the fact that fluid motion is not effective in the layer. In the case of *Ge*, dT/dx_s has been shown to be linear and independent of dx/dt for rates less than 3.6 cm/hr. The gradient is about 150°/cm. In Si (Dash, 1958; Edwards, 1960) in spite of the higher melting point (1412°C vs. 937°C for Ge) the gradient is similar 125–150°C/cm.

Slichter and Burton (1958) calculated dT/dx_l assuming $dT/dx_s = 150°$ and a rotation speed of 100 rpm. Their results are plotted in Fig. 5.10. The amount of heat the crystal must dispose of will depend on the rate of growth because the amount of latent heat of crystallization is rate dependent. The

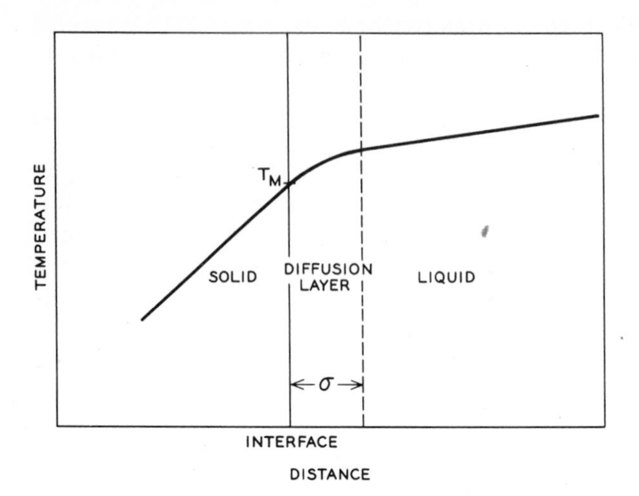

Fig. 5.9 Temperature profile in Czochralski growth (after Slichter and Burton, 1958). (Reproduced from *Transistor Technology*, Vol. 1, by H. E. Bridgers, J. H. Scaff, and J. N. Shive, eds., by permission of Van Nostrand Reinhold Company, a division of Litton Educational Publishing, Inc., Litton Industries, Princeton, N. J., 1958.)

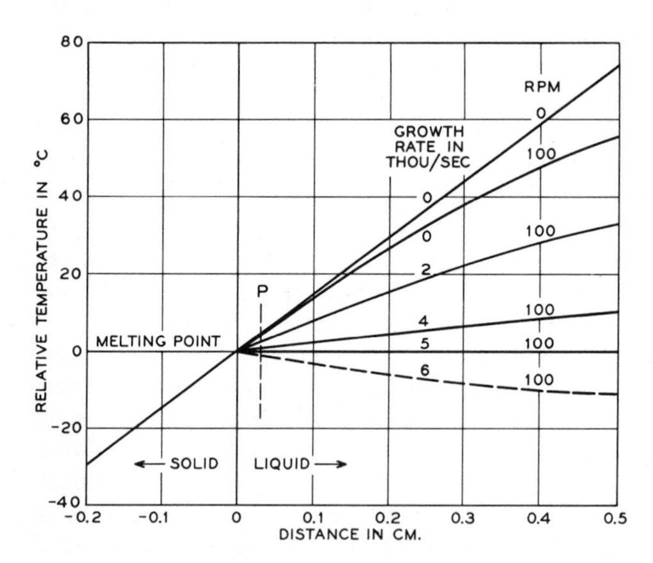

Fig. 5.10 Temperature profile in Czochralski growth at various growth rates (after Slichter and Burton, 1958). (Reproduced from *Transistor Technology*, Vol. 1, by H. E. Bridgers, J. H. Scaff, and J. N. Shive, eds., by permission of Van Nostrand Reinhold Company, a division of Litton Educational Publishing, Inc., Litton Industries, Princeton, N. J., 1958.)

greater the rate, the smaller the gradient permitted in the melt. The maximum rate will thus occur as the melt gradient approaches 0. Crystal rotation, of course, aids in stirring, and Cochran (1934) showed that a rotating disc acts like a fan, drawing up liquid in the center (see Fig. 5.11) so that the flow near the interface is laminar. Goss and Adlington (1959) indicate that stable vortices will occur at from 30–60 rpm. It might be pointed out that in some cases greater rates can be accommodated as the crystal diameter increases (Goss and Adlington, 1959) because the stirring increases at a given rpm as the diameter increases. However, it is usually harder to control interface profiles at larger diameters and consequently more difficult to get good-quality growth unless the rate is moderate.

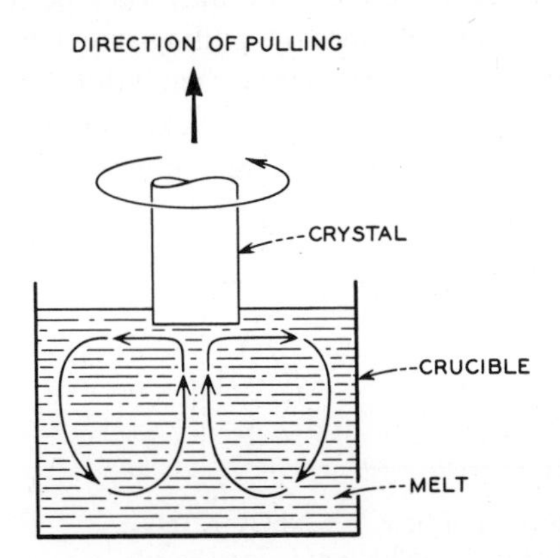

Fig. 5.11 Fluid flow during stirring in Czochralski growth.

Thermal stability during growth is essential to the production of good-quality crystals. Changes in gradient lead to changes in instantaneous growth rate as the interface seeks the melting-point isotherm. Thus it is important to achieve the best possible temperature control. Similar changes in conditions occur as the melt level falls in the crucible and as the crystal diameter and length change.

In addition to these changes there are short-term fluctuations in temperature of the melt caused by nonuniform (unstable) thermal convection, as pointed out by Müller and Wilhelm (1964). Wilcox and Fullmer (1965) have analyzed experimental data obtained during the pulling of CaF_2, and their results are probably applicable to most puller geometries. Hurle (1966) and Cole and Winegard (1962) also noted temperature fluctuations in molten materials and analyzed them in terms of fluid circulation.

Figure 5.12 (Barns, 1965) shows the temperature recorded by an unshielded Pt vs. Pt–20% Rh thermocouple located 1 mm below the surface in the center of the crucible in a melt heated in a resistance furnace. Wilcox and Fullmer and others have obtained similar temperature–time plots in rf-heated melts. The fluctuations are not caused by fluctuations in control temperature because, in both Wilcox and Fullmer's experiments and Barns' experiments, rather sophisticated control equipment was used, and temperature fluctuations of the control point were an insignificant fraction of the observed fluctuations of melt temperature close to the interface. Wilcox and Fullmer have shown that the thermal fluctuations are reduced as the distance from the melt interface increases. The magnitude of the fluctuations increases as the growing temperature increases, increases as the temperature gradient in the melt increases, and increases as the temperature gradient above the melt increases. All experimental observations of the phenomenon are consistent with the explanation that it is caused by irregular convection.† Fluctuations in melt temperature close to the interface cause fluctuations in growth rate. Thus, under such conditions, the true rate may vary between a small fraction of the pulling rate and several times the pulling rate. Changes in rate cause changes in distribution constant for impurities and dopants. Resistivity striations in semiconductors grown by the Czochralski technique have long been known, and Müller and Wilhelm (1964) have comprehensively surveyed this problem.

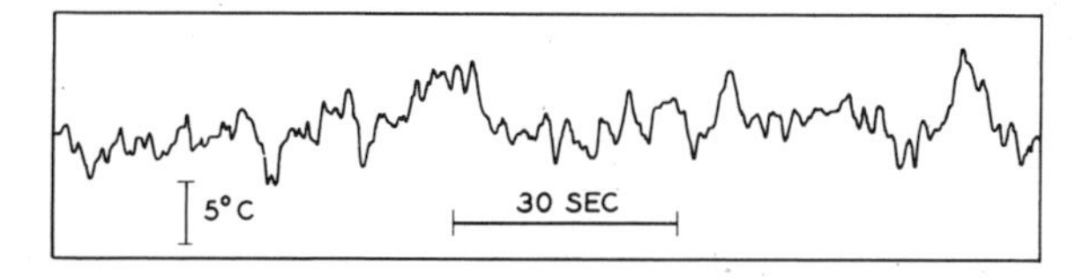

Fig. 5.12 Temperature as a function of time in $KNbO_3$–$KTaO_3$ melt with temperature difference of 10° from bottom to top (Barns, 1965).

It would seem that many of the imperfections associated with pulled crystals have their origin in the short-term thermal fluctuations that are almost always present in such systems. This problem is deserving of further study and, as a minimum, information regarding thermal stability close to the interface should be reported as a part of all papers dealing with melt pulling. Müller and Wilhelm (1964) have also observed similar fluctuations during crystal growth of InSb by zone-melting, and it is appropriate to suspect similar thermal fluctuations caused by irregular convection in most mono- and polycomponent liquid–solid growth techniques and even in crystal growth from the vapor phase.

Figure 5.13 illustrates schematically the mechanism of irregular convection responsible for thermal fluctuations. Convection occurs in the heated melt with liquid tending to rise near the heated walls and fall near the center.

†Wilcox and Fullmer used the term *turbulent convection*, but *irregular convection* is probably a more appropriate term because Reynolds numbers in the turbulent range have not been proved and indeed are not required to explain the observed effects.

The larger ΔT_L, the temperature difference from bottom to top of the liquid, the greater this convection. One can think of small volumes of liquid V_1, V_2, V_3, . . . , rising to the surface. At the surface these volumes cool principally by radiation to the environment. The rapidity with which they cool will depend on the thermal conditions above the melt. The thermal gradient, the presence of drafts, the rate of gas flow through the system, etc., will all effect the cooling of a given volume. When the volume is cooled so that its density is sufficiently large, it sinks and is replaced by another volume of liquid. Eliminating or reducing the gradient in and above the melt and preventing drafts, etc., will thus reduce thermal fluctuations. Another means of reducing fluctuations is to stir the melt violently enough that the stirring overrides

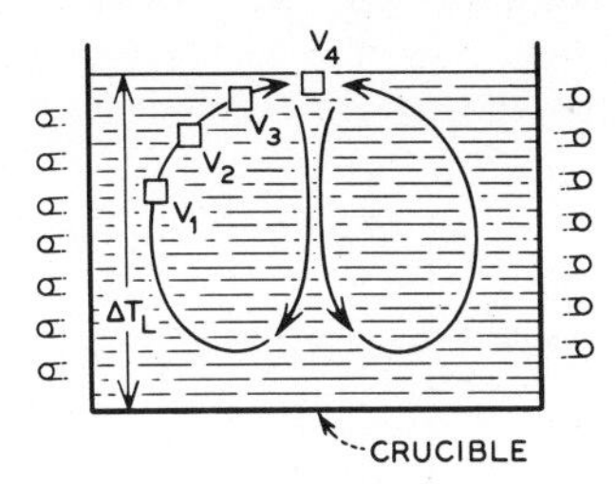

Fig. 5.13 Mechanism of irregular convection.

random convective variations. However, as Fig. 5.14 (after Wilcox and Fullmer, 1965) shows, stirring at rates normally used is generally not enough to override convection. Thus altering the thermal environment by the use of carefully designed furnaces and after-heaters offers the best solution of this problem.

The presence of regular inhomogeneities of a planar nature of index of refraction, resistivity, or other properties is usually called *banding*. Banding may even be caused by changes in stoichiometry, when no impurities or dopant are present. When it is caused by irregular convection, the bands are parallel to the growing face (perpendicular to the growth direction) as shown

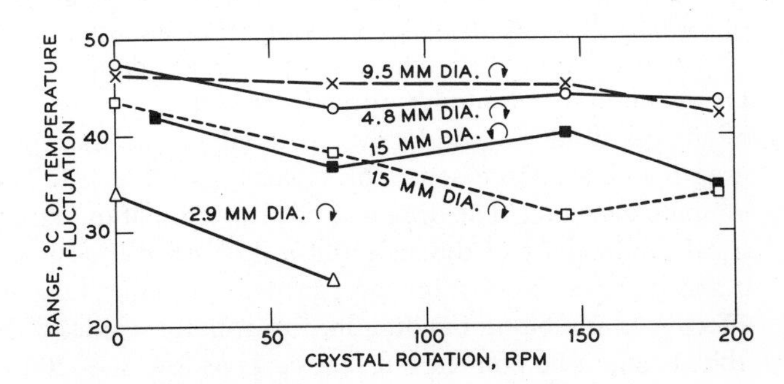

Fig. 5.14 Effect of stirring on irregular convection (after Wilcox and Fullmer, 1964).

in Fig. 5.14A(a). When constitutional supercooling is a problem, facets, cellular interfaces, and, in the limit, dendritic growth result (see Secs. 3.11, 3.12, 3.13 and 5.8). The banding is generally parallel to the growth direction, as shown in Fig. 5.14A. To repress constitutional supercooling, a large thermal gradient at the interface is called for. In contrast, to repress irregular convection, minimal thermal gradients are required. Thus in a system where both irregular convection and constitutional supercooling are problems, regulation of thermal gradient may be ineffective in producing band-free crystals. However, if the system is arranged so that the upper region of the melt is hotter than the lower, convection cannot take place regardless of the magnitude of the temperature gradient. A system where this geometry is possible is Bridgman–Stockbarger growth and for this reason when banding is a problem this technique should be considered.

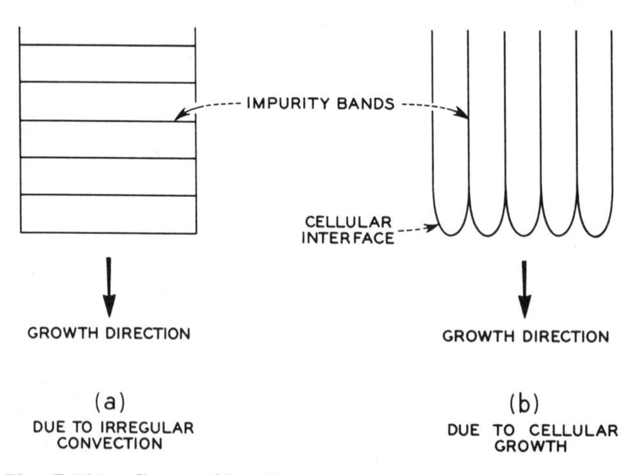

Fig. 5.14A Causes of banding.

Utech and Flemings (1965) have shown that magnetic fields of the order of 1000 G decrease irregular convection to the point where banding is not observable in tin, aluminum–copper, and tellurium-doped Czochralski-grown indium antimonide provided thermal gradients are also kept to a minimum. When a conductive material moves through a magnetic field, drag is induced. In a sense the viscosity is increased. The result then is to inhibit convection. The drag will be proportional to $(\sigma\mu)$, where σ is the electrical conductivity of the melt and μ is its magnetic susceptibility. Values of σ and μ are not known for most melts, so whether magnetic fields would be effective in inhibiting banding in, for instance, oxides is not readily predictable. It might be pointed out that fields as low as $\sim$200 G produced effects in indium antimonide so that the technique surely bears further investigation and may be of very general utility.

Another cause of imperfections during pulling arises as the crystal is cooled to room temperature. If cooling is rapid, strains may be set up. These strains can be caused by temperature gradients in the pulling geometry used or because the crystal is too rapidly removed from the system and brought to room temperature. Under such conditions one region of the crystal will be at a temperature different from another. The result will be dimensional differences among different regions of the crystal. If these differences introduce elastic strain it will be removed when the crystal reaches isothermal conditions. If plastic deformation results at high temperature and is nonuniform, the whole crystal may possess nonuniform elastic strains when it is cooled to room temperature. If the strain results in dislocation or grain-boundary initiation, such imperfections will remain at room temperature, and a repeat anneal of the crystal after it has once been at room temperature may not be especially effective. It is thus preferable to anneal from growing temperature without first cooling to room temperature.

5.4.2-1 Crystal Growth of Semiconductors by the Czochralski Technique

Czochralski (1917) first used crystal pulling and applied the technique to metals. From 1917 until the work of Teal and Little (1950) the technique was applied mainly to metals. However, in many ways the triumph of the technique and certainly the great impetus to its widespread use was provided by Teal and Little's pioneering work. Indeed, it is fair to say that the whole semiconductor industry (which has itself provided great impetus to solid-state research in other areas and in turn to crystal growth) would not have developed and could not exist today without the high-quality single crystals provided by crystal pulling. We will discuss the growth of Ge and Si in some detail as prototype materials. Other semiconductors and their conditions of growth are listed in Table 5.2.

Germanium (mp 937°C) is easier to grow than silicon (mp 1412°C) principally because of its lower melting point. Germanium melts are usually contained directly in a graphite crucible that acts as the susceptor in rf heating because germanium carbides are not formed and the solubility of C in Ge at the melt temperature is negligible. Most growth is with rf heating often because resistance furnaces are more likely to cause melt contamination. For high-purity Ge the highest-purity starting material prepared by zone melting is used. Boron is a particularly deleterious impurity in group-IV semiconductors where it acts as an electrical acceptor. Because its distribution constant in Si is close to one, it is not removed by ordinary zone melting or by crystal pulling. B contamination from graphite crucibles can be a serious problem. Fortunately, essentially B-free graphite has been prepared for nuclear applications and is available in crucible form. Boron originally present in starting

Table 5.2

SOME REPRESENTATIVE CRYSTALS GROWN BY CRYSTAL-PULLING AND KYROPOULOS TECHNIQUES

Compound	*Formula*	*Melting Point (°C)*	*Crucible Material*	*Pulling Rate*	*Growth Direction*	*Atmosphere*	*Comments*	*References*
Germanium	Ge	937					See text	
Silicon	Si	1412					See text	
Zinc	Zn	419	Pyrex	1.2 cm/min	Various	N_2	2.7-mm-diameter crystal, air-cooled	Hoyem and Tyndall, 1929
Gallium arsenide	GaAs	1240	Vitreous silica		Various	As	over-pressure of As	
Potassium chloride	KCl	770	Pt or porcelain			Air	Method really Kyropoulos; air-cooled cold seed immersed in melt used; a variety of alkali halides grown	Kyropoulos, 1930
Water	H_2O	0	Glass			Air	9-cm diameter × 6 cm high Kyropoulos method	Jona and Scherrer, 1952
Calcium tungstate	$CaWO_4$	1535	Rh	0.5–2 cm/hr	Various	Air	See text	
Lithium niobate	$LiNbO_3$	1260	Pt	0.5–2 cm/hr	Various	Air	See text	
Sapphire	Al_2O_3	2050	Ir	0.5–2 cm/hr	Various	Air	See text	
Yttrium–aluminum garnet	$Y_3Al_5O_{12}$	~1900	Ir	0.5–2 cm/hr	Various	Air	See text	

material Si can be removed by zone melting in the presence of water vapor (Theuerer, 1956), which selectively oxidizes B. The oxide is removed by vaporization. Figure 5.15 shows a pulling setup suitable for Ge or Si growth. Furnace energy is supplied by a 10-kW, 450-cycle oscillator that heats a graphite susceptor. Temperature is measured by a Pt vs. Pt–10% Rh thermocouple in a Mo shield placed at a predetermined point in the susceptor. Atmosphere control is provided by flowing gases through a fused silica tube that is sealed into water-cooled brass plates. The seed crystal is held in a chuck in a stainless-steel shaft that enters the chamber through a compressed Teflon O ring. The shaft and the motor that provides rotation are mounted on a platform above the puller. The platform is raised and lowered

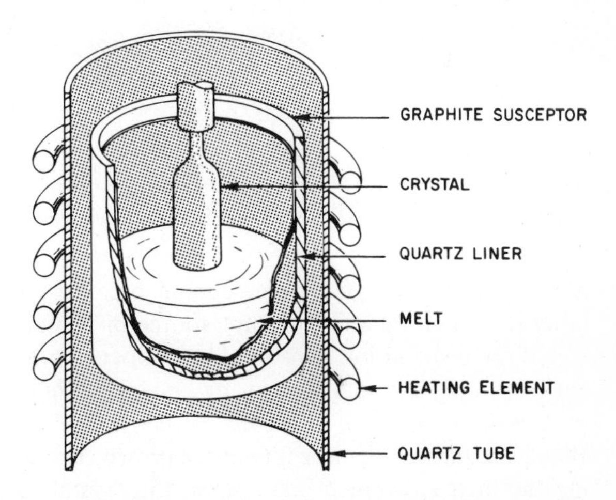

Fig. 5.15 Crystal pulling arrangement for growth of Si. (Courtesy of J. R. Carruthers.)

by a screw that is driven by a rotating nut whose speed is controlled by a motor-and-control system. The shaft to which the seed is fixed is also rotated by a motor–control system. The melt volume is about 100 cm^3. Crystals can be grown at rates from about 10^{-4} cm/sec to 10^{-2} cm/sec. At rates lower than 10^{-4} cm/sec growth-rate fluctuations caused by puller and temperature-control inadequacies and irregular convection became comparable to the imposed growth rate. At rates approaching 10^{-2} cm/sec single crystals do not result because dT/dx_s (the gradient in the solid) is not large enough. Cooling of the crystal would probably permit larger rates. Indeed, dendrites are deliberately grown under somewhat similar conditions (see Sec. 5.8) by imposing large rates. The gas atmosphere in which the crystal is grown is usually supplied by an inlet at the top and leaves by an outlet at the bottom. Thus cool gas flows over the growing crystal tending to increase dT/dx_s. Germanium can be grown in H_2, He, Ar, or N_2 but high-purity gases free of oxidizing agents, hydrocarbons, and water are required.

The distribution constants for a number of dopants and impurities of electronic interest are shown in Table 5.3. It is important that semiconductor

Table 5.3

DISTRIBUTION COEFFICIENTS FOR VARIOUS IMPURITIES†

Solute	*Si*	*Ge*	*Conductivity Type*
B	0.8	17	*p*
Al	.002	.073	*p*
Ga	.008	.087	*p*
In	4×10^{-4}	.001	*p*
P	.35	.08	*n*
As	.30	.02	*n*
Sb	.023	.003	*n*

†Thurmond, 1959; Trumbore, 1960; J. R. Carruthers (private communication).

purity reagents be used when doping because the dopants themselves will otherwise often be an important source of contamination.

If the distribution constant for a dopant (or an impurity) is much different from one, a crystal grown by crystal pulling will have a concentration gradient in it because, as growth proceeds, the concentration of dopant will change in the melt. Uniformity of the grown crystal may be improved by making the melt size much larger than the crystal to be grown. The floating-crucible technique of Leverton (1958) is one modification of this procedure. Constant composition in the melt may be maintained by melting a rod of appropriate composition into the melt as the crystal is pulled and zone leveling (see Sec. 5.5) is another means of homogenizing dopant concentration throughout a grown crystal.

Silicon is more difficult to grow than germanium because a truly inert crucible is not available. Si reacts to form silicon carbide in graphite and it alloys with noble metals so that these materials cannot be used. The best crucible is vitreous silica. Heating may be in a resistance furnace, but it is more common to use rf heating and a graphite susceptor. Silicon slowly reacts with vitreous silica:

$$Si_{(l)} + SiO_{2(s)} \rightleftharpoons 2\,SiO_{(g)} \tag{5.14}$$

and SiO has an appreciable volatility (~10 Torr at 1412°C, Schäfer and Hörnle, 1950). In addition a common contaminant of vitreous SiO_2 is B, which dissolves in the melt during attack and acts as an acceptor when it enters the Si lattice. Oxygen has an appreciable solubility in molten silicon, and it, too, can enter the Si lattice and adversely affect the electrical properties. To obtain high-resistivity (intrinsic) Si it is necessary to use crucibleless techniques (see Sec. 5.5). Nevertheless, Si of a resistivity suitable for many applications can be obtained by pulling and, wherever the resistivity, lifetime, and mobility requirements on the Si permit growth by pulling, this technique is preferred over crucibleless methods because of its greater ease. With the exception of conditions imposed by the higher temperature and reactivity the growth conditions and technique for Si growth are otherwise similar to those for Ge. However, in Si growth, crucible rotation is especially helpful to decrease the velocity difference between the crucible wall and the melt and reduce the rate at which Eq. (5.14) proceeds.

Dislocation-free crystals of Si and Ge were first grown by Dash (1960) and recently compounds such as GaAs have been grown as dislocation-free crystals. These are probably the most perfect crystals yet prepared. It might be more accurate to state that the crystals are dislocation-free insofar as dislocations that can be revealed by etching and decoration are concerned. It is possible (but not probable) that other techniques might reveal imperfections. The crystals, of course, are not free of vacancies, and, although of unusual purity, not free of chemical impurities. Nevertheless, the work of Dash should be clearly recognized as a milestone in our control of the properties of solid matter. In order to achieve dislocation-free material the following procedure was used.

1. The initial seed diameter (1–2 mm) must be smaller than is ordinarily used in pulling so that it contains a small number of dislocations. The small-diameter seed also results in fewer thermal stresses being generated in the seed when it makes initial contact with the melt. Reduced thermal stresses lead to less likelihood of dislocation generation. As an alternative a growing crystal can be necked down to 1–2 mm and then carefully enlarged. In the necked-down region the number of dislocations will be small. It is important that the necked-down region be of considerable length so that dislocations have a maximum chance of terminating at the surface before the diameter is enlarged.

2. Growth should be such that the $\{111\}$ glide planes make a large angle with the growth direction. Dislocations generated during the plastic deformation of Si and Ge concentrate in the $\{111\}$ glide planes. If the growth axes can be chosen at a large angle to the $\{111\}$ planes, any dislocations present in the crystal will tend to grow out to the surface and disappear as the crystal lengthens.

3. To favor climb of dislocations to the surface Dash suggested that an adequate supply of vacancies be available to help propagate the dislocation. It is thus useful to supersaturate the crystal with vacancies and prevent their diffusion to the surface. Quenching will effect this. In growth this is not pract-

ical but the next best thing is rapid growth in a large temperature gradient, especially during initial stages where necking down is being performed. This procedure has been found helpful in assisting dislocation climb.

It is interesting to point out that once dislocation-free material is obtained, when the bonding in the material is covalent, it is quite difficult to nucleate new dislocations in it. Dash showed that the thermal stresses generated in a dislocation-free Ge crystal produced by withdrawing it from the melt and then reinserting it did not create any dislocations. Apparently the generation of dislocations from already existing dislocations is considerably easier. For instance, when a crystal containing about 10^3 dislocations/cm^2 was withdrawn from the melt and then reinserted, the dislocation density increased to more than about 10^6/cm^2. Some semiconductors that have been grown by crystal pulling together with their conditions of growth are listed in Table 5.2.

5.4.2-2 Growth of Insulators and Metals by the Czochralski Technique

Although crystal pulling was first applied to metals and has had its commercial triumph with semiconductors, perhaps the area of most rapid advance in recent years has been in the growth of refractory oxide and salt crystals that are dielectrics. This work has in large part been stimulated by the need for host crystals for lasers and for ancillary laser materials for modulators and harmonic generators.

The first material of this sort grown extensively was $CaWO_4$ (calcium tungstate, scheelite), which has been studied widely as a laser host. Nassau and Van Uitert (1960) first reported the growth of scheelite-type structures by pulling, and Nassau and coworkers (Nassau and Broyer, 1962; Nassau and Loiacono, 1963; Nassau, 1963) in a series of papers reported the conditions for growth and rare-earth doping. The culmination of this work was an extremely low-threshold continuous solid-state laser where the active ion was Nd^{3+} and charge compensation was achieved by controlled Na^+ doping (Johnson et al., 1962).

Calcium tungstate can be grown in a rf crystal puller similar to that described for semiconductor growth. Its melting point is 1570°C and the most satisfactory crucible is iridium. With such a setup it can be pulled in air, so no atmosphere control is required. Appropriate rates are from $\frac{1}{4}$–$\frac{3}{4}$ in./hr. Useful growth directions are normal to the (110) and (100) faces.

It is necessary to use high-purity starting materials and better-quality crystals occur when the starting materials are zone-refined or when previously pulled crystals are remelted and used for growth. If after-heaters are not used, the crystal–melt interface is concave [like Fig. 5.8(a)]. If after-heaters are used, the melt interface can be made planar with resultant improvement in quality.

The distribution coefficients for rare earths and alkali metals were studied extensively by Nassau (1963). Because Nd^{3+} resides at a Ca^{2+} site, charge compensation is required (see Sec. 2.5) and Nassau determined the dependence of $k_{Nd^{3+}}$ and $k_{Na^{+}}$, the distribution constants for Nd^{3+} and Na^{+} on the concentration of Nd^{3+} and Na^{+}. The results of this work allowed control of Na^{+} and Nd^{3+} concentrations and their ratio in the grown crystal.

Many of the scheelite isomorphs and related compounds were grown by Van Uitert and coworkers (Van Uitert et al., 1963; Van Uitert and Preziosi, 1962; Preziosi et al., 1962).

The discovery that $LiNbO_3$ could be pulled with comparative ease (Ballman, 1965) and that its large nonlinear optical coefficients and birefringence made it useful for harmonic generator and parametric experiments led to an intense interest in this material. The fact that the ferroelectric Curie temperature is close to the melting temperature complicates poling procedures but single domain material is now easy to obtain (Nassau et al., 1966). Lerner et al. (1968) were the first to report a phase diagram in the Li_2O–Nb_2O_5 system which gave accurate information on the liquid and solid compositions near $LiNbO_3$. The maximum in melting point occurs at 48.6 mole % Li_2O; at any other composition the solid which freezes will not have a composition equivalent to the melt (the material melts incongruently but without a peritectic). Stoichiometric $Li_{1.00}Nb_{1.00}O_3$ will only be crystallizable from a melt whose composition is very rich in Li_2O. In addition, crystals pulled from melts other than 48.6% will have a gradient in composition. This can be understood if we realize that the distribution constants for Nb and Li are not equal to one except at 48.6 mole % Li_2O. Moreover, compositional fluctuations caused by temperature fluctuations, growth-rate fluctuations and irregular convection will be a problem. The compositional variations are especially troublesome because Curie temperature, birefringence, phase-matching temperature and other properties of device interest are sensitive functions of composition. Growth at 48.6% and growth in low thermal gradients do much to minimize these inhomogeneities. Present growth procedures use rf heating, growth in Pt crucibles in air and growth rates from 0.5–2 cm/hr.

Barium sodium niobate ($Ba_2NaNb_5O_{15}$) like $LiNbO_3$ can be pulled with comparative ease (Van Uitert et al., 1968) and its nonlinear optic coefficients and resistance to laser damage make it superior to $LiNbO_3$. Growth procedures and stoichiometry problems are similar to $LiNbO_3$.

Lithium tantalate is also readily pullable (Ballman, 1965) and of special interest as a laser modulator and piezoelectric.

The extension of crystal pulling to higher-melting compounds would be possible if one can devise a crucibleless technique. Van Uitert (Monforte et al., 1961) has had some success with a technique where the melt is contained in a frozen shell of itself. A typical apparatus is shown in Fig. 5.16. In a typical modification of this technique (Monforte et al., 1961) a closely spaced rf coil was used to heat a manganese-ferrite charge placed within it. Alundum cement was pasted between the coils and the charge was placed inside. It was necessary to premelt the charge with an oxyhydrogen torch but, once it was melted, coupling to the rf coil was sufficient to provide the desired temperature. The water-cooled coils kept a shell-shaped layer of charge close to them from melting. This shell contained the melt.

One of the recent triumphs of crystal pulling has been the growth of materials with melting points above 2000°C by F. R. Charvat and coworkers at the Linde Division of Union Carbide. Both sapphire (mp 2015°C) and yttrium–aluminum garnet (mp 1970°C) are now routinely grown (Charvat et al., 1965; Paladino and Roiter, 1964, 1966; Linares, 1964). Both rf heating and oxyhydrogen flames have been used to heat Ir crucibles. Rubin and Van Uitert (1966a) have recently shown that the lifetime of Ir crucibles is markedly prolonged when they are flame-sprayed on the exterior with refractories such as zirconia. Bardsley and Cockayne (1967) have discussed the role of constitutional supercooling and the effect of after-heaters in reducing crystalline imperfections in the Czochralski growth of high-melting-point oxides.

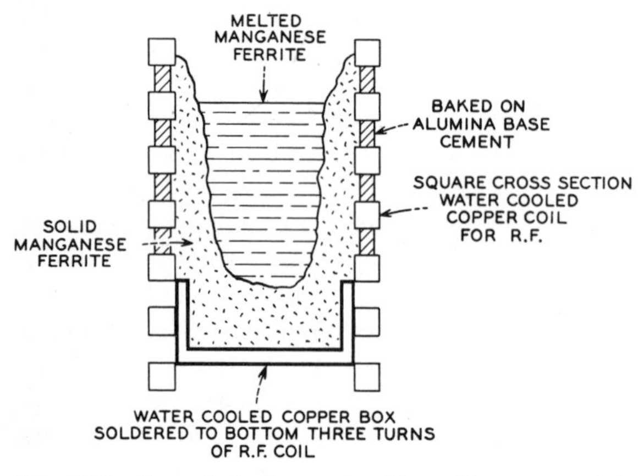

Fig. 5.16 A method of crystal pulling without a crucible (after Monforte, Swanekamp, and Van Uitert, 1961).

Volatility of the melt causes considerable difficulties in crystal pulling. One solution of this problem is a heated-wall crystal puller of the sort used by Gremmelmaier (1956). Metz et al. (1962) have pointed out that if an immiscible nonvolatile material that floats on the melt can be found, it will be extremely efficient in inhibiting volatilization. They succeeded in using B_2O_3 for this purpose in pulling PbTe and PbSe. This technique, called *liquid encapsulation*, has been extended to the growth of GaAs, GaP, and other materials by workers at the Royal Radar Establishment, Great Malvern,

England (Mullin et al., 1968), where an inert gas at substantial pressures has been used over the B_2O_3 melt. The inert gas keeps the liquid encapsulate in contact with the melt. The encapsulate thus acts as a "piston" and an equilibrium pressure of the volatile constituents from the melt can be thought of as existing between the melt and its piston.

Many authors have suggested inert gasses in the absence of liquid encapsulates as volatility repressants. The effects to be expected here at moderate pressures will usually be marginal once the inert gas pressure exceeds the pressure of the volatile constituent. At pressures where Dalton's law describes the situation or where the volume change of the condensed phases is small, only inhibition of the kinetics of volatilization can be expected.

Another variation of a crucibleless technique is to use focused radiation, plasmas, or arcs as the heat source and the material itself as the container. Some of these heat sources are discussed in Sec. 5.6.2. A combination of these heat sources and Van Uitert's crucibleless technique (Monforte et al., 1961) might be useful. In particular, various *cold-hearth* techniques might be advantageous for holding the melt. Cold hearths are water-cooled platforms that hold a melt while a layer of the material in direct contact with the hearth remains unmelted. They have been used extensively in metallurgical processing and have so far been applied to crystal growth in only one instance (Reed, 1968).

Table 5.2 lists a number of other materials, including metals, that have been pulled. Table 5.2 is meant to be indicative and not inclusive.

5.5 Zone-Melting Techniques

Zone melting is probably mainly thought of as a purification technique. It was first used by Pfann for this purpose in 1952 (Pfann, 1966) and has been a mainstay of semiconductor purification techniques. We shall not attempt to discuss the use of the technique for purification here. Numerous papers, reviews, and a fine book (Pfann, 1966) discussing this aspect of the technique are readily available. However, zone melting may be used as a single-crystal-growth technique and, indeed, single crystals often result when the technique is used for purification. Kapitza (1928) and later Andrade and Roscoe (1937) used a traveling zone to produce single crystals. The purpose of the zone was (and still is when the object is primarily crystal growth) to produce a temperature gradient near the growing interface. However, it remained for Pfann to realize and exploit the potential of such a zone as a distributor of solutes. It is, of course, an advantage of crystal growth by zone melting that control of impurities can usually be obtained at the same time.

Figure 5.17(a) and (b) are schematics of the commonly employed zone-melting configurations. Figure 5.17(a) shows horizontal zone melting. The zone is started at the far left. If single crystals are desired a single-crystal seed may be placed in the left end of the boat. The seed is partially melted to provide a clean surface for growth and then the zone is moved to the right. If a material crystallizes readily, a seed may not be required. The heat source may be radio-frequency heating coupling to the melt, the boat, or a susceptor. Other heat sources include radiative heating from a resistance element, electron bombardment, and focused radiation from a high-intensity lamp or the sun. In horizontal zone melting the container, of course, must be compatible with the melt. Even if the melt does not react with the boat, enough wetting may take place that the grown crystal sticks to the boat. This may induce strains because of differential contraction during cooling and will

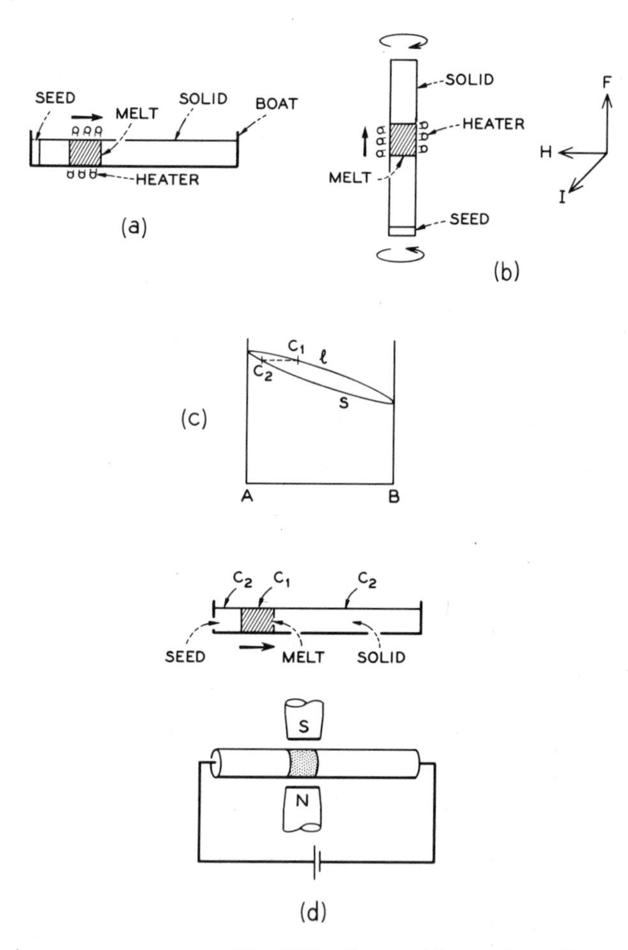

Fig. 5.17 Zone-melting configurations.

often make the crystal difficult to remove from the boat. Deformable or soft boats can sometimes be used to prevent these difficulties. If the left end of the boat is pointed, nucleation of a single crystal without need of a seed is often possible. Other crucible geometries, such as those used in Bridgman–Stockbarger growth to induce nucleation of a single crystal, have been used.

Many passes are, of course, conveniently made and, if one desires, zone travel may be arranged not to melt a "seed" section at the left. It is often observed that, as a material is purified by subsequent passes, the grain size tends to increase so that single crystals result as a happy "accident" of ultrapurification.

Figure 5.17(b) shows the float-zone process. This technique was first described by Keck and Golay (1953) and appears to have been developed independently by several others (Emeis, 1954; Theuerer, 1956). Its first application was to the purification of silicon. In this technique surface tension holds a molten zone of liquid in a sample whose axis is vertical. The technique does away with the container—it is "crucibleless"—so reactivity with the boat is no longer a problem.

In the absence of a boat, rf coupling to the boat is not possible, so heating must be accomplished by direct coupling to the melt (provided it is conductive enough) or by radiant heating from a resistance heater, an rf-heated susceptor, or by focusing a radiant source. If rf coupling is a problem, in some cases better coupling in resistive materials may be accomplished by rf heating at high frequencies. Some stirring may be effected by independently rotating the two ends of the sample in opposite directions. Unless the sample is quite dense the melt may tend to penetrate voids by capillary action, and it will be difficult to keep the zone width under control. To prevent this, material previously prepared by horizontal melting, casting, sintering, or hot pressing is generally required.

The conditions for zone stability have been studied by Heywang and Ziegler (1954) and by Heywang (1956). Heywang assumed that the only active forces are surface tension and the gravitational field, that the melt completely wets the solid, that the volume change on melting is negligible, and that the interfaces are planes perpendicular to the sample axis and the gravitational field. Figure 5.18 shows Heywang's results where λ is a parameter proportional to the maximum zone length, l, so that

$$\lambda = l\sqrt{dg/\sigma} \tag{5.15}$$

and ρ is proportional to the rod diameter, r, so that

$$\rho = r\sqrt{dg/\sigma} \tag{5.16}$$

In Eqs. (5.15) and (5.16) d is the density of the liquid, g is the gravitational constant, and σ is the surface tension. As can be seen, as the rod diameter increases, λ approaches $\sim$2.7. Thus if the zone length could be controlled at small values of l there would be no limit to r. In practice when $l > r$, control

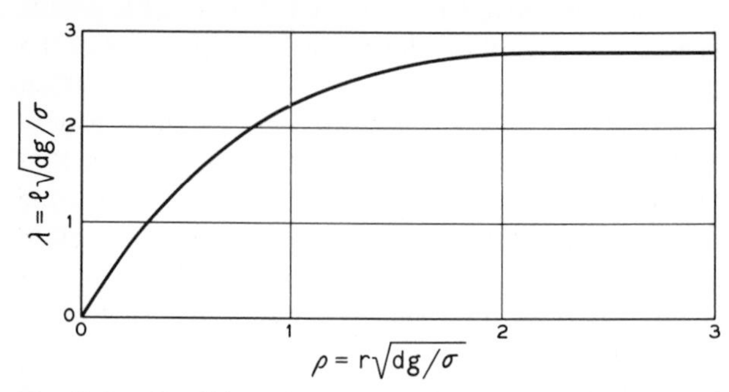

Fig. 5.18 Conditions for zone stability (after Heywang, 1956). (Courtesy of *Verlag der Zeitschrift für Naturforschung*, Tübingen.)

is difficult. If we assume $l \simeq r$ for stability then the only other requirement (Heywang, 1956) is that

$$l = \frac{1}{2.7}\sqrt{\sigma/dg} \tag{5.17}$$

If the ratio σ/d is sufficiently large, as in light metals such as Al where d is small and denser high-melting metals such as iron and titanium where σ is large, l will be large enough to make float-zone melting practical. Many materials where σ/d is small, such as low-melting-point dense inorganics, give such a small l that they are not conveniently float-zone-melted.

By the use of low-frequency electromagnetic fields and the proper design of the rf induction coil it is possible to produce a substantial upward force on the liquid region and thereby support greater l's than Eq. (5.17) would predict. The technique of supporting a material in a gravitational field without mechanical connection to it is called *levitation*. Radio-frequency levitation results because of the interaction of the applied low-frequency electromagnetic field and the field caused by induced eddy currents in the liquid. In the case of Si some levitation results at 5 Mc but 500 kc give a much stronger levitating force. The levitating force F is given by

$$F \propto HI \tag{5.18}$$

where H is the magnetic field produced by the rf frequency in the melt and I is the current in the melt. F, H, and I are mutually perpendicular [Fig. 5.17(b)] according to the right-hand rule. The analysis of the situation with an rf current is rather complicated because the exact position and magnitude of the eddy currents is affected by the fluid currents that they cause in the melt and may be influenced by convection. Indeed, these fluid currents tend to stir the melt and are an important source of stirring in melting by rf heating. It is also possible to cause levitation by producing current flow in the melt by imposing a voltage across the end of the sample and producing a magnetic field in the melt by placing it between the poles of a permanent magnet or electromagnet, as shown in Fig. 5.17(d). This arrangement might be used to effect crucibleless horizontal zone melting. One variant of zone leveling, a technique for uniformly doping, is shown in Fig. 5.17(c). Concentration is kept uniform at C_2 along the sample by starting with a zone of concentration C_1. For other methods applicable to crystal growth see Pfann (1966). Table 5.4 lists a variety of materials that have been grown by zone melting together with conditions for growth. Thermal-gradient zone melting is covered in Chapter 7.

Electrón-beam melting sometimes provides a convenient method of growing high-melting-point metals in the float zone. Because electrons are easily scattered, the beam must be generated and melting must take place in vacuo. For successful melting, it is generally necessary to have an initial vacuum in the working chamber of about 10^{-6} Torr. The necessity for vacuum operation is the principal disadvantage of float-zoning by electron-beam heating. However, considerable advantage results because higher energy density inputs into the specimen are possible than by rf heating and thus very refractory materials may be melted with relative ease. Generally, unless the specimen is conductive, experimental complications are introduced, so this is an additional constraint on the technique. More work on zone refining than on crystal growth has been reported but there is every expectation that with a modicum of care in control the technique can readily be extended to the growth of many of the materials where purification has been successful.

A schematic of a typical electron-beam melting apparatus is shown in Fig. 5.18A.

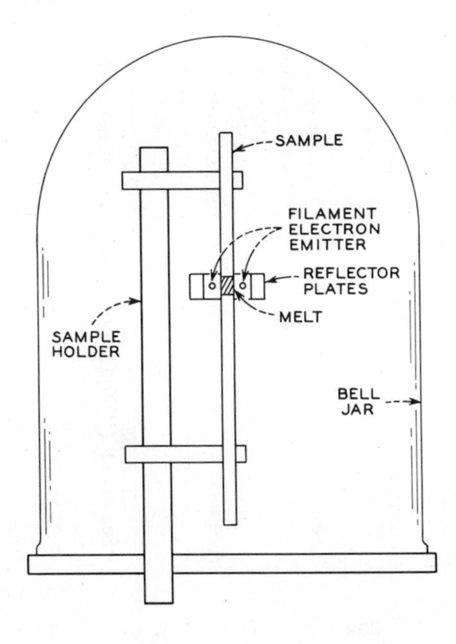

Fig. 5.18A Electron-beam melting configuration.

Table 5.4

SOME CRYSTALS GROWN BY ZONE-MELTING TECHNIQUE†

Material	*Melting Point (°C)*	*Boat*	*Rate of Zone Movement*	*Comments*	*References*
Germanium	942	Vitreous silica, graphite-coated	0.1–5 cm/hr		Pfann and Olsen, 1953
Germanium	942	Vitreous silica, graphite-coated	~0.1 cm/hr	After-heater and slow rate improve perfection	Bennett and Sawyer, 1956
Gallium arsenide	1240	Vitreous silica or silica coated with pyrolytic graphite	1.25–5 cm/hr	Dislocation densities $\approx 100/cm^2$ (As pressure)	Richards, 1960
Tungsten	3370	Water-cooled copper reactor and float zone	0.2–4 cm/min	Arc melting and electron beam	Carlson, 1959; Schadler, 1960; Geach and Jones, 1959

†See Pfann, (1966) for other materials that have been zone-melted for purification and in some cases crystal growth.

The electron emitter is commonly a simple ring filament surrounding the specimen. In some cases several electron emitters replace a single filament. The electron-emitter material (1) should have a convenient work function (i.e., the electron flux should be reasonable at practical voltages), (2) should not react with the ambient at the conditions of use, (3) should have a low enough vapor pressure that specimen contamination is not a problem, and (4) should be strong enough to support itself at the temperature of operation. An ideal material is tungsten and an ideal situation is when the emitter material and specimen are identical. The reflection plates are made of a refractory conductor, often tantalum, and are grounded or biased to aid in focusing the beam on the specimen. Beam deflection can be accomplished by either electron or magnetic fields, and the equations and techniques are well known (Bakish, 1962). A bias voltage is applied to the sample to aid in focusing the electrons. Control of temperature is achieved by controlling the sample bombardment current by adjusting the filament temperature or bombardment voltage (potential difference between anode–sample and cathode–filament). Emission temperature control is more common. Emission current is easily measured by passing it through a known resistor and measuring the voltage drop. Bombardment equipment is commercially available and is described in Bakish (1962). If the solid sample is not conductive but the melt is, electrical contact with the melt may be made and melting is possible. If the solid sample is not conductive at moderate temperatures, a conductive strip may be deposited along its length to permit bombardment to proceed until the temperature is reached where the material becomes conductive.

The materials where crystals have been grown include tungsten, molybdenum, and rhenium and alloys of tantalum–niobium and of molybdenum–rhenium (Bakish, 1962). It is useful if polycrystalline rods of high density are available as the starting sample. If the rod is not dense or if high gas evolution takes place, maintaining a zone is difficult.

The extensive potential for electron-beam melting of refractory materials has yet to be fully realized.

An interesting modification of electron-beam melting that allows the growth of insulators is the hollow-cathode or so-called "cold-cathode" technique (Class et al., 1967). This method depends on a self-sustaining dc gas discharge emanating from an annular-shaped hollow cathode that encircles the workpiece. At pressures of the order of a few millimeters of Ar, O_2, and other gases when the voltage of the cathode is a few kilovolts, electrons are emitted that ionize the gas and form a conducting plasma. A current of several hundred milliamperes is produced in this conducting plasma and current flow is to any convenient ground in the system. Thus the workpiece need not be conductive to complete the circuit for current flow, as in conventional electron-beam melting. If the interior of the cathode is properly shaped, the electrons and produced ions are focused on the workpiece and their energy causes melting. A suitable cathode material is stainless steel and, although the cathode is partly cooled by water circulating within it, best results seem to be obtained when the cathode temperature is only somewhat below red heat. Thus *cold* cathode is somewhat of a misnomer for the method. Materials other than stainless steel may be required for melting materials with melting points much above 2200–2500°C. To maintain the plasma, low gas pressure of a moderately easily ionizable gas is required (in contrast to the high vacuum needed in ordinary electron-beam melting). This gas may be useful in repressing decomposition of the material to be melted. The technique has been used successfully (Class et al., 1967) to grow single crystals or sapphire, yttrium–aluminum garnet, and other materials by float zoning. Its potential for very high-melting insulators such as carbides, borides, and nitrides should be considerable (see Sec. 5.6.2 for a discussion of plasma techniques).

5.6 Other Crucibleless Techniques

As we have seen, an important constraint on the application of a liquid–solid equilibrium to crystal growth is melt reactivity with the container and container melting. This constraint becomes especially severe if high-melting-point crystals are desired because, where reactivity is not a problem, crucible melting sets the upper limit. Tungsten (mp 3370°C) and carbide or nitride containers (mp $>$ 4000°C) offer partial solutions to this problem. However, perhaps the highest-melting-point material that has been successfully grown from a container is sapphire (mp 2015°C) pulled from an Ir (mp 2554°C) crucible, where reactivity with the melt presents no serious problem (see Sec. 5.4.2-2). If successful growth without reactivity is the criterion of success, it remains to be seen whether containers stable at higher temperature can be made practical. One solution to the container problem, as we have seen, is float zoning. However, it is often difficult to get enough thermal energy into the zone to melt high-melters. This is especially true if the melt is not conductive (rf heating is impractical) and does not absorb appreciably in the infrared and visible regions (radiant heating from a resistive heater or focused radiation from a lamp, arc, or the sun is impractical). One means of improving thermal-energy input is to increase the surface/volume ratio of the melt. This can be done easily if a molten puddle is produced on top of a seed or a polycrystalline mass of the material as shown in Fig. 5.19(a). The puddle is held on the seed or polycrystalline mass by surface tension. The remaining crucibleless techniques are variants of this procedure.

5.6.1 FLAME FUSION

In flame fusion the molten puddle is produced on top of a seed crystal and the thermal source is a flame [Fig. 5.19(b)]. The material for growth

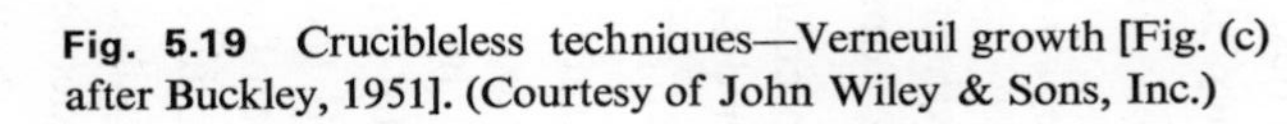

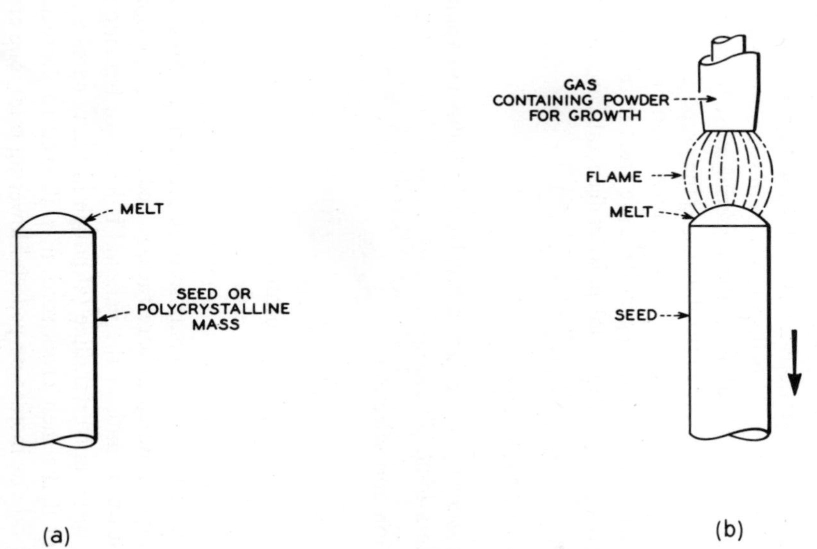

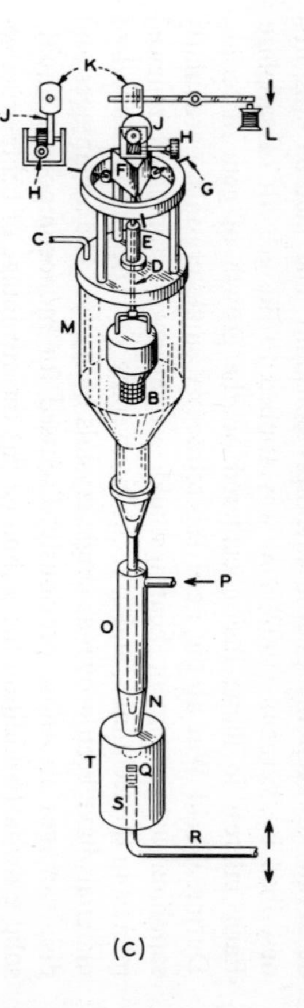

Fig. 5.19 Crucibleless techniques—Verneuil growth [Fig. (c) after Buckley, 1951]. (Courtesy of John Wiley & Sons, Inc.)

is added to the gas supply as a powder and the seed crystal is withdrawn at a rate to keep the puddle thickness constant. The method was first described by Verneuil (1902) and was applied to the growth of sapphire. Early interest in flame fusion centered on the growth of gem materials. During World War II the need for refractory single crystals, particularly sapphire, as instrument bearings and watch jewels stimulated renewed interest in the technique, and over the intervening years perhaps one hundred materials have been grown as single crystals by its use. The advantages and disadvantages are those previously discussed for monocomponent liquid–solid growth. In addition the technique has the advantage of being crucibleless, although with some constraints imposed by the necessity of supporting a liquid layer on the seed (the analysis of liquid-layer height, although never conducted in detail, should be analogous to the calculation of zone length in floating-zone technique). The fact that more surface area is exposed to thermal energy makes it easier to melt refractories. However, the thin melt layer with its resultant small heat capacity means that growth rates are severely influenced by thermal fluctuations and it is difficult to get uniform doping without striations. Similarly, growth-rate variations produce strain and index-of-refraction variations. In addition the material must be compatible with the chemical conditions present in the flame, and the preparation of powder for addition to the gas and its feeding into the melt often present experimental difficulties. In particular, thermal gradients are great in the technique, and consistency of flame temperature and powder feed rate are difficult to control.

5.6.1-1 Apparatus and Technique

The growth of crystals by flame fusion involves the following apparatus and technique considerations:

1. Seed preparation and selection
2. Burner design
3. Powder preparation and feed

Because apparatus and technique are central to success in this method we shall examine considerations 1–3 in some detail.

Seeds can be obtained from natural sources or by means of another technique, but it is more common to prepare seeds in preliminary flame fusion experiments. If the flame is aimed at a refractory pedestal it is possible to form a puddle of melt on it, and, as the pedestal is lowered, freezing takes place. The product is often polycrystalline (although in many cases a single crystal will result), but if the melt cross section is allowed to decrease (by lowering powder feed rate or increasing pedestal-lowering rate), one crystallite will often dominate the growing interface. The melt cross section may then be increased and single-crystal growth will result. This procedure is analogous to "necking down" in crystal pulling. The grown crystal may then be used as a seed or cut into smaller seeds for subsequent runs. Seeds are selected for perfection. Low-angle grain boundaries and strains are important to avoid if highly perfect crystals are desired. In the present state of the technique it is impractical to be concerned with dislocations. Crystals are oriented before use as seeds because in most materials orientation influences perfection and in many cases will even determine whether single crystals can be grown.

Burner design has evolved considerably since the time of Verneuil. All practical work employs the oxyhydrogen flame. Acetylene–oxygen and illuminating gas–oxygen have been investigated with limited success because clean flames are hard to obtain. If higher temperatures are required, the use of fluorine or other strong oxidants is possible, although so far there has been little work in this direction. Purity of the gas and constancy of gas flow is important because poor crystals will result if temperature fluctuations are severe. Oxygen and hydrogen should not be mixed before their ignition, so all burner designs depend on separate O_2 and H_2 tubes leading to the ignition region. In the simplest case [Fig. 5.19(c)] there are two concentric tubes and the inner tube carries oxygen and feed powder, while the outer tube carries hydrogen. In one popular modification a number of small tubes are nested inside a larger tube. Oxygen and feed powder are supplied through the small tubes and hydrogen is supplied through the interstices between the tubes. This design produces a large flame front with a fairly uniform flame temperature, which is useful in the growth of large-diameter crystals. This design is often referred to as the multitube burner. Perhaps the most important innovation in burner design since Verneuil's original work is the tricone burner developed by Merker (1947). Three concentric tubes are used, with the inner tube carrying hydrogen and the two outer tubes oxygen. This produces a pointed flame useful for starting growth when a seed crystal is not available and useful for routinely growing small-diameter rods. Small-diameter sapphire rods are of great commercial interest because they can be conveniently cut into watch jewels and instrument bearings. The burner produces a strongly oxidizing flame that is necessary in growing easily reducible materials such as nickel oxide. The tricone burner tends to produce cooler flames than other burners because the flame is oxygen-rich. This is advantageous if lower-melting materials are being grown but prevents the growth of materials with melting points much above sapphire.

Popov (1959) and Bauer and Field (1963) have extensively discussed Verneuil growth. For the growth of sapphire, seed powder must be pure and of uniform particle size in the range of a few microns. Carefully calcining alum at about 1000°C provides powder of the requisite properties. As a general rule a powder of desired properties is best prepared by use of a calcining reaction where chemically bonded water or some other volatile constituent is driven off during a reaction at a low enough temperature that sintering of the feed powder does not take place. For sapphire, the best growth takes place at about 60° from the *c* axis and rates of a few inches per hour are possible.

5.6.2 TECHNIQUES RELATED TO FLAME FUSION —PLASMA AND ARC HEATING

There are several techniques similar to flame fusion in which the heat source is not a flame. Some authors have called these techniques *modified Verneuil growth.* These heating methods in addition to flames have been used in modified Verneuil growth:

1. Radiation heating
2. Plasma heating
3. Arc-image heating

Radiation heaters have been heated by rf coupling and by resistance heating. Keck et al. (1954) inductively heated a metal susceptor. They found it necessary to use inert gases in order to prevent oxidation of the susceptor at its operating temperature. Dickinson and Field (1956) heated their susceptor resistively. The experimental arrangement is thus greatly simplified because rf equipment and controls are not required. Dickinson and Field also found that inert atmospheres were required to prevent heater oxidation. Radiation heaters partially circumvent the constraint of flame fusion, that the crystal grown be compatible with the flame. However, if very high temperatures are required, then oxidizing atmospheres are not permitted because heater oxidation is a problem. It might be possible to place the heaters in an envelope separate from the growing crystal and thus independently control their environment. In a sense this approach is realized in arc-image growth. Several ferrite compositions have been grown by radiation heating in the Verneuil geometry.

Plasma heating has been pioneered as a technique by Reed (1961). (Hollow-cathode heating, which also involves a plasma, is discussed in Sec. 5.5) Commercial dc plasma torches had been applied to crystal growth previously (Bauer, 1960) but Reed was the first to use an inductively coupled plasma as a heat source. The plasma state can be considered a fourth state of matter in which some or all of the electrons have been stripped from gaseous atoms. Plasma temperatures can be very high, ranging up to many thousands of degrees. Plasmas are produced by the ionization of atoms in flames and in electrical discharges. The arc that is produced when current discharges between two electrodes, as in an arc welder, is a common example of a plasma. Heating by means of arcs has been known since the ready availability of substantial electric currents. Direct-current plasma torches became commercially available in the mid-1950's and techniques for introducing powdered feed material into torches are well known. Commonly, the torch is allowed to impinge on a cool surface where the feed material freezes as a fine-grained ceramic. This technique is called flame spraying and it is discussed in an extensive literature. Such a torch can be modified to replace the flame in a Verneuil crystal-growth setup. Figure 5.20 shows a dc plasma torch. Essentially, a dc arc is formed between the electrodes and rapid gas flow through the arc directs the plasma ahead of the electrodes. In welding the workpiece is one of the electrodes, and material cannot be melted unless the workpiece is conductive. The plasma torch removes this constraint. Direct-current plasma torches commonly run at 10–100 V and at currents from several hundred to several thousand amperes. Temperatures up to about 15,000°C have been reported. There is often considerable difficulty in gas stabilization. At the worst, the plasma blows from between the electrodes and is extinguished; at the best the plasma tends to fluctuate in intensity and wander in position. Plasmas have been stabilized by introducing liquid water into the torch. This produces an inherently "dirty" plasma that negates the primary plasma advantage—that the atmosphere is clean and can be varied

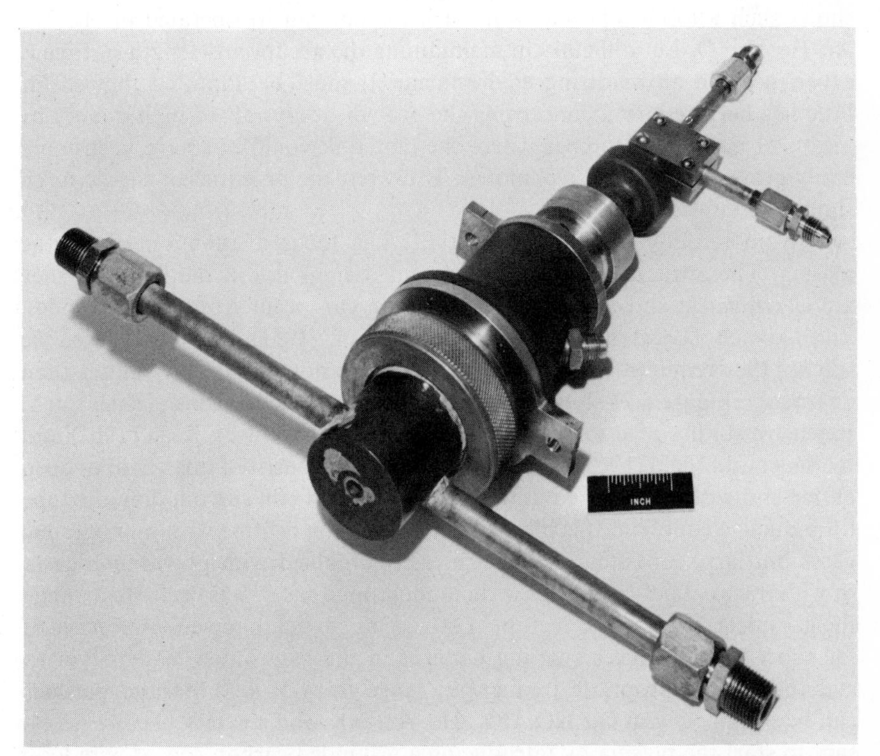

Fig. 5.20 D.C. plasma torch (after Bauer and Field, 1963). (Courtesy of John Wiley & Sons, Inc.)

at will. Plasmas are attractive principally because, in contrast to flames, oxidizing, reducing, and inert conditions can be obtained. A more satisfactory method of stabilization is the use of magnetic fields. Because the plasma is a stream of gas conducting current between the electrodes it is subject to a force by a magnetic field and a suitably shaped field can "shape" the discharge. The details of such interactions are considered by the field of magnetohydrodynamics. The present impetus for such studies is the desire to contain high-temperature plasmas for controlled fusion research. Although magnetic stabilization is inherently attractive, it has been little applied to crystal growth. Plasma stabilization in crystal growth has been mainly by controlling velocity and direction of the gases swept between the electrodes. If the flow can be arranged to have a circular motion so that a low-pressure region froms in its center, i.e., to form a vortex, then the plasma will stabilize in this region. *Vortex stabilization* is particularly useful in rf plasmas. *Gas sheath stabilization* is most commonly used in dc torches. In this design the arc goes from a tungsten cathode to a hollow water-cooled anode. The arc remains within the chamber and is prevented from striking the wall by a sheath of gas that is considerably larger than the wall diameter. Figure 5.20 shows such a torch. Plasmas with such torches can be operated in H_2, N_2, Ar, He, and O_2 but difficulty in maintaining the arc for any length of time is experienced in an oxidizing environment. It must be admitted that so far little has been reported concerning the growth of crystals of high quality by means of dc plasma torches where the material would not have been more easily growable by other techniques. However, the potential of the dc torch should be substantial.

More recently, Reed's work (1961) has focused attention on the ac plasma. The attractiveness of these recent designs lies in the fact that they are *electrodeless* and thus no contamination can occur from the electrodes. Zilitinkovich (1928) and Cobine and Wilbur (1951) previously used ac torches that required electrodes, but they were not significantly better than dc torches. Figure 5.21 shows a typical ac coupled electrodeless plasma torch. Extensive studies of torch design have been carried out by Reed (1961) and by Bauer and Field (1963). Essentially all torches consist of quartz tubes open at one end with gas injected at the other with an rf coil surrounding the tube for inductive coupling. It is more difficult to obtain heating with monoatomic gases but large amounts of heat are easily obtained with polyatomic gases or with traces of polyatomic gases in monoatomic gases. It is useful to arrange the gas inlets so that the direction of flow is tangential to the circumference of the tube. This produces swirling patterns in the gas, which lead to vortex stabilization. Appropriate frequencies range from 5–3000 Mc and plasmas can be sustained with O_2, NO, CO_2, He, Ar, SO_2, and air. It is usually necessary to start the plasma by introducing a grounded carbon rod or wire loop

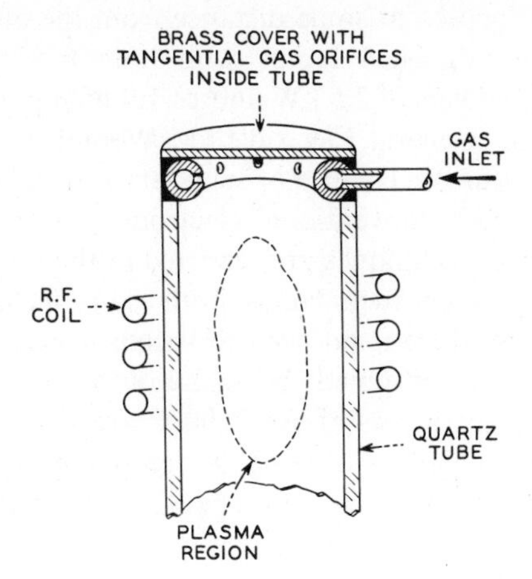

Fig. 5.21 R.F. plasma torch (courtesy of R. L. Barns).

within the rf field. This introduces sufficient ions to make the gas conducting. The plasma forms and then it is hot enough to be self-sustaining. Bauer and Field (1963) have discussed various geometries, including modifications that are capacitively coupled. Comparatively high-melting oxides such as sapphire can be grown in the rf plasma but their perfection so far has not been as good as that of sapphire made by more conventional techniques.

If the material to be melted absorbs sufficiently at wavelengths where a powerful lamp is available, it may be melted by focusing radiation onto it. This family of techniques is usually called *arc-image growth.* Until recently the best lamps were carbon arc lamps and the main difficulty with the technique was the instability and the relatively short lifetime of these heat sources. As an arc lamp burns, it consumes the carbon electrode at rates of the order of inches per hour. The most convenient lamps available are theater projection lamps of the kind used in projectors in "drive-in movies." Such lamps provide for automatic feed of the carbon and LaRue and Halden (1960) had considerable success in focusing the output of such lamps for crystal growth (see Fig. 5.22). However, eventually the carbon is consumed and, while a new electrode is inserted, growth ceases with a resultant severe inhomogeneity in the final crystal. One solution of this problem is to provide two arc sources to heat the crystal and to switch from one source to the other while electrodes are replaced. This reduces the inhomogeneity but does not eliminate it because there are bound to be some thermal fluctuations during such a change-over. A convenient geometry for the two source heaters is the so-called "clam-shell arrangement" described by Ploetz (Bauer and Field, 1963). A typical system consists of two clam-shell arrangements both fixed to focus radiation on the sample. Each of the clam shells consists of two parabolic reflectors, with a hole cut from the center of each reflector. A source of light is placed just outside one reflector and focused at some distance from the other.

Recently high-temperature tungsten lamps with long lifetimes have become available. Lamps with powers of 2.5 kW and useful life up to 1500 hr are now available from Osram (General Electric Ltd., Wembley, England) and other major lamp manufacturers. The use of such lamps should obviate most of the difficulties found with conventional electrode sources. Recent developments in thermal imaging techniques are reported in the book edited by Glaser and Walker (1964). With such lamps both clam shells can be operated simultaneously because there is no need to switch from one to the other when carbon electrodes are consumed. With the output of two clam shells focused on a sample, temperatures above 2500°C are possible. It is, of course, easy to control the atmosphere around the sample by enclosing it

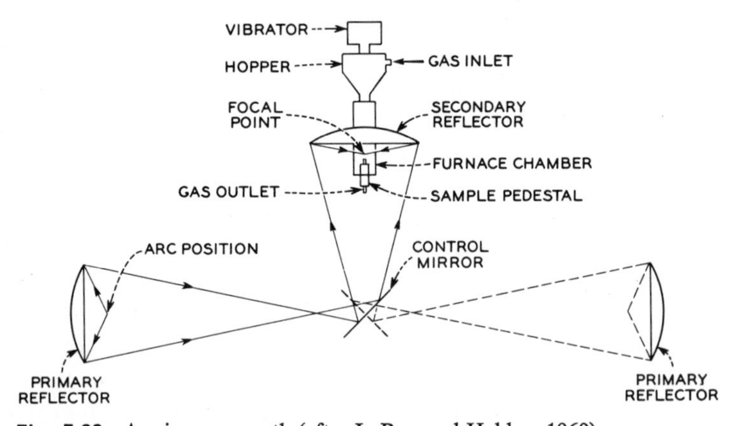

Fig. 5.22 Arc-image growth (after LaRue and Halden, 1960).

in a suitable transparent envelope. This technique provides perhaps the most promise of all the Verneuil-type methods for providing a clean high-temperature environment for crystal growth with reasonable experimental simplicity. Table 5.5 lists some arc-image-grown crystals.

Crystals are subject to high thermal gradients during growth by Verneuil-type methods. Postgrowth annealing helps greatly to reduce strains. Commercially grown crystals† are strained and shatter easily before annealing in gas-fired furnaces. If heater arrangements can be utilized that permit annealing without cooling to room temperature, the strains are further reduced. If heater arrangements that reduce gradients are used, even lower strain results. Heat sources such as arcs and lamps can, of course, be adapted to growth by float zoning.

---

†Verneuil and other crystals are often called boules when they lack planar morphology (F. *boule* = ball).

Table 5.5

SOME CRYSTALS GROWN BY VERNEUIL AND ARC-IMAGE TECHNIQUES

Material	Melting Point (°C)		Comments	References
Al_2O_3	2040	Corundum, sapphire	Growth in a variety of directions; best growth in a cone of directions 60° from *c*	Popov, 1946; Alexander, 1946; Verneuil, 1904, 1911; Ikornikova and Popova, 1956; Popov, 1946
$Al_2O_3: Cr$		Ruby	Verneuil growth	
$MgAl_2O_4$	2130	Magnesium–aluminum spinel	Verneuil growth	Farben, 1945
$3Al_2O_3 \cdot 2SiO_2$	1810	Mullite	Verneuil growth	Bauer et al., 1950
$CaWO_4$	1530	Scheelite	Verneuil growth	Gillette, 1950; Zerfoss et al., 1949
TiO_2	1830	Rutile	Verneuil growth	Alexander, 1949; Merker, 1955
ZrO_2	2700	Zirconia	Verneuil growth	Halden and Sedlacek, 1963
Y_2O_3	2400	Yttria	Verneuil growth	Lefever, 1962
$MgFe_2O_4$	Above 1200	Magnesium (nonstoichiometric) spinel	Arc-image growth	Halden, 1964
$NiFe_2O_4$	Above 1200	Nickel ferrite	Arc-image growth	Poplawsky, 1964

5.7 Other Liquid–Solid Methods

Differential pulling might be considered either as a modification of pulling, float zoning, or Verneuil growth. The technique is illustrated in Fig. 5.23 and is useful where a crucibleless technique is required. It provides a greater solid angle for energy input than float zoning and the possibility of better control than in Verneuil-related techniques.

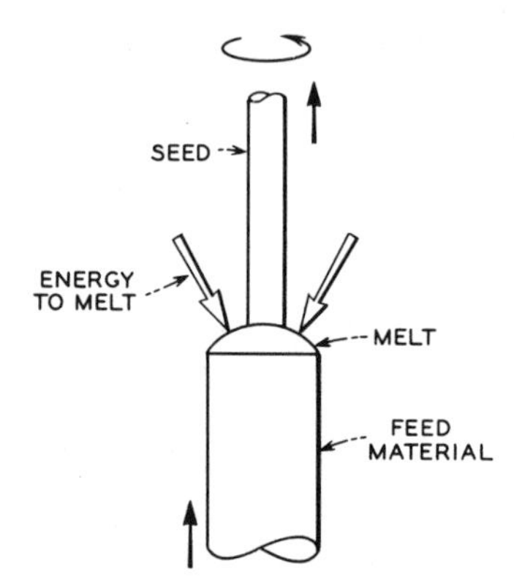

Fig. 5.23 Differential pulling.

5.8 Dendrite Growth

In some cases it is deliberately desired to grow crystals with elongated-protrusion and branched-protrusion morphology. Such crystals are generally called dendrites. As discussed in Secs. 3.11, 3.12, 3.13 and 3.14, dendrite growth can occur under conditions of high constitutional supercooling. Generally, whenever the temperature in front of the growing interface is considerably below the equilibrium temperature, dendrites are likely to grow. This can occur because

1. Diffusion problems are severe and the melt is greatly enriched with respect to an impurity, dopant, or solvent close to the growing interface. This is constitutional supercooling and has been discussed in Sec. 3.11.

2. The growth rate is so rapid, the heat of crystallization is so large, or the thermal coupling to a heat sink is so poor that the crystal is not able to dispose of its heat of crystallization rapidly. Under such conditions, the interface will be considerably warmer than the nearby liquid and dendrite growth will result.

Figure 5.24(a) shows the temperature distribution during normal growth and Fig. 5.24(b) the distribution when dendrites are formed because of this mechanism. In Fig. 5.24(a) supercooling in front of the interface is negligible;

in Fig. 5.24(b) it is severe and any protrusion in the interface can grow more easily than nearby areas. The result is that such protrusions (caused by growth steps, facets, the edges of cells if growth is cellular, etc.) grow rapidly into the melt. Dendritic growth is not confined to metals or to liquid–solid growth but can occur under any circumstances where condition 1 or 2 is fulfilled. For completeness we have shown the temperature profiles in constitutional supercooling in Fig. 5.24(c). Here the concentration of impurity, dopant, or solvent is greater near the interface than in the bulk of the liquid. Impurities lower the melting point so that the region in front of the interface is greatly supercooled.

Wagner (1960), Bolling and Tiller (1961), and Temkin (1962) have discussed the conditions for dendrite formation in some detail, and Billig (1955) and Bennett and Longini (1959) were the first to grow controlled dendrites.

Probably Ge has been most studied. For the growth of a diamond-type lattice such as Ge:

1. Usually, a seed from a previously grown dendrite is used. This is because such a crystal, if properly grown, will be twinned. A properly oriented

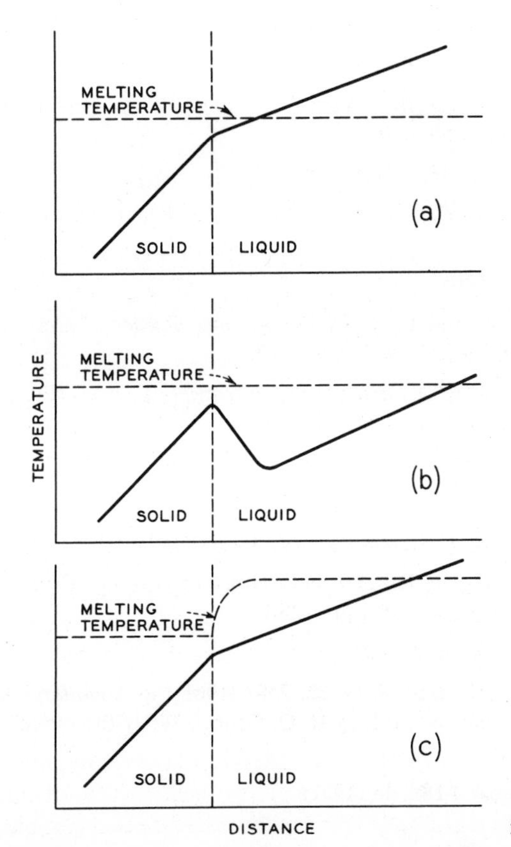

Fig. 5.24 Temperature profile in dendrite growth.

twin face provides a permanently stepped interface that allows rapid growth (see, for instance, Holmes, 1961; John and Faust, 1961; Seidensticker and Hamilton, 1963).

2. After thermal equilibrium has been obtained, the temperature of the melt is lowered $\sim$15–20°.

3. After a few seconds, pulling is commenced at a rapid rate (10 cm/sec).

A typical product is a tapelike crystal with smooth faces and jagged edges several meters long, a few tenths of a millimeter thick, and a fraction of a centimeter wide. Such crystals may have some advantages for semiconductor devices (e.g., cutting of bulk crystals is eliminated) and much work is described in an extensive patent literature.

REFERENCES

Alexander, A. E., *J. Chem. Educ.* **23**, 418 (1946).

Alexander, A. E., *J. Chem. Educ.* **26**, 254 (1949).

Andrade, E. N. da C., and R. Roscoe, *Proc. Phys. Soc.* (*London*) **49**, 152 (1937).

Aust, K. T., and B. Chalmers, *Can. J. Phys.* **36**, 977 (1958).

Bakish, Robert, Ed. *Introduction to Electron Beam Technology*, Wiley, New York, 1962. See especially pp. 21ff, 168ff, and 184ff.

Ballman, A. A., *J. Am. Ceram. Soc.* **48**, 112 (1965).

Bardsley, W., and B. Cockayne, in *Crystal Growth*, Ed. by S. Peiser, Pergamon, New York, 1967, pp. 109ff.

Barns, R. L., unpublished work, 1965.

Bauer, W. H., Reported in Bauer and Field, in *The Art and Science of Growing Crystals*, Ed. by J. J. Gilman, Wiley, New York, 1963.

Bauer, W. H., and W. G. Field in *The Art and Science of Growing Crystals*, Ed. by J. J. Gilman, Wiley, New York, 1963, pp. 398ff.

Bauer, W. H., I. Gordon, and C. H. Moore, *J. Am. Ceram. Soc.* **33**, 140 (1950).

Bennett, A. I., and R. L. Longini, *Phys. Rev.* **116**, 53 (1959).

Bennett, D. C., and B. Sawyer, *Bell. System Tech. J.* **35**, 637 (1956).

Berry, C., W. West, and F. Moser, in *The Art and Science of Growing Crystals*, Ed. by J. J. Gilman, Wiley, New York, 1963, pp. 226ff.

Billig, E., *Proc. Roy. Soc.* (*London*) **A229**, 346, 1955.

Bolling, G. F., and W. A. Tiller, *J. Appl. Phys.* **32**, 2587 (1961); in *Metallurgy of Elemental and Compound Semiconductors*, Ed. by R. O. Grubel, Wiley-Interscience, New York, 1961.

Bowles, J. S., *Trans. AIME* (*J. Met.*) **189**, 44 (1951).

Bridgman, P. W., *Proc. Am. Acad. Arts Sci.* **60**, 303 (1925).

Brown, H. G., C. N. Hoyler, and R. A. Bierwith, *Theory and Applications of Radio Frequency Heating*, Van Nostrand, Princeton, N.J., 1947.

Buckley, H. E., *Crystal Growth*, Wiley, New York, 1951, pp. 71ff, 99ff.

Carlson, R. G., *J. Electrochem. Soc.* **106**, 49 (1959).

Carruthers, J. R., and K. Nassau, *J. Appl. Phys.* **39**, 5205 (1968).

Chalmers, B., *Proc. Roy. Soc.* (*London*) **A175**, 100 (1940).

Champier, G., *Compt. Rend.* **243**, 657 (1956).

Charvat, F. R., O. H. Nestor, and J. C. Smith, Semiannual Technical Report, Contract No. 4132(00), ARPAN 306–62, Union Carbide, Linde Division, Speedway Laboratories, Indianapolis, 1965.

Class, W., H. R. Nestor, and G. T. Murray, in *Crystal Growth*, Ed. by H. S. Peiser, Pergamon, New York, 1967, pp. 75ff.

Cobine, J. D., and D. A. Wilbur, *J. Appl. Phys.* **22**, 835 (1951).

Cochran, W. G., *Proc. Cambridge Phil. Soc.* **30**, 365 (1934).

Cole, G. S., and W. C. Winegard, *Can. Met. Quart.* **1**, 29 (1962).

Czochralski, J., *Z. Physik. Chem.* (*Leipzig*) **92**, 219 (1917).

Dash, W. C., in *Growth and Perfection of Crystals*, Ed. by R. H. Doremus, B. W. Roberts, and David Turnbull, Wiley, New York, p. 31, 1958.

Dash, W. C., *J. Appl. Phys.* **31**, 736, 2275 (1960).

Dickinson, S. K., and W. G. Field, 1956, reported in Bauer and Field, 1963.

Edmund, J. T., R. F. Brown, and F. A. Cunnell, *Services Electron. Res. Lab. Tech. J.* **6**, 123 (1957).

Edwards, W. D., *Can. J. Phys.* **38**, 439 (1960).

Elam, C. F., *Proc. Roy. Soc.* (*London*) **112**, 289 (1926).

Emeis, R., *Z. Naturforsch.* **9a**, 67 (1954).

Farben, I. G., *FIAT, Rev. Ger. Sci. Rept.* No. 655 (1945).

Fleischer, R. L., and R. S. Davis, *Trans. AIME* **215**, 665 (1959).

Fong, Francis K., *J. Chem. Phys.* **41**, 2291 (1964a).

Fong, Francis K., *J. Chem. Phys.* **41**, 245 (1964b).

Geach, G. A., and F. O. Jones, *J. Less Common Metals* **1**, 56 (1959).

Gillette, R. H., *Rev. Sci. Instr.* **21**, 294 (1950).

Gilman, J. J., Ed., *The Art and Science of Growing Crystals*, Wiley, New York, 1963. See pp. 214ff, 314ff, 343ff, 365ff, 381ff, and 410ff.

Glaser, P. E., and R. F. Walker, Eds., *Thermal Imaging Techniques*, Plenum Press, New York, 1964.

Goss, A. J., and R. E. Adlington, *Marconi Rev.* **22**, 18 (1959).

Goss, A. J., in *The Art and Science of Growing Crystals*, J. J. Gilman, Ed., Wiley, New York, 1963, pp. 314ff.

Gremmelmaier, R., *Z. Naturforsch.* **11a**, 511 (1956).

Guggenheim, H., *J. Phys. Chem.* **64**, 938 (1960).

Guggenheim, H., *J. Appl. Phys.* **32**, 1337 (1961); **34**, 2482 (1963).

Guggenheim, H. J., and J. V. Kane, *Appl. Phys. Letters* **4**, 172 (1964).

Halden, F. A., in *Thermal Imaging Techniques*, Ed. by P. E. Glaser and R. F. Walker, Plenum Press, New York, 1964, pp. 193ff.

Halden, F. A., and R. Sedlacek, *Rev. Sci. Instr.* **34**, 622 (1963).

Heywang, W., *Z. Naturforsch.* **11a**, 238 (1956).

Heywang, W., and G. Ziegler, *Z. Naturforsch.* **9a**, 561 (1954).

Holmes, P. J., *Metallurgy of Elemental and Compound Semiconductors*, Ed. by R. O. Grubel, Wiley-Interscience, 1961.

Hoyem, A. G., and E. P. T. Tyndall, *Phys. Rev.* **33**, 81 (1929).

Hurle, D. T. J., *Phil. Mag.* **13**, 305 (1966).

Ikornikova, N., and A. A. Popova, *Dokl. Akad. Nauk SSSR* **106**, 460 (1956).

John, H. F., and J. W. Faust, in *Metallurgy of Elemental and Compound Semiconductors*, Ed. by R. O. Grubel, Wiley-Interscience, New York, 1961.

Johnson, L. F., and K. Nassau, *Proc. IRE* **49**, 1704 (1961).

Johnson, L. F., G. D. Boyd, K. Nassau, and R. R. Soden, *Phys. Rev.* **126**, 1406 (1962).

Jona, F., and P. Scherrer, *Helv. Phys. Acta* **25**, 35 (1952).

Kapitza, P., *Proc. Roy. Soc.* (*London*) **A119**, 358 (1928).

Keck, P. H., and M. J. E. Golay, *Phys. Rev.* **89**, 1297 (1953).

Keck, P. H., S. B. Levine, J. Broder, and R. Lieberman, *Rev. Sci. Instr.* **25**, 298 (1954).

Kyropoulos, S., *Z. Anorg. Chem.* **154**, 308 (1926).

Kyropoulos, S., *Z. Physik* **63**, 849 (1930).

LaRue, R. E. De, and F. A. Halden, *Rev. Sci, Instr.* **31**, 35 (1960).

Lawson, W. D., *J. Appl. Phys.* **22**, 1444 (1951); **23**, 495 (1952).

Lawson, W. D., and S. Nielsen, *Preparation of Single Crystals*, Butterworth, London, 1958, pp. 14ff.

Lefever, R. A., *Rev. Sci. Instr.* **33**, 769, 1470 (1962).

Lerner, P., C. Legras, and J. P. Dumas, *J. Cryst. Gr.* **3, 4**, 231 (1968).

Leverton, W. F., *J. Appl. Phys.* **29**, 1241 (1958).

Linares, R. C., *Solid State. Commun.* **2**, 229 (1964).

Lovell, L. C., and J. H. Wernick, *J. Appl. Phys.* **30**, 234 (1959).

McAfee, K. B., *J. Chem. Phys.* **28**, 218 (1958).

Merker, L., *FIAT Rept.* No. 1001 (1947).

Merker, L., *Trans. AIME* **202**, 645 (1955).

Metz, E. P. A., R. C. Miller, and J. Mazelsky, *J. Appl. Phys.* **33**, 2016 (1962).

Monforte, F. R., F. W. Swanekamp, and L. G. Van Uitert, *J. Appl. Phys.* **32**, 959 (1961).

Müller, A., and M. Wilhelm, *Z. Naturforsch.* **19a**, 254 (1964).

Mullin, J. B., R. J. Heritage, C. H. Holliday, and W. B. Stranghan, *J. Cryst. Gr.* **3, 4**, 281 (1968).

Nash, H. C., and C. S. Smith, *J. Phys. Chem. Solids* **9**, 113 (1959).

Nassau, K., *J. Phys. Chem. Solids* **24**, 1511 (1963).

Nassau, K., and A. M. Broyer, *J. Am. Ceram. Soc.* **45**, 474 (1962); *J. Appl. Phys.* **33**, 3064 (1962).

Nassau, K., and G. M. Loiacono, *J. Phys. Chem. Solids* **24**, 1503 (1963).

Nassau, K., H. J. Levenstein, and G. M. Loiacono, *J. Phys. Chem. Solids* **27**, 983 (1966).

Nassau, K., and L. G. Van Uitert, *J. Appl. Phys.* **31**, 1508 (1960).

Nix, F. C., *Rev. Sci. Instr.* **9**, 426 (1938).

Obreimov, J., and L. Schubnikov, *Z. Physik* **25**, 31 (1924).

Paladino, A. E., and B. D. Roiter, *J. Am. Ceram. Soc.* **47**, 465 (1964); **49**, 51 (1966).

Pearson, R. F., *Brit. J. Appl. Phys.* **4**, 342 (1953).

Pfann, W. G., *Trans. AIME* **194**, 747 (1952).

Pfann, W. G., *Zone Melting*, 2nd ed., Wiley, New York, 1966.

Pfann, W. G., and K. M. Olsen, *Phys. Rev.* **89**, 322 (1953).

Poplawsky, P. P., in *Thermal Imaging Techniques*, Ed. by P. E. Glaser and R. F. Walker, Plenum Press, New York, 1964, pp. 269ff.

Popov, S. K., *Bull. Acad. Sci. USSR Div. Chem. Sci. English Transl.* **10**, 504 (1946).

Popov, S. K., in *Growth of Crystals*, Vol. II, Ed. by A. V. Shubnikov and N. N. Sheftal, translated from Russian by Consultant's Bureau, New York, 1959, pp. 103ff.

Preziosi, S., R. R. Soden, and L. G. Van Uitert, *J. Appl. Phys.* **33**, 1893 (1962).

Reed, T. B., *J. Appl. Phys.* **32**, 821 and 2534 (1961).

Reed, T. B., in *Crystal Growth*, Ed. by H. Peiser, Pergamon, New York, 1967, pp. 39ff.

Reed, T. B., *J. Cryst. Growth*, **6**, 243 (1968).

Rhodes, R. G., *Imperfections and Active Centers in Semiconductors*, Macmillan, New York, 1964, pp. 125ff.

Richards, J. L., *J. Appl. Phys.* **31**, 600 (1960).

Rowland, P. R., *Trans. Faraday Soc.* **47**, 193 (1951).

Rubin, J. J., and L. G. Van Uitert, *J. Appl. Phys.* **37**, 2920 (1966).

Rubin, J. J., and L. G. Van Uitert, *Mater. Res. Bull.* **1**, 211 (1966a).

Schadler, H. W., *Trans. AIME* **218**, 649 (1960).

Schadler, W. H. in *The Art and Science of Growing Crystals*, J. J. Gilman, Ed., Wiley, New York, 1963, p. 344.

Schäfer, H., and R. Hörnle, *Z. Anorg. Allgem. Chem.* **263**, 261 (1950).

Seidensticker, R. G., and D. R. Hamilton, *J. Phys. Chem. Solids* **24**, 1585, 1963.

Siegel, S., *Phys. Rev.* **57**, 537 (1940).

Simpson, P. G., *Induction Heating*, McGraw-Hill, New York, 1960.

Slichter, W. P., and J. A. Burton, in *Transistor Technology*, Vol. I, Ed. by H. E. Bridgers, J. H. Scaff, and J. N. Shive, Van Nostrand, Princeton, N. J., 1958, pp. 107ff.

Smakula, A., *Einkristalle*, Springer, Berlin, 1962, pp. 214ff and 238ff.

Stansfield, D., *Phil. Mag.* **1**, 934 (1956).

Stepanov, V., and P. P. Feofilov, in *Rost Kristallov*, Ed. by A. V. Shubnikov and N. N. Sheftal, Translated by Consultant's Bureau, New York, 1958, pp. 181ff.

Stockbarger, Donald C., *Rev. Sci. Instr.* **7**, 133 (1936).

Stockbarger, Donald C., *Discussions Faraday Soc.* **5**, 294, 299 (1949).

Tammann, G., *Metallography*, translated by Dean and Swenson, Chemical Catalog Co., New York, 1925, p. 26.

Tanenbaum, M., in "Solid State Physics", Vol. 6, Part A, of *Methods of Experimental Physics*, Ed. by K. Lark-Horovitz and Vivian A. Johnson, Academic Press, New York, 1959, pp. 86ff.

Teal, G. K., and J. B. Little, *Phys. Rev.* **78**, 647 (1950).

Temkin, D. E., *Soviet Phys. Cryst. English Transl.* **7**, 354 (1962).

Theuerer, H. C., *Trans. AIME* **206**, 1316 (1956).

Thurmond, C. D., in *Conference on Properties of Elemental and Compound Semiconductors*, Ed. H. C. Gatos, Wiley-Interscience, New York, 1959, p. 121.

Trumbore, F. A., *Bell System Tech. J.* **39**, 205 (1960).

Utech, H. P., and M. C. Flemings, *J. Appl. Phys.* **37**, 2021 (1966).

Van Uitert, L. G., and S. Preziosi, *J. Appl. Phys.* **33**, 2908 (1962).

Van Uitert, L. G., J. J. Rubin, and W. A. Bonner, *J. Am. Ceram. Soc.* **46**, 512 (1963).

Van Uitert, L. G., J. J. Rubin, and W. A. Bonner, *IEEE J. Quantum Electronics* **QE4**, 622 (1968).

Verneuil, A., *Compt. Rend.* **135**, 791 (1902); **151**, 1063 (1911).

Verneuil, A., *Nature* **32**, 177 (1904); *Ann. Chim. Phys.* **3**, 20 (1904).

Wagner, R. S., "Factors Influencing Dendritic Growth in Germanium" in *Metallurgy of Elemental and Compound Semiconductors*, Ed. by R. O. Gruble, Wiley-Interscience, New York, 1961.

Wernick, J. H., and E. Buehler, unpublished work, 1959.

Wernick, J. H., and H. M. Davis, *J. Appl. Phys.* **27**, 149 (1956).

Wernick, J. H., K. Benson, and D. Dorsi, *J. Metals* **9**, 996 (1957).

Wilcox, W. R., and L. D. Fullmer, *J. Appl. Phys.* **36**, 2201 (1965).

Yamamoto, M., and R. Miyasawa, *Sci. Rept. Res. Inst. Tohoku Univ.* **A6**, 333 (1954).

Zerfoss, S. L. R., L. R. Johnson, and O. Imber, *Phys. Rev.* **75**, 320 (1949).

Zilitinkovich, S. J., *Tecegr. i Telef. bex. Provod.* **9**(6), 51 (1928).

6

CRYSTAL GROWTH BY VAPOR–SOLID EQUILIBRIA

6.1 Introduction

Growth by vapor–solid equilibria in a monocomponent system is of relatively limited application and is in a number of ways so similar to growth in a polycomponent system that both mono- and polycomponent growth (sometimes called growth by vapor-phase reaction) will be considered in this chapter. Vapor growth has been employed to produce bulk crystals and to produce thin films on crystals prepared by other techniques. The latter is usually called vapor-phase epitaxial growth (see Sec. 6.5.2). One of the advantages of vapor growth is that very often the equilibrium constants of proposed reactions and their temperature dependence can be estimated beforehand (see Sec. 2.11). The techniques of vapor–solid growth may be summarized as follows:

I. Monocomponent Techniques
  A. Sublimation–condensation techniques
  B. Sputtering techniques
  C. Ion-implantation techniques
II. Polycomponent Techniques
  A. Growth by reversible reactions
  B. Growth by irreversible reactions

Of the monocomponent techniques only sputtering techniques have found wide applicability to crystal growth and they are useful only for the epitaxial growth of films. Sublimation–condensation is limited because relatively few materials have a high enough vapor pressure for growth. It is generally useful to add an additional component to complex involatile materials into volatile species, in which case growth is by reaction. Ion implantation is not, strictly speaking, a growth technique. Rather, it is useful for doping crystals of previously grown materials.

Vapor growth as a technique has been reviewed by Brenner (1963), who discussed growth of metals; Bradley (1963), who discussed organic materials; Reynolds (1963), who discussed sulfides; Glang and Wajda (1963), who discussed silicon; O'Connor (1963), who discussed SiC†; Mason (1963), who discussed ice; and Antell (1962), who discussed III–V compounds. In addition, semiconductor vapor growth is extensively discussed in the book edited by Grubel (1961) and general vapor growth in the book edited by Powell et al. (1966). The monograph of H. Schäfer (1964) is essential for general background, the book edited by Francombe and Sato (1964) has much useful information on the preparation of single-crystal films, and the author has made considerable use of Ellis' review (1968) in writing this chapter.

Vapor growth can be either monocomponent or polycomponent. Monocomponent growth avoids impurity and diffusion problems associated with the additional component but generally requires higher temperatures. Polycomponent growth usually permits growth at lower temperatures and thus is indicated for the growth of low-temperature polymorphs, materials that volatilize incongruently, and where the temperatures required for monocomponent growth are experimentally not realizable.

6.2 Sublimation–Condensation

Two types of systems for sublimation–condensation growth are generally used. These are closed-tube systems and open-tube systems. In the closed-tube system a single tube or a vacuum evaporator is used. In a typical version of the closed-tube technique [Fig. 6.1(a)] the material to be transported is placed within a tube (usually vitreous silica) and the tube is evacuated (if the substance is unreactive with air, evacuation is sometimes not necessary). Sometimes the tube is backfilled to a desired pressure with a gas that represses decomposition or with an inert gas. If reaction between the gas and the transported material is important, the technique should probably be

†In addition, the book edited by O'Connor and Smiltens (1960) has several papers devoted to SiC vapor growth.

properly considered reactive transport. The inert gas may act to improve transport, because its convection may help to move the subliming species toward the growth region. However, because the mean free path of the subliming species is a function of pressure, better transport is usually achieved in high vacuum. Experimentally it is observed that sometimes inert carrier gases help growth kinetics and sometimes they hinder them. Carrier gases also often alter morphology.† Vapor pressures have been extensively tabulated in the literature. Table 6.1 lists some crystals grown by sublimation–condensation techniques, using both closed and open tubes. Conventional horizontal and vertical tube furnaces and the usual controllers and recorders are used to establish the desired temperature and gradient.

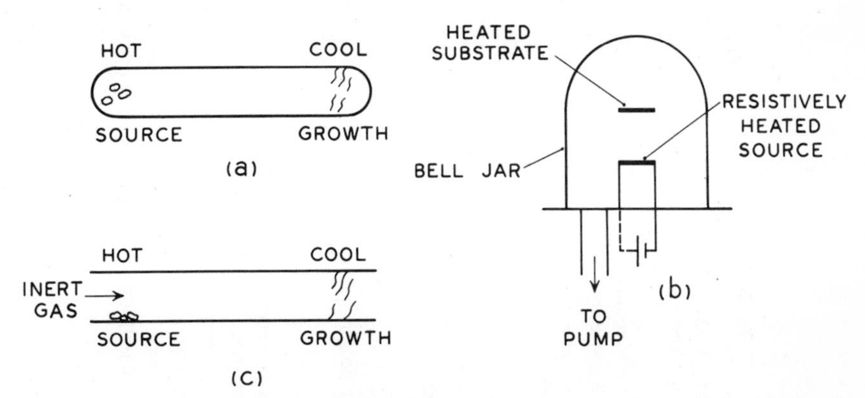

Fig. 6.1 Sublimation–condensation growth methods: (a) closed-tube technique; (b) vacuum evaporation; (c) open-tube technique.

Perhaps the most technologically interesting crystal grown by monocomponent transport is SiC. The Lely technique (Lely, 1955; Hamilton, 1960) for the preparation of SiC is essentially sublimation. Lumps of SiC are heated to temperatures of about 2500°C in an inert atmosphere. Masses of α-SiC plates form near the center of the charge. The use of a hollow cylindrical charge of SiC (Hamilton, 1960) lined with a porous graphite sleeve [see Fig. 6.1A] increases the size of the grown crystals. The graphite sleeve helps to maintain the shape of the charge and serves to restrict initial nuclea-

†It is probable that the vapor-phase analogue of constitutional supercooling is responsible for some of these effects. Constitutional supercooling can take place in vapor growth in both poly- and monocomponent systems. In monocomponent systems, because of concentration gradients caused by diffusion or convection, the partial pressure in some parts of the system may exceed the equilibrium partial pressure over the solid for the temperature under consideration. Such a situation can be considered the vapor analogue of constitutional supercooling. Reed and LaFleur (1964) have discussed constitutional supercooling in iodine vapor crystal growth.

Table 6.1

REPRESENTATIVE CRYSTALS AND FILMS GROWN BY SUBLIMATION–CONDENSATION AND SPUTTERING

Material	*Sublimation Temperature* (°C)	*Condensation Temperature* (°C)	*Gas*	*References*
Cd	320–330 (open or closed tube)	250–290	Vacuum, Ar	Brenner, 1963; Sears, 1955; Price, 1961
Zn	375–475 (open or closed tube)	350–~380	Ar, He	Sears, 1955; Coleman and Sears, 1957
CdS*	1150–1200 (open or closed tube)	1100	Ar	Piper and Polich, 1961; Reynolds, 1963
ZnS*	1550–1600 (open or closed tube)	1475–1500	Ar	Piper and Polich, 1961; Reynolds, 1963
SiC	> 2500 (open or closed tube)	< 2500	Inert atmosphere	Lely, 1955; Hamilton, 1960
Ta	Electron beam (vacuum evaporator) heated probably > 2000	700 Mainly polycrystalline film	10^{-6} Torr	Marcus, private communication, Bell Telephone Laboratories, Murray Hill, N. J.
Al	> 1000 (mp 660)	< 600	Vacuum	Holland, 1963
Be	1100–1200 (sometimes electron beam)	Mainly polycrystalline film below mp	Vacuum	Holland, 1963
Co, Ni, Fe	> 1500	Mainly polycrystalline film below mp	Vacuum	Holland, 1963

*Considerable dissociation in vapor phase; growth might be polycomponent by reversible reaction (See Sec. 6.4).

Table 6.1—*Cont.*

Material	*Volts*	*Current*	*Condensation Temperature* (°C)	*Gas*	*References*
Pb	1000 sputtering	1000 mA 2.5-cm cathode	Mainly polycrystalline film −195	~30 × 10^{-3} Ar	Theuerer and Hauser, 1964, 1965
Al	1500 sputtering	2–10 mA 2.5-cm cathode	−195 ± 300	27–73 × 10^{-3} Ar	Theuerer and Hauser, 1965
Cu	1500 sputtering	9.5 mA 2.5-cm cathode	−195	59 × 10^{-3} Ar	Theuerer and Hauser, 1965
V_3Ga	2250 sputtering	10.5 mA 2.5-cm cathode	1000	100 × 10^{-3} Ar	Theuerer and Hauser, 1964, 1965

Material	*Cathode*	*Reactive Gas*		*References*
TaN	Ta	N	(reactive sputtering)	Schwartz, 1963
TaC	Ta	CO, CH_4	(reactive sputtering)	Schwartz, 1963
Ta_2O_5	Ta	O_2	(reactive sputtering)	Schwartz, 1963
SnO_2	Sn	O_2	(reactive sputtering)	Sinclair et al., 1965
SiO_2	Si	O_2 or H_2O	(reactive sputtering)	Sinclair et al., 1965
NbO	Nb	O_2	(reactive sputtering)	Schwartz, 1963
ZrO_2	Zr	O_2	(reactive sputtering)	Schwartz, 1963
Al silicates (Mullite)	Al–Si alloy	O_2	(reactive sputtering)	Schwartz, 1963; Williams et al., 1963

tion. The crystals that do grow are nucleated by tendrils of SiC that propagate through the graphite tube. Hamilton has discussed the morphology and electrical properties of crystals grown by this method. The problems associated with polymorphism, stacking faults, and polytypism in SiC are discussed in Sec. 1.3.5.

Figure 6.1(b) shows a schematic of a vacuum evaporator. Vacuum evaporation has been especially useful in preparing epitaxial films of metals. The desired temperature in Fig. 6.1(b) is often obtained by means of resistance heating of a strip of the metal to be evaporated. Heating, of course, can be accomplished by many other techniques, e.g., resistance heating of a nearby heater (a filament), rf heating, focused radiation, electron beam, etc. Often provision is made for heating the substrate. Very high vacuums (10^{-9}–10^{-12} Torr) are now possible and clean conditions (less-than-monolayer absorption on surfaces) have been developed for surface-physics studies. These techniques should provide further impetus to vacuum evaporation. Presently, pressures of about 10^{-6} Torr are generally used. Table 6.1 lists some crystals and films prepared by vacuum evaporation.

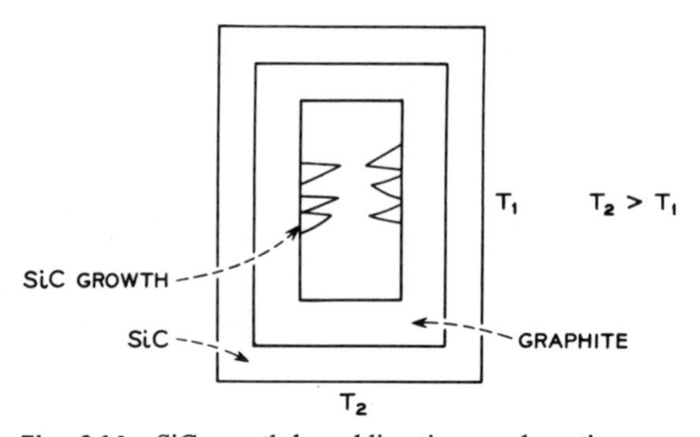

Fig. 6.1A SiC growth by sublimation condensation.

In open-tube techniques [Fig. 6.1(c)], perhaps more properly called flow systems, an inert carrier gas sweeping through the system carries the volatilizing species to the growth region. Table 6.1 lists some crystals grown in open tubes. Conventional tube furnaces, etc., are used. Growth on a seed has so far been hard to achieve in the tube techniques but is the principal object of vacuum-evaporator techniques. In vacuum-evaporator techniques the flux of material to the substrate is rather small, so the growth of bulk crystals is generally too time-consuming.

6.3 Sputtering

Sputtering has been widely used for the preparation of polycrystalline and amorphous films but, with suitable control, single-crystal films are possible.

Its principal advantage is that film growth can take place at lower temperatures than in ordinary sublimation–condensation growth because volatilization is brought about by an electric field instead of thermally. Three main techniques have been used: cathode sputtering (and a modification called getter sputtering), reactive sputtering, and ion implantation (which is a technique for doping grown crystals rather than for crystal growth).

Figure 6.2 is a schematic of a cathode sputtering apparatus (Berry, 1963). A dc potential of several thousand volts is maintained between the cathode and anode. The cathode is made of the material to be sputtered and the anode is the substrate. Suitable masks may be used to deposit the film in a desired orientation and the anode may include a sample holder that holds the work piece. Provision for the independent control of anode and cathode temperature is provided.

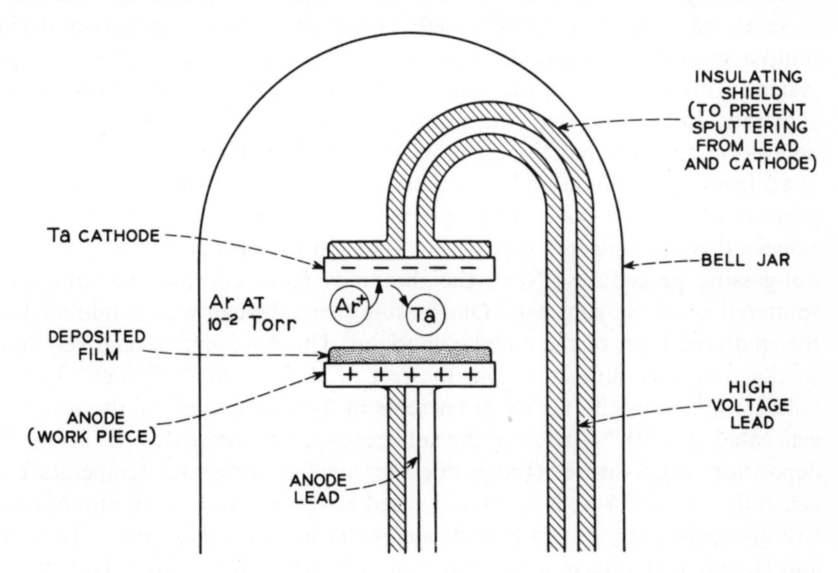

Fig. 6.2 Cathode sputtering apparatus.

The system is filled with an inert gas such as Ar at 10^{-1}–10^{-2} Torr. Ar^+ ions created in the gas impinge on the cathode and detach tantalum species, which are propelled to the substrate and form a film. Alloy films can be prepared by using alloy cathodes. High temperature at the substrate favors single-crystal growth. Films from a few angstrom to a few microns are possible. Impurities are a problem, so highly purified gases, good vacuum technique, and care to keep the sputtering chamber clean are necessary. The parameters of importance in establishing growth rate include cathode-current density, voltage, cathode nature, and geometry of cathode and anode.

In reactive sputtering (Schwartz, 1963; Sinclair et al., 1965; Williams et al., 1963), which, of course, is a polycomponent method, a reactive gas is added to a system of the same geometry as in cathode sputtering and the reactive gas combines with the species from the cathode, so that the deposit on the substrate is formed from components of the cathode material and the reactive gas. Details of the mechanism of the reaction and even where it takes place are obscure. Voltages of a few kilovolts are required and pressures in the 10^{-1}–10^{-3}-Torr range are used. The atmosphere consists of a mixture of inert and reactive gas. Multicomponent films may be made by simultaneous sputtering cathodes of different materials. Reactive sputtering provides the means of depositing many compounds, because the metallic element is usually sputterable and the nonmetallic-element ion may be provided as the reactive gas.

In getter sputtering, as described by Theuerer and Hauser (1964, 1965), reactive constituents are removed from the gas (gettered) by reactive sputtering before film deposition is initiated by cathode sputtering. The result is sputtered films of especially high purity. In a typical getter-sputtering setup a second anode (besides the normal substrate anode) is present in the system. This anode is in the shape of a can surrounding the cathode and substrate. In the first step, the substrate is covered by a shutter to prevent deposition, and any reactive gases in the can are gettered by the metal sputtered from the cathode and deposited on the can wall. By this process the pressure of reactive gases in the system can be reduced to $\sim 10^{-10}$ Torr. To achieve this pressure in a conventional system requires elaborate pump and out-gassing procedures. Next, the shutter is removed, and the cathode is sputtered onto the substrate. Out-gassing from the can wall is inhibited by the sputtered layer of the metal compound. The can is made relatively tight so that impurity diffusion from the rest of the system is difficult. Typical voltages are from 1.0–1.5 kV at currents of 2–10 mA. Initially the system is evacuated at $\sim 10^{-6}$ Torr, and the substrates are out-gassed $\sim 50°$ above the deposition temperature. (Independent control of substrate temperature is generally essential.) Ar is commonly used as the gas and 15–30 min of reactive sputtering are generally sufficient to clean the atmosphere. Theuerer and Hauser (1965) have used Ar pressures from $31–185 \times 10^{-3}$ Torr during the decomposition step. No special emphasis has so far been placed on single-crystal growth, and higher substrate temperatures and the use of single-crystal substrates should prove helpful. Getter sputtering provides a possible means of studying crystal-growth mechanisms under ultrapure conditions as well as a means for the preparation of ultrapure films.

In ion implantation an ion propelled by a dc field bombards the substrate, which is the cathode. The ion damages the substrate and thereby accelerates the diffusion rate in the damaged layer. If a gas is introduced (or already present) in the system, components from the gas can diffuse into the substrate. Thus doping levels ordinarily not obtained by thermally driven diffusion processes are obtainable. In some cases the ions themselves may be involved in producing the change in electrical properties desired in the substrate.

Table 6.1 lists some films prepared by sputtering techniques.

6.4 Growth by Reversible Reactions

6.4.1 THEORY

As discussed in Sec. 2.11, provided the equilibrium constant for a reaction is approximately 1 ($\Delta G^\circ \approx 0$) and its temperature dependence is appropriate (a calculation of ΔH° will allow an estimate of this) a reversible transport reaction is possible. Such growth is polycomponent and may be at temperatures considerably below those required in a monocomponent process.

The basic requirements of a successful transport reaction are

1. A chemical reaction where the desired phase is the only stable solid product in the temperature range of interest at the partial pressures of gas species chosen.

2. Free energy close to zero. This is for ease in reversibility and to ensure that significant quantities of reactants and products are present at equilibrium. If the concentration of reactants and products is too low, it will be difficult to cause a reasonable flow of material from source to crystallization region. This will be especially true in the closed-tube systems usually employed, where convection and diffusion are the driving forces for transport. In many cases the conventional problems associated with polycomponent growth, such as constitutional supercooling, facet effects, and dendrite formation, may appear. Such effects are commonly observed in polycomponent vapor growth but have been little discussed.

3. A nonzero ΔH°, so that the equilibrium can be shifted toward crystal formation in the growth zone and reversed in the volatilization zone by means of a temperature differential between zones. ΔH° will thus fix the temperature differential, ΔT. ΔT should not be too small or temperature control will be a problem. If ΔT is too large (rarely the case with the usual ΔH° values) such low temperatures may be required in the cool region that kinetics will be inhibited (not exceeding the softening temperature of SiO_2 in the hot region, i.e., ~1100°C, is experimentally very helpful). A large ΔT will, however, aid convective transport.

4. Nucleation control and rapid enough kinetics to grow a good quality crystal in reasonable time are, of course, essential. Hence predictions are not possible from equilibrium thermodynamic data alone. High temperature (to increase surface mobility) at the seed is usually helpful, and a large enough ΔT to enable reasonable convection is generally required.

Choice of a reaction is made on the basis of such compilations of thermodynamic data as Brewer's (1950), Kelley's (1935), Coughlin's (1954), and Glassner's (1957). Prior knowledge (intuition often will suffice) of the general chemistry and volatility of possible reactants, products, etc., is helpful. H. Schäfer's monograph (1964) summarizes much information on transport reactions of general inorganic interest that may be readily transposed to crystal-growth terms. If the ΔG° of the first postulated reaction is unsatisfactory, it may often be modified by subsequent addition of other reactions. It is often necessary to approximate thermodynamic data from analogous compounds. The following example from Sagal (1966) is perhaps informative.

Suppose that one wants to grow Y_2O_3. Consider the reaction:

$$Y_2O_{3\,(s)} + 3\,Cl_{2\,(g)} \rightleftharpoons 2\,YCl_{3\,(g)} + \tfrac{3}{2}\,O_{2\,(g)} \tag{6.1}$$

Making use of appropriate thermodynamic data (Brewer, 1950) for Eq. (6.1) we can calculate ΔG° at $1000^\circ K$ as

$$\Delta G^\circ_{1000^\circ K} = 59.4 \text{ kcal/mole} \tag{6.2}$$

which corresponds to the equilibrium constant:

$$\log K_{1000^\circ K} = -13.0 \tag{6.3}$$

The equilibrium position of the reaction of Eq. (6.1) lies much too far to the left for practical crystal growth. If we replace Cl_2 with Br_2 or I_2, the situation will be even worse because ΔG° for YBr_3 and YI_3 are larger than ΔG° for YCl_3 (the negative free energies are smaller) with the result that ΔG° for the reactions analogous to reaction (6.1) will be even larger when B_2 or I_2 are substituted for Cl_2. This should be intuitively obvious even in the absence of thermodynamic data if we are aware that YBr_3 and YI_3 are less stable than YCl_3. We must look for a reaction with a negative free energy that can add to reaction (6.1). Such a reaction is

$$CO + \tfrac{1}{2}O_2 \rightleftharpoons CO_2 \tag{6.4}$$

$$\Delta G^\circ_{1000^\circ K} = -46.7 \tag{6.5}$$

Thus a possible transport reaction is

$$Y_2O_{3\,(s)} + 3\,CO_{(g)} + 3\,Cl_{2\,(g)} \rightleftharpoons 2\,YCl_{3\,(g)} + 3\,CO_{2\,(g)} \tag{6.6}$$

$$\Delta G^\circ_{1000^\circ K} = 59.4 + 3(-46.7) = -80.7 \tag{6.7}$$

The equilibrium point now lies too far to the right. However, if we remember that the negative free energies of YBr_3 and YI_3 are smaller, we realize that the analogous Br_2 and I_2 reactions for Eq. (6.6) will have larger ΔG°'s. It turns out that for

$$Y_2O_{3\,(s)} + 3\,CO_{(g)} + 3\,Br_{2\,(g)} \rightleftharpoons 2\,YBr_{3\,(g)} + 3\,CO_{2\,(g)} \tag{6.8}$$

$$\Delta G^\circ_{1000^\circ K} = 27 \text{ kcal/mole}$$

Although not as close to zero as might be desired, a ΔG° of 27 leads to a partial pressure of YBr_3 of $\sim 10^{-2}$ atm, when the total pressure is ~ 1 atm. Transport has been successful in other systems with partial pressures in this range. A number of complications have been neglected in the analysis. For instance, data on the partial pressure of YBr_3 should be investigated because the formation of solid YBr_3 would be a complication. Similarly the possibility of YOBr would present difficulties. Further data at other temperatures or estimates of ΔH in order to estimate the ΔT required would also be necessary. However, the foregoing treatment illustrates the considerable predictive power one can bring to bear because of the availability of thermodynamic data. Such predictions are virtually impossible in any other polycomponent growth.

H. Schäfer (1964) has discussed many aspects of the kinetics of vapor growth in some detail. In the case of growth by a reversible reaction, the transport process can be divided into three separate steps:

1. The heterogeneous reaction on the starting solid.
2. The transport of volatile species in the gas.
3. The heterogeneous reverse reaction at the place where crystals are formed.

The processes associated with transport are of special importance in vapor-growth techniques, while the processes associated with step 3 are common in many other techniques and have been discussed previously (see Chapter 3), and the processes associated with volatilization are often trivial in affecting rate. Consequently, we shall discuss transport in some detail. It is interesting to point out that in all polycomponent growth where a dissolving step takes place in one part of the system and a growth step in another the dissolved species must be transported to another part of the system. As we shall see in Chapter 7, processes akin to those to be discussed here take place in aqueous solvents, molten salt, and hydrothermal solutions.

There are essentially three possible regimes for the gas-transport process:

1. At pressures $< 10^{-3}$ atm the mean free path of atoms in the gas phase is comparable with or greater than the dimensions of typical apparatus, and atomic or molecular collisions are negligible. The transport rate will be established by the atomic velocity, which by simple kinetic theory is

$$\mu = \sqrt{\frac{3RT}{M}} \tag{6.9}$$

where μ is the root mean square velocity, R is the gas constant, T is the absolute temperature, and M is the molecular weight. If transport is rate limiting, an idealized schematic of the situation is a simple tube, as shown in Fig. 6.3. Because the gas can be assumed to obey the perfect-gas law at low pressures, the rate of transport, $\mathscr{R}$, in atoms per second per unit cross-section area of the tube is given by

$$\mathscr{R} = \frac{p\mu}{RT} \tag{6.10}$$

where p is pressure. Combining Eqs. (6.9) and (6.10),

$$\mathscr{R} = p\sqrt{\frac{3}{RTM}} \tag{6.11}$$

Typical conditions where Eq. (6.11) might be applicable would include a collimated molecular beam used as the means of producing growth.

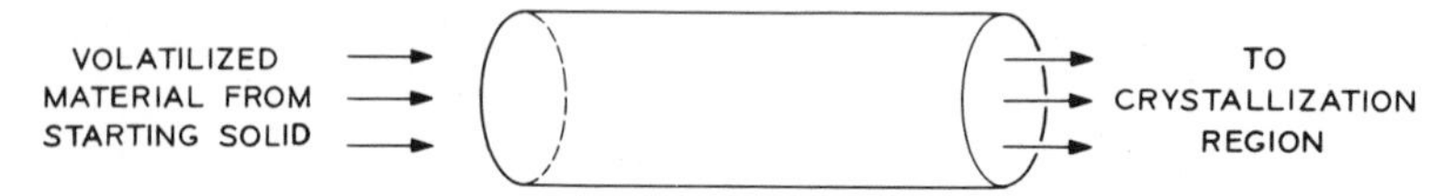

Fig. 6.3 Schematic of transport-limited growth.

2. If one works at pressures between 10^{-3} and 3 atm then gas motion is essentially determined by diffusion. Fick's laws (see Sec. 3.2) will describe the situation and for a constant concentration gradient the diffusion constant will decrease with increasing total pressure. Diffusion constants for common gases are well known, but the concentration gradient may be difficult to estimate.

3. At pressures $\gtrsim$ 3 atm thermal convection will be of overriding importance in determining gas motion. As H. Schäfer has pointed out, the transition range from diffusion to convection-controlled transport will often be determined by the details of apparatus geometry.

In most practical vapor-phase crystal-growth situations transport is by the diffusion mechanism and is rate limiting. Thus calculation of transport rate, assuming diffusion and comparing with actual transport rate, is often used as a test of whether a system is "well behaved." The following is based on Schäfer's discussion (1964).

An idealized closed tube apparatus that is useful in an analysis of diffusion-limited transport is shown in Fig. 6.4. Diffusion along the axis of the tube is assumed to be rate limiting in Fig. 6.4. Radial diffusion such as

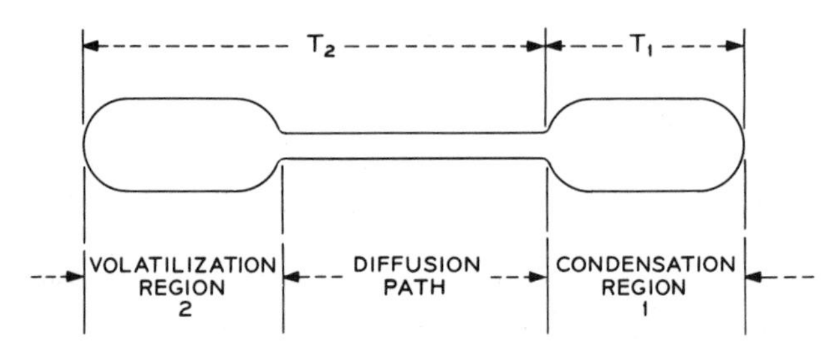

Fig. 6.4 Schematic of diffusion-limited growth.

occurs in growth on hot wires is considered by analogous methods and is discussed by Oxley and Blocker (1961) and Shapiro (1955). Consider a generalized reaction

$$iA_{(s)} + kB_{(g)} \rightleftharpoons jC_{(g)} \qquad (6.12)$$

where increasing temperature favors the reaction to the right, i.e., volatilization of A occurs at a higher temperature (T_2) than condensation (T_1). If increasing temperature favors reaction to the left, then the volatilization and condensation regions of Fig. 6.4 are reversed. If the number of moles of gas reactants does not equal the number of moles of gaseous products ($k \neq j$) in Eq. (6.12), then the requirement of equal pressure throughout the tube will give rise to a flow of the entire gas mass. This flow combined with diffusion is responsible for the transport in the tube. It is easy to show (H. Schäfer, 1964) that the number of moles of transported substance n_A is given by the equation

$$n_A = \frac{i}{j}\frac{Dqt}{sRT}\Delta P_c \qquad (6.13)$$

where D is the diffusion constant of the mixture, q is the cross-section area of the diffusion path, s is its length, t is the time of the experiment, and $\Delta P_c = P_{c_2} - P_{c_1}$ where P_{c_2} is the partial pressure of c in region 2 and P_{c_1} is the partial pressure in region 1. Equation (6.13) has been derived where flow caused by $k \neq j$ can be neglected. It can be shown that this approximation is justified under almost all conditions (H. Schäfer, 1964).

D_0, the diffusion coefficient, at total pressure 1 atm and $T = 273°$, for a number of gas pairs is known (Schäfer, 1964) and D can be estimated at other temperatures and pressures by means of the semiempirical equation

$$D = \frac{D_0}{\Sigma P}\left(\frac{T}{273}\right)^{1.8} \qquad (6.14)$$

Lacking other data, a good value for D_0 in most hydrogen-free systems is 0.1 cm^2/sec.

H. Schafer has discussed reactions where more than two volatile species are involved and has described the appropriately modified equations analogous to Eq. (6.13).

When thermal convection is important, the situation is considerably more complicated. A simple case where analysis is possible is shown in Fig. 6.5. All of the tube is at T_1 except the region labeled T_2.

Schäfer (1964) has shown that

$$n_A = \frac{i}{j}\Delta P_c P_{B\,init}\left\{\frac{4.7r^4 l_w M_B t}{RT_1\eta l}\left(\frac{1}{T_1} - \frac{1}{T_2}\right)\right\} \qquad (6.15)$$

where $P_{B\,init}$ is the initial partial pressure of B, r is the tube radius, l_w is the length of the zone at T_2, M_B is the molecular weight of B, η is the gas viscosity, t is time, and l is the total length of the loop. Clearly the analysis of a conventional horizontal transport tube will be much more complicated. Under such conditions the braced expression in Eq. (6.15) may be determined empirically and used as a "tube constant." Comparison of Eqs. (6.14) and (6.15) shows that transport rate depends on $1/P$ in the case of diffusion and on P in the case of convection. Thus our earlier statement that at low pressure transport was diffusion limited, while at high pressure it is convection limited is borne out. Indeed, a measure of transport rate as a function of pressure can be used to establish whether diffusion, convection, or a combination of both is the rate-determining mechanism.

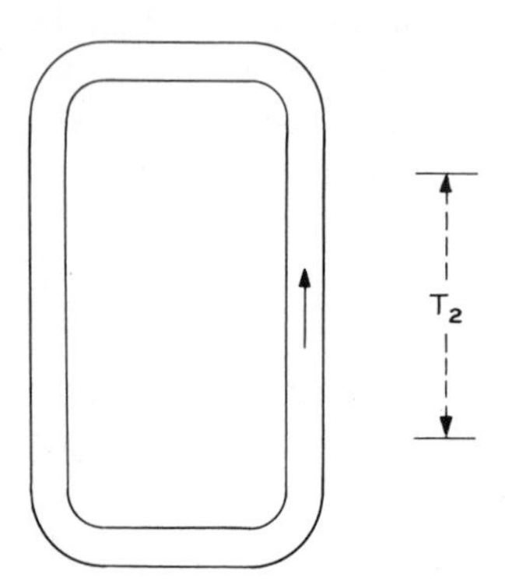

Fig. 6.5 Schematic of convection-limited growth.

Nucleation control in vapor-phase growth is usually a problem, especially when bulk crystals are desired. Epitaxial films may usually be grown without too much difficulty in nucleation control. This is because the thickness of the film required is not great, the growth rate may be slow, and the supersaturation need not be large enough to cause nucleation anywhere but on the substrate (seed). When bulk crystals are desired, growth rates must be higher (to produce reasonable volumes of crystal in a reasonable time), the required supersaturation is higher, and homogeneous and wall nucleation ("spontaneous nucleation") becomes a problem. Scholz and Kluckow (1967) have shown that such nucleation can be greatly reduced in vapor growth by temperature-cycling during growth. Consider vapor transport from a zone at temperature T_2 to a zone at temperature T_1 where $T_2 > T_1$. If no seed is present, $\Delta T = T_2 - T_1$ must be great enough initially to wall or homogeneously nucleate a particle on which further growth takes place. Ideally, for the largest final crystal, one such nucleus is best. At least the number of growing nuclei should be small. Often at ΔT's large enough to cause practically useful growth rates, additional nuclei are formed. If, however, ΔT is varied with time throughout the growth so that for some of the time $T_1' > T_2'$ small nuclei will be revolatilized with the result that fewer but larger crystals will result. If $T_2 - T_1 > T_1' - T_2'$ the time-integrated supersaturation will still be positive and growth will result. The time-integrated supersaturation may also be made positive even when $T_2 - T_1 \leqq T_1' - T_2'$ if the time interval over which $T_1' > T_2'$ is less than the time interval over which $T_2 > T_1$. Scholz and Kluckow (1967) discuss various apparatus for carrying out temperature cycling and describe temperature and time cycles that have been useful in increasing remarkably the size of vapor-grown αFe_2O_3 crystals. The technique should also be applicable to other polycomponent growth methods, especially to flux and hydrothermal growth where spontaneous nucleation is quite troublesome.

6.4.2 TYPICAL SYSTEMS

6.4.2-1 Hot-Wire Process

Van Arkel and de Boer (1925) may perhaps rightly be considered the fathers of closed-tube reversible transport processes. Their interest was in preparing refractory metals in a pure state, but the technique that bears their names is often capable of producing very good single crystals. The technique is sometimes called the hot-wire process and is illustrated in Fig. 6.6 by the preparation of Zr. The reaction

$$Zr_{(s)} + 2I_{2\,(g)} \underset{1200}{\overset{450}{\rightleftharpoons}} ZrI_{4\,(g)} \tag{6.16}$$

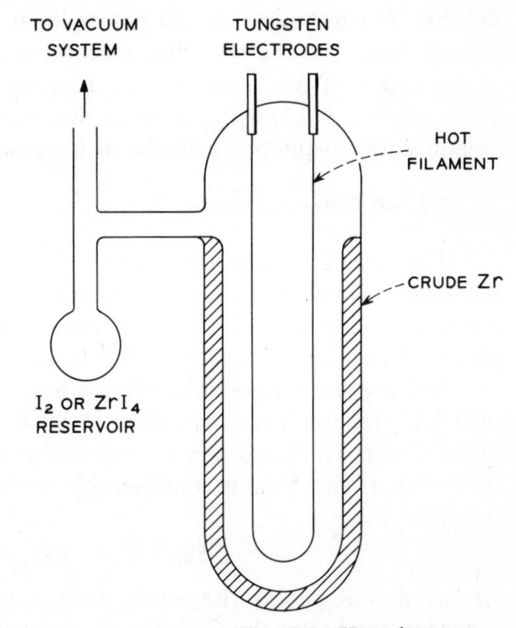

Fig. 6.6 Hot-wire process.

is favored to proceed to the right as temperature is lowered. The filament (often of W) is heated resistively to 1200°C. To prevent filament oxidation and the formation of ZrO_2, air must be excluded from the system. The partial pressure of I_2 and ZrI_4 is regulated by controlling the temperature of the reservoir. Radiant heating from the filament (plus auxiliary heaters if necessary) brings the crude Zr lining of the reaction vessel to 450°C, where Eq. (6.16) proceeds to the right. In the pressure range of usual operation ZrI_4 moves by convection and diffusion to the hot filament, where Eq. (6.16) proceeds to the left with the deposit of Zr. The deposit is often polycrystalline but is always purer than the crude Zr starting materials and usually some large crystals are formed. A similar process was used by van Arkel and de Boer for the formation of Ti. The literature (see, for instance, van Arkel and de Boer, 1925; van Arkel, 1934; Shapiro, 1955; Campbell et al. 1948; Schäfer et al., 1956; Chizhikov and Grin'ko, 1959) on the preparation of pure metals by hot-wire processes is too voluminous to summarize here, but it should be pointed out that W, Mo, Zr, Ni, CuFe, Hf, Nb, Ta, Th, V, Pt, Ir, Au, and U are among the metals prepared by this technique (Schäfer, 1964; Rolsten, 1961). Often the crystals are small, but improved control would no doubt improve their size. Using a single-crystal wire of the desired material (prepared, for instance, by strain annealing) would be advantageous, as would careful control of temperatures and gas pressures. Creating deliberate thermal gradients along the wire or programming the temperature of the wire to cause one nucleus to be the site of all subsequent growth would be sensible. The technique should not be limited to metals and might be considered useful for any reversible reaction where the desired solid is formed at the higher of two temperatures. Thermal cycling (Scholz and Kluckow, 1967; see Sec. 6.4.1) of the wire so that unwanted nuclei are destroyed by volatilization might prove useful in increasing crystalline size.

6.4.2-2 Other Processes

The more conventional technique for growth by reversible reaction is the use of a closed horizontal tube as shown in Fig. 6.1(a). The volatilization region is, of course, not hotter than the growth region, when transport is by means of a reaction where growth is favored at the higher temperature.

Perhaps the best examples of growth by this modification are the growth of II–VI compounds such as ZnS and CdS. CdS is often considered an example of a material that is prepared by sublimation. However, dissociation at the transport condition is considereable so that a more realistic reaction for the process is

$$CdS_{(s)} \rightleftharpoons Cd_{(g)} + S_{(g)} \tag{6.17}$$

Czyzak and Reynolds (Reynolds, 1963; Czyzak et al., 1952) suggest the following conditions for CdS growth: growth temperature—1000°C; atmosphere—H_2S (~1 atm); apparatus—closed quartz tube and horizontal tube furnace; and reaction time—48–72 hr. The initial powder charge serves as the nucleation site for growth, so that the driving force is caused by temperature fluctuations and the size differences between the finely divided powder starting material and the large grown crystals. It is surprising that this technique routinely yields 1-cm^3 CdS crystals (Czyzak et al., 1952). A similar technique, where the growth temperature was 1150°C, yielded ZnS crystals.

Greene et al. (1958) modified Czyzak and Reynolds' procedure by first subliming CdS (~950°C) or ZnS (~1160°C) in H_2S and then subliming in vacuo. This double sublimation improves the purity of the grown crystals. Growth takes place in ~1 atm Ar or H_2S in a temperature differential of ~50°. The hot (volatilization) zone is about 1575°C, and a vitreous-silica plate serves as a substrate for nucleation in the cooler region.

CdS crystals are invariably wurtzite-structure. ZnS is sphalerite if prepared below ~1020°C and wurtzite if prepared above. Stacking faults and polytypism are problems in ZnS. KCl is reported to stabilize the sphalerite structure (Samelson and Brophy, 1961).

Piper and Polich (1961) succeeded in growing very large crystals of CdS by a vapor-phase technique containing some elements of the geometry used in Bridgman–Stockbarger melt growth. This technique is illustrated in in Fig. 6.7. Figure 6.7 shows the conditions used by Kaldis and Widmer

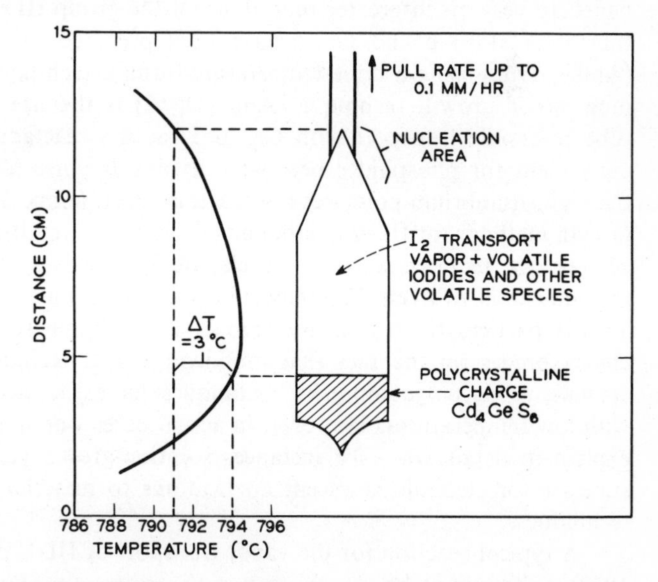

Fig. 6.7 Kaldis and Widmer's (1965) method for Cd_4GeS_6 growth.

(1965), who have used iodide transport for the growth of Cd_4GeS_6 by a reaction of the sort

$$Cd_4GeS_{6\,(s)} + 6\,I_{2\,(g)} \rightleftharpoons 4\,CdI_{2\,(g)} + GeI_{4\,(g)} + 6\,S_{(g)} \qquad (6.18)$$

In addition to CdS and Cd_4GeS_6, a variety of compounds, including ZnSe (Kaldis, 1965), have been grown in apparatus of similar geometry. The advantages of the geometry are that the temperature difference between the source material and the seed remains relatively constant throughout growth and that initial nucleation takes place in the conical tip at the top of the growth capsule. Thus if several nuclei form several crystallites initially, there is a good chance that one of them will dominate the growing interface by the time that interface has grown down to the point of maximum capsule diameter. If this is the case, because ΔT is nearly constant, the chance of the remainder of the growth taking place as a single crystal is quite good. By the use of this geometry Kaldis and Widmer (1965) have routinely grown crystals of ZnSe by the reaction

$$ZnSe_{(s)} + 2\,I_{2\,(g)} \rightleftharpoons ZnI_{4\,(g)} + S_{(g)} \qquad (6.19)$$

and crystals of Cd_4GeS_6 up to lengths of 12 mm and 24 mm, respectively. ZnSe crystals as large as 20 × 15 × 15 mm were grown in 36 days.

III–V compounds are often grown by closed-tube reversible reactions using halogens as the transport reagent. In the case of III–V compounds direct sublimation is generally not practical because the compounds decompose at their melting points and the vapor pressures of the group-V components are very much greater than those of the group-III elements. However, thin films of InSb and GaAs have been prepared by vacuum deposition (Antell, 1962) using a three-temperature-furnace technique. The more common vapor growth technique (Antell, 1962) is the use of closed reaction tube reversible transport with Cl_2 or I_2 as the reactant. Vapor growth is convenient for phosphides because it avoids the necessity of maintaining the high equilibrium pressures needed at the melt temperature. The choice of growth methods for III–V compounds depends on whether thin films or bulk crystals are required and on the electrical properties desired. The growth method can be of overriding importance in determining electronic properties. This is particularly true in the case of electroluminescent properties. The choice hinges on the fact that monocomponent techniques lead to purer crystals, while polycomponent techniques have the advantages associated with low temperatures. However, in actual cases one is not usually able to explain in detail why, for instance, solution-grown gallium phosphide is superior for electroluminescent applications to material prepared by other techniques.

A typical reaction for the vapor transport of III–V compounds (Antell, 1962) is thought to be

$$3\,InCl_{(g)} + 2\,As_{(g)} \underset{T_2}{\overset{T_1}{\rightleftharpoons}} InCl_{3\,(g)} + 2\,InAs_{(s)} \quad T_2 > T_1 \qquad (6.20)$$

Probably analogous reactions are involved in the case of most of the other III–V compounds, and iodine can replace chlorine in at least some cases. Typical growth conditions are

Material (*Transport Agent*)	*Growth Temperature* (°C)
InAs (Cl)	725
InAs (I)	650
GaP (I)	1050

Halide pressures are reported as "not critical."

In exploratory crystal growth by reversible reaction transport it is sensible to employ a geometry such as Fig. 6.1(a) in preliminary experiments and to use one such as Fig. 6.7 when preliminary experiments have indicated that transport of the desired phase has been achieved and larger crystals are desired.

For exploratory work if thermodynamic data is inadequate as a guide post, experience has indicated that high temperatures and iodide pressures are more likely to enhance transport and that a rather large ΔT is appropriate. Thus recommended exploratory conditions are T—1000°C (safely below the softening point of quartz) and concentration of I_2—$\sim$5 mg/cm³ tube volume. This is equivalent to an I_2 pressure well under 3 atm, which quartz can easily contain. An appropriate ΔT is $\sim$100°C or less. It is convenient to observe exploratory experiments in furnaces with an open slit permitting a view along the length of the capsule and a high-intensity lamp for illumination. Such an arrangement permits observation of transport as the experiment progresses and allows the temperature and ΔT to be determined during the experiment. Material may be back-transported to the source if one wants to begin anew at any time. An especially convenient furnace arrangement for direct observation can be made by winding a grooved vitreous-silica tube with resistance wire. In order to decrease convective cooling of such a furnace, the tube is often placed within a larger diameter silica tube. To minimize radiative losses the outer tube is sometimes partially covered with a thin (vacuum evaporated) Au film. Once proper conditions have been obtained, experiments are conducted in opaque furnaces without slits so that axial thermal uniformity is obtained. A representative list of materials, including a number of ternary calcogenides that Nitsche, Kaldis, and Widmer have prepared by iodide transport, is included in Table 6.2. Their success in forming good crystals of these complex materials attests to the usefulness of halide transport in particular and reversible vapor transport in general. Table 6.2 includes both materials prepared under closed-tube conditions by reversible reactions and those prepared under open-tube reac-

Table 6.2

CRYSTALS GROWN BY VAPOR REACTION (ELLIS, 1968)

Crystal	*Reaction*	*System*†	*Reference*
Iron	$Fe_{(s)} + 2\,HCl_{(g)} \underset{730°C}{\overset{1000°C}{\rightleftharpoons}} FeCl_{2\,(g)} + H_{2\,(g)}$	Closed; open	Schäfer, 1964; Brenner, 1956 a & b
Copper	$Cu_{(s)} + 2\,HI_{(g)} \underset{650°C}{\rightleftharpoons} CuI_{2\,(g)} + H_{2\,(g)}$	Closed; open	Schäfer, 1964
Cobalt	$Co_{(s)} + 2\,HBr_{(g)} \underset{650°C}{\overset{900°C}{\rightleftharpoons}} CoBr_{2\,(g)} + H_{2\,(g)}$	Closed; open	Schäfer, 1964; Brenner, 1956 a & b
Nickel	$Ni_{(s)} + 2\,HBr_{(g)} \underset{740°C}{\overset{1000°C}{\rightleftharpoons}} NiBr_{2\,(g)} + H_{2\,(g)}$	Closed; open	Schäfer, 1964; Brenner, 1956 a & b
Silicon	$Si_{(s)} + 4\,HCl_{(g)} \underset{1200°C}{\rightleftharpoons} SiCl_{4\,(g)} + 2\,H_{2\,(g)}$	Open	Theuerer, 1961; Mark, 1961
Gallium arsenide	$2\,GaAs_{(s)} + 6\,HCl_{(g)} \rightleftharpoons 2\,GaCl_{3\,(g)} + As_{2\,(g)} + 3\,H_{2\,(g)}$	Closed	Moest and Shupp, 1962
Sapphire, Al_2O_3	$Al_{(l)} + H_2O_{(g)} \overset{1400°C}{\longrightarrow} AlO_{(g)}‡ + H_{2\,(g)}$ $Al_2O_{3\,(s)} + Al_{(l)} \rightleftharpoons 3\,AlO_{(g)}$	Open (see Sec. 6.5.2 for additional reactions)	Webb and Forgeng, 1957; Schaffer, 1965, 1966
	$Al_2O_{3\,(s)} + 2H_{2\,(g)} \underset{<1900°C}{\overset{1900°C}{\rightleftharpoons}} Al_2O_{(g)}‡ + 2\,H_2O_{(g)}$	Closed	DeVries and Sears, 1959
	$Al_2O_{3\,(s)} + 6\,HCl_{(g)} \overset{1150°C}{\rightleftharpoons} 2\,AlCl_{3\,(g)} + 3\,H_2O_{(g)}$	Open	Kerrigan, 1963
Beryllia, BeO	$BeO_{(s)} + H_{2\,(g)} \overset{1500°C}{\longleftarrow} Be_{(g)} + H_2O_{(g)}$	Closed	Edwards and Happel, 1962

†Closed-tube reactions are reversible; open-tube reactions often are not (see Secs. 6.4 and 6.5).
‡The composition of the volatile oxide of aluminum is not known with certainty.

Table 6.2—*Cont.*

Crystal	*Reaction*	*System*	*Reference*
Silica, SiO_2	$SiO_{2(s)} + 2\,CO_{(g)} + 4\,HCl_{(g)} \overset{1200°C}{\longleftarrow} SiCl_{4(g)} + 2\,H_{2(g)} + 2\,CO_{2(g)}$	Closed	Tung and Caffrey, 1965; Steinmaier and Bloem, 1964
Gallium phosphide, GaP	$2\,GaP_{(s)} + H_2O_{(g)} \underset{1050°C}{\overset{1100°C}{\rightleftharpoons}} Ga_2O_{(g)} + P_{2(g)} + H_{2(g)}$	Open	Frosch, 1964; Thurmond and Frosch, 1964
Gallium arsenide, GaAs	$GaAs_{(s)} + \frac{3}{2}\,I_{2(g)} \underset{900°C}{\overset{1100°C}{\rightleftharpoons}} GaI_{3(g)} + As_{(g)}$	Closed	Holonyak, Jillson, and Bevacqua, 1962; Holonyak, 1964
	$2\,GaAs_{(s)} + 3\,ZnCl_{2(g)} \underset{900°C}{\overset{1100°C}{\rightleftharpoons}} 2\,GaCl_{3(g)} + 2\,As_{(g)} + 3\,Zn_{(g)}$ Exact reactions not known, may involve disproportionation, as in Eq. (6.20). Material may be doped by means of cation in transport reagent. $HgCl_2$, $CaCl_2$, $SnCl_2$, $AlCl_3$, $MgCl_2$ also used as transport reagents.		
Gallium phosphide, GaP	Similar to GaAs reactions.		Holonyak, Jillson, and Bevacqua, 1962; Holonyak, 1964
Gallium phosphide–arsenide, GaP_xAs_{1-x}	Procedures similar to GaAs. Starting material is GaP_xAs_{1-x}		Holonyak, Jillson, and Bevacqua, 1962; Holonyak, 1964

tions by (usually) irreversible reactions (see Sec. 6.5). A number of other materials prepared by iodide transport have been discussed by Nitsche (1967).

Carbon monoxide is a convenient reactant for the transport of many of those metals that form volatile carbonyls. For instance, nickel crystals can be grown by the reaction

$$Ni_{(s)} + 4\,CO_{(g)} \underset{200^\circ C}{\overset{80^\circ C}{\rightleftharpoons}} Ni(CO)_{4\,(g)} \tag{6.21}$$

The decomposition of $Ni(CO)_4$ in an open tube is the Mond–Langer process (Thorpe and Whiteley, 1961), and is used for the preparation of ultrapure Ni. In the closed-tube process (generally favored for crystal growth involving carbonyls because of their toxicity) finely powdered Ni is placed in one end of a tube (Schäfer, 1964) that contains CO in a gradient furnace. The source temperature is about 80°C, and the crystals form at the tube walls or on a suitable substrate at about 200°C. Carbonyl iron is made by a similar process.

Reversible disproportionation reactions are often useful in effecting transport. Perhaps their most important use is in preparing thin single-crystal layers of the semiconductors Si and Ge on semiconductor substrates.† This technique is generally called epitaxial growth, and it takes advantage of several of the inherent features of vapor growth. In semiconductor technology it is often important to grow a crystal with a sharp (or definitely shaped) concentration profile for one (or more) of its impurities. For instance, in a *p–n* junction the object is to have as sharp a change from *p*- to *n*-type material as possible along a flat interface. Thus, for instance, because, in Si, B-doped material is *p*-type and As-doped material is *n*-type, the idealized concentrations desired at the junction would be those of Fig. 6.8. It is generally not practical to produce the sharp idealized gradients shown but for most real devices profiles of the sort labeled *actual concentration* are acceptable. However, for many applications sharper profiles are advantageous. Indeed, for some devices some exact shape of the profile (other than those of Fig. 6.8) is desired. Vapor growth presents several advantages in profile control:

1. The temperature can be low so that solid-state diffusion after growth will not alter the "as grown" profile.

2. Concentration of dopants in the gas phase can be altered abruptly (if the dopants or their compounds are volatile) by simple expedients such as opening a valve or changing the temperature of a reservoir containing dopant.

3. Growth can be made quite slow and can often be observed so that very thin layers are possible. Indeed, the fact that growth is slow usually dictates that the substrate (seed on which doped growth takes place) be prepared by some other technique.

---

†There is a possibility that halide transport of III–V compounds of the sort described in Eq. (6.20) also involves disproportionation.

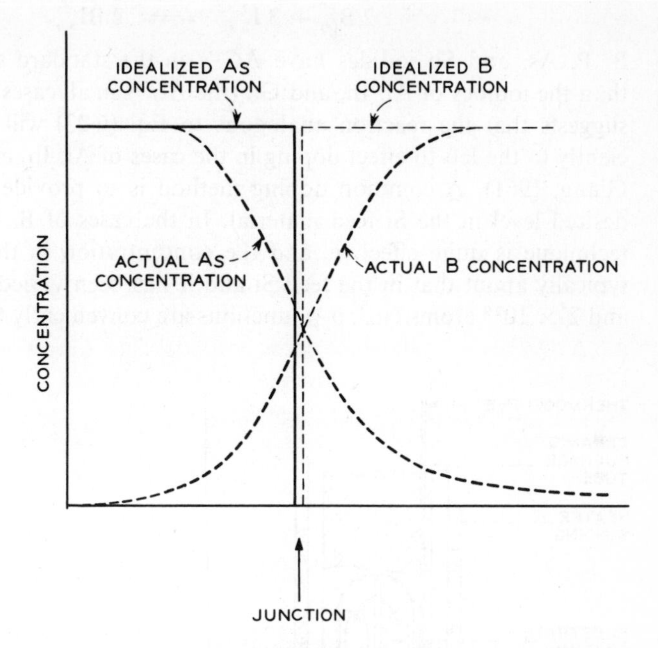

Fig. 6.8 Impurity profiles at a *p–n* junction.

4. Suitably designed masks can enable one to grow differently doped material in different regions of the substrate, and the application of masks and their removal in various sequences could allow the preparation of elaborate three-dimensional arrays of single-crystal material with complicated concentration profiles. Such techniques, where masking is used during diffusion processes, are central to the preparation of monolithic circuits.

Marinace (1960) described a system where iodine reacts in a closed quartz tube, transports germanium as an iodine vapor, and deposits it epitaxially via disproportionation reactions. Similar methods have been applied to a variety of semiconductors, including Ge, GaAs, GaP, and alloys of GaAs and GaP [Eq. (6.20) is a typical example].

For the epitaxial growth of silicon the reaction is

$$SiI_{4\,(g)} + Si_{(s)} \underset{900^\circ C}{\overset{1100^\circ C}{\rightleftharpoons}} 2\,SiI_{2\,(g)} \tag{6.22}$$

The SiI_4 may be made directly in the system by the reaction of Si with I_2. Figure 6.9 (Wajda and Glang, 1961) shows a typical experimental setup. Useful I_2 concentrations (Glang and Wajda, 1963) are from 12 to 48 mg/cm^3.† Impurities are transported by reactions such as

†Care should be exercised at such high fills of I_2 because their pressures will be close to the rupture point of the quartz ampoules.

$$2\,B_{(s)} + 3\,I_{2\,(g)} \rightleftharpoons 2\,BI_{3\,(g)} \tag{6.23}$$

B, P, As, and Sb iodides have ΔG°'s in the standard state closer to zero than the iodides of Al, In, and Ga (the ΔG°'s in all cases are negative). This suggests that the reaction analogous to Eq. (6.23) will not proceed sufficiently to the left to effect doping in the cases of Al, In, and Ga (Wajda and Glang, 1961). A common doping method is to provide the dopant at the desired level in the Si feed material. In the cases of B, P, As, and Sb this technique is quite effective, and the concentration in the grown crystal is typically about that in the feed Si and it has been varied between 5×10^{16} and 2×10^{20} atoms/cm³. *p–n* junctions are conveniently formed from doped

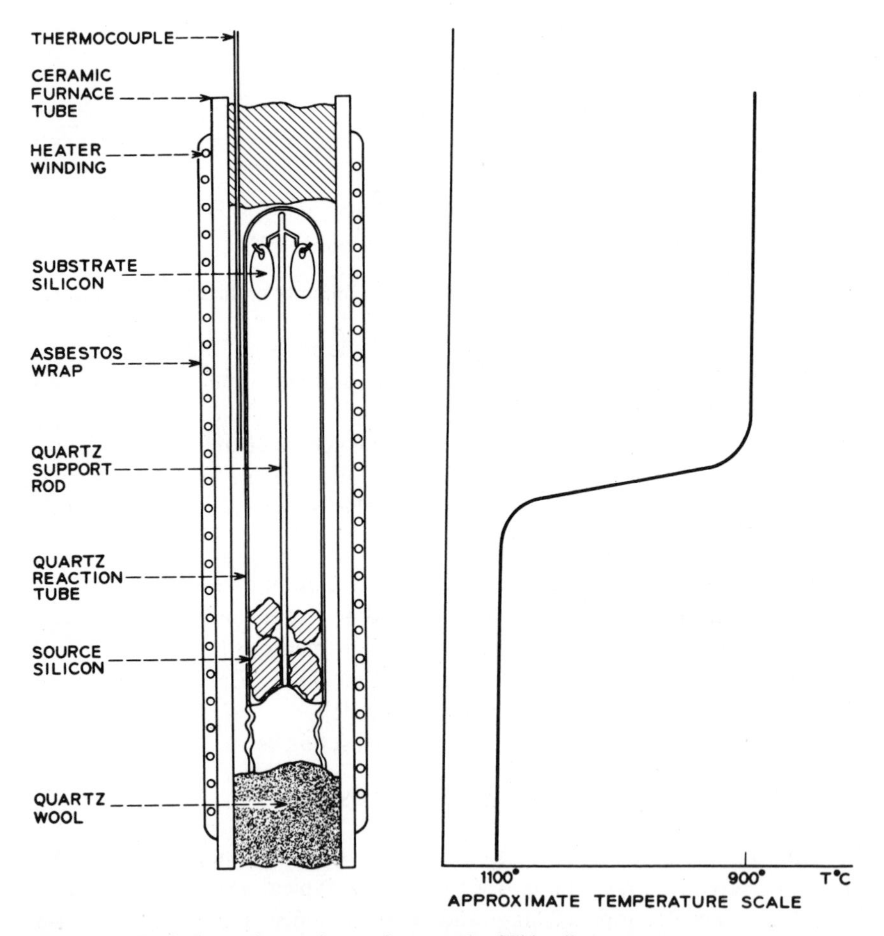

Fig. 6.9 Typical experimental setup for growth of Si by disproportionation of SiI_2 (after Wajda and Glang, 1961). (Courtesy of John Wiley & Sons, Inc.)

feeds by a number of methods. The following is typical: n-type feed material, p-type material, and substrate (seed) are placed in a tube in an appropriate furnace. The temperature profile is arranged so that deposition is from the p-type material to the substrate (the p-type material is at T_2, the substrate at T_1, and the n-type material at T_0; $T_2 > T_1 \gg T_0$ and transport is from T_2 to T_1). Then the tube is moved so that the n-type material and the substrate are at T_1 and the p-type material is at T_2. Transport will be from the n-type material (a small thickness of p-type material deposited on the n-type material will first be volatilized) to the substrate.

Germanium epitaxial layers are prepared by a similar technique to that used for Si where the reaction is (Maunail et al., 1961; Marinance, 1960)

$$Ge_{(s)} + GeI_{4\,(g)} \underset{400^\circ C}{\overset{500^\circ C}{\rightleftharpoons}} 2\,GeI_{2\,(g)} \tag{6.24}$$

in both open- and closed-tube systems, and the rate has been shown to be dependent on the I_2 density. Growth rates from 5–10 μ/day occur at I_2 densities of 0.03 mg/cm^3 and growth (although of poor quality) rates as high as 24,000 μ/day occur at 30 mg/cm^3. Acceptable quality at rates of $\sim$100 μ/day can be obtained at I_2 densities of $\sim$5 mg/cm^3. As is probably the case with most vapor-grown materials, among the undesirable impurities incorporated into the crystal is the transporting reagent. In the case of Ge about 10^{14}–10^{15} atoms/cm^3 of I_2 are found in the crystal. Growth rates (as well as distribution coefficients for I_2 and other impurities), as might be expected, are directionally dependent with $\langle 211 \rangle \sim \langle 110 \rangle > \langle 111 \rangle \sim \langle 100 \rangle$. Doping has been achieved by the use of doped Ge feed material, the addition of dopants in elemental form to the system, and the addition of dopant iodides to the system.

The literature concerning the preparation of Ge, Si, and other semiconductors by epitaxial processes is too extensive for summarization here, especially because excellent summaries are available elsewhere (Grubel, 1961; Willardson and Goering, 1962).

6.5 Growth by Irreversible Reactions

6.5.1 GENERAL

Perhaps a better classification for this section would be *growth by open-tube processes* because very often the thermodynamics are not known well enough to assure that a given process, although practiced in one direction only, might not be reversible. Thus those processes where the reactants are not recycled, either for thermodynamic or practical reasons, fall naturally together and will be discussed here.

6.5.2 TYPICAL SYSTEMS

Conditions for the growth of a number of materials by the open-tube process are listed in Table 6.2. The simplest sort of open-tube process is illustrated by the growth of Pd by the decomposition reaction (Riebling and Webb, 1957; Webb, 1965)

$$PdCl_{2\,(g)} \longrightarrow Pd_{(s)} + Cl_{2\,(g)} \tag{6.25}$$

Solid $PdCl_2$ is placed in a vitreous-silica tube in a gradient furnace and heated to a temperature where its vapor pressure is appreciable (~900°C, which is above the mp). A slow stream of Ar is passed over the $PdCl_2$ and the reaction to form Pd takes place in a higher-temperature region of the furnace. The resultant crystals include hexagonal and triangular platelets and long filamentary single crystals or whiskers, some of which are straight, while others are helical.

Vapor growth is one of the favorite techniques for the formation of whiskers. A whisker is a long filamentary crystal often containing a single-screw dislocation (see Sec. 7.5 for a discussion of the V–L–S method, which produces crystals without a screw dislocation).† The core of the dislocation usually follows the whisker axis. Whiskers are thus single crystals with a very low dislocation density, and hence they are of special interest for studies of the influence of dislocations on properties. For instance, whiskers have unusually high tensile properties because of their low dislocation density (Brenner, 1956 a & b, and references therein).

Open-tube processes are important in epitaxial growth. Silane, SiH_4, is pyrolyzed according to the reaction (Kagdis, 1962; Joyce and Bradley, 1963)

$$SiH_{4\,(g)} \longrightarrow Si_{(s)} + 2\,H_{2\,(g)} \tag{6.26}$$

on a heated Si substrate at temperatures from 950 to 1250°C. Unless the silane pressure is below about 1 Torr, homogeneous decomposition in the gas phase takes place, but at low pressures the heterogeneous reaction on the substrate can be quite easily controlled. Use of reactive transport species such as SiH_4 should, of course, be made with caution. Their reactivity (in the case of SiH_4, especially with oxidants) poses explosion hazards.

Probably the process enjoying the widest commercial use in Si epitaxy in open-tube growth is the hydrogen reduction of halides. The generalized reaction is (Sangster et al., 1957; Allegretti et al., 1961; Litton and Anderson, 1954; McCarty, 1959; Herrick and Krieble, 1960)

$$SiH_{4-x}Y_{x\,(g)} + (x-2)H_{2\,(g)} \longrightarrow Si_{(s)} + xHY_{(g)} \tag{6.27}$$

where Y is a halogen and x has values from 1 to 4. Crystals over 1 cm^3 have

†Indeed, recent work (see Sec. 7.5) indicates that a good case can be made that most whiskers grow by V–L–S without a dislocation.

been produced by hydrogen reduction. Doping is accomplished by introducing, for example, $BCl_{3\,(g)}$ and $PCl_{5\,(g)}$ into the reaction region.

Theuerer (1961) has described the method for producing epitaxial layers of Si by the reaction

$$SiCl_{4\,(g)} + 2\,H_{2\,(g)} \longrightarrow Si_{(s)} + 4\,HCl_{(g)} \tag{6.28}$$

The furnace arrangement is shown in Fig. 6.10.

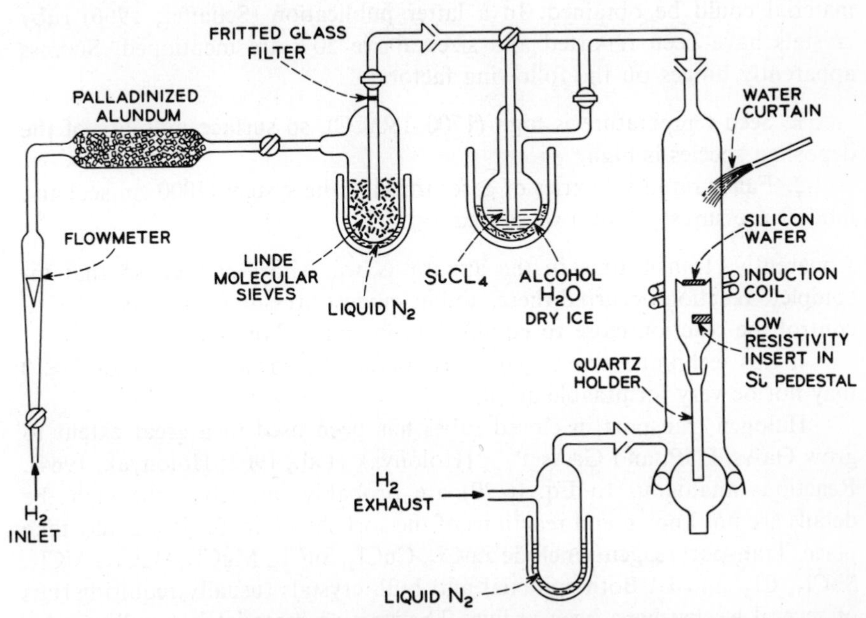

Fig. 6.10 Typical experimental arrangement for growth of Si by halide reduction (after Theuerer, 1961). (Courtesy of the Electrochemical Society.)

The substrate is a single crystal of Si and its temperature is about 1270°C. At lower temperatures low surface mobility of species on the substrate leads to polycrystalline growth. Under typical conditions of a $SiCl_4$–H_2 mole fraction of 0.02 and a hydrogen flow of 1 liter/min an epitaxial film of 8 μ was obtained in a 6.3-in^2 substrate in five minutes. Doping is effected by introducing appropriately complexed dopants with satisfactory volatility.

The mechanism and kinetics of several vapor-phase reactions have been studied (Theuerer, 1961; Monchamp et al., 1964; Miller and Grieco, 1963; Steinmaier, 1963), and, while the simple reactions we have written are valid *overall reactions*, the step-wise reactions actually taking place in many instances are considerably more complicated.

The growth of refractory materials by open-tube techniques has received considerable impetus because of the work of Schaffer (1965, 1966), who has grown sapphire in an open tube. The reaction employed was

$$2\,AlCl_{3\,(g)} + 3\,H_{2\,(g)} + 3\,CO_{2\,(g)} \longrightarrow Al_2O_{3\,(s)} + 3\,CO_{(g)} + 6\,HCl_{(g)} \tag{6.29}$$

and the reaction was run under conditions where K, the equilibrium constant, was very large.† It is interesting that, in spite of this, nucleation control sufficient to produce controlled growth on a seed of low-dislocation-density material could be obtained. In a latter publication (Schaffer, 1966) ruby crystals have been reported and sizes above 50 g are mentioned. Success apparently hinges on the following factors:

1. Seed temperature is high (1700–1800°C), so surface mobility of the deposited species is high.
2. Fairly rapid velocities of gases through the system (1000 cm/sec) and modest pressures (9 Torr) were used.

Apparently, transit time in the apparatus was short enough so that incomplete reaction occurred there, and at modest pressures it was possible to control the reaction close to equilibrium in spite of the large K.

If this technique proves generally useful, the constraint that $\Delta G^\circ \approx 0$ may not be very formidable at all.

Halogen transport in closed tubes has been used to a great extent to grow GaAs, GaP, and $GaAs_xP_{1-x}$ (Holonyak et al., 1962; Holonyak, 1964). Reactions analogous to Eq. (6.20) are probably operative, although the details are not known and reactions of the sort shown in Table 6.2 may take place. Transport reagents include $ZnCl_2$, $CuCl_2$, $SnCl_2$, $MgCl_2$, $HgCl_2$, $AlCl_3$, $SbCl_3$, Cl_2, and I_2. Both epitaxial and bulk crystals (usually requiring runs of several weeks) have been grown. The starting material is usually GaAs, GaP, or $GaAs_xP_{1-x}$, transport is from the hot to the cold portion of the tube, and the crystals may be made *n*- or p-type depending on the cation in the transport reagent. Laser-quality crystals have been produced.

Silicon carbide has been grown by an open-tube process employing the reaction of toluene and $SiCl_4$ on a hot wire (O'Connor and Smiltens, 1960) and techniques of this sort provide real hope of producing large controlled crystals of this recalcitrant material. Kendall (1960) used an open-tube hot-wire process for the preparation of SiC by gaseous cracking. The filament was graphite, and its temperature was ~2000°C. $SiCl_4$–toluene with about a 20:1 excess of $SiCl_4$ seemed to provide the best crystals. On a 10-mil filament, crystals up to 25 mils long of the low-temperature modification of SiC could be readily obtained.

---

†The possibility exists that K for the relevant *rate-determining* reaction is not large. That is, the *rate-determining* reaction is not the reaction assumed.

REFERENCES

Allegretti, J. E., D. J. Schombert, E. Schaarschmidt, and J. Waldman, in *Metallurgy of Elemental and Compound Semiconductors*, Vol. 12, Ed. by R. O. Grubel, Wiley-Interscience, New York, 1961, pp. 255ff.

Antell, G. R., in *Compound Semiconductors*, Vol. I, Ed. by R. K. Willardson and H. L. Goering, Reinhold, New York, 1962, pp. 288ff.

Berry, R. W., *Bell Lab. Record* **41**, 46 (1963).

Bradley, R. S., in *The Art and Science of Growing Crystals*, Ed. by J. J. Gilman, Wiley, New York, 1963, pp. 55ff.

Brenner, S. S., *J. Appl. Phys.* **27**, 1484 (1956); *Acta Met.* **4**, 62 (1956).

Brenner, S. S., *The Art and Science of Growing Crystals*, Ed. by J. J. Gilman, Wiley, New York, 1963, pp. 30ff.

Brewer, L., Papers 6 and 7 in *The Chemistry and Metallurgy of Miscellaneous Materials*, Ed. by L. I. Quill, National Nuclear Energy Series, Division IV, Vol. 19B, McGraw-Hill, New York, 1950.

Campbell, I. E., R. I. Jaffee, J. M. Blocker, J. Gurland, and W. B. Gonser, *Trans. J. Electrochem. Soc.* **93**, 271 (1948).

Chizhikov, D. M., and A. M. Grin'ko, *Russ. J. Inorg. Chem.* (*English Transl.*) **4**, 446 (1959).

Coleman, R. V., and G. W. Sears, *Acta Met.* **5**, 131 (1957).

Coughlin, J. P., *U.S. Bur. Mines Bull.* 542 (1954).

Czyzak, S. J., D. J. Craig, C. E. McCain, and D. C. Reynolds, *J. Appl. Phys.* **23**, 932 (1952).

DeVries, R. C., and G. W. Sears, *J. Chem. Phys.* **31**, 1256 (1959).

Edwards, P. L., and R. J. Happel, Jr., *J. Appl. Phys.* **33**, 943 (1962).

Ellis, W. C., in *Techniques of Metals Research*, Ed. by R. F. Bunshah, Vol. I, Part 2, Interscience, New York, 1968.

Francombe, M. H., and H. Sato, Eds., *Single Crystal Films*, Macmillan, New York, 1964.

Frosch, C. J., *J. Electrochem. Soc.* **111**, 180 (1964).

Glang, R., and E. S. Wajda, in *The Art and Science of Growing Crystals*, Ed. by J. J. Gilman, Wiley, New York, 1963, pp. 80ff.

Glassner, Alvin, *The Thermochemical Properties of Oxides, Chlorides and Fluorides to 2500°K*, Argonne National Laboratory Rept. 5750, 1957.

Greene, L. C., D. C. Reynolds, S. J. Czyzak, and W. M. Baker, *J. Chem. Phys.* **29**, 1375 (1958).

Grubel, O., Ed., *Metallurgy of Elemental and Compound Semiconductors*, Vol. 12, Wiley-Interscience, New York, 1961.

Hamilton, D. R., in *Silicon Carbide—A High Temperature Semiconductor*, Ed. by J. R. O'Connor and J. Smiltens, Pergamon, New York, 1960, pp. 67ff.

Herrick, C. S., and J. G. Krieble, *J. Electrochem. Soc.* **107**, 111 (1960).

Holland, L., *Vacuum Deposition of Thin Films*, Chapman and Hall, London, 1963, pp. 671–735.

Holonyak, N., Jr., *Trans. AIME* **230**, 276 (1964).

Holonyak, N., Jr., D. C. Jillson, and S. F. Bevacqua, in *Metallurgy of Elemental and Compound Semiconductors*, Ed. by J. B. Schroder, Vol. 15, Wiley-Interscience, New York, 1962, pp. 49ff.

Joyce, B. A., and R. A. Bradley, *J. Electrochem. Soc.* **110**, 1235 (1963).

Kagdis, W. A., *J. Electrochem. Soc.* **109**, 71C (1962).

Kaldis, E., *J. Phys. Chem. Solids* **26**, 1701 (1965).

Kaldis, E., and R. Widmer, *J. Phys. Chem. Solids* **26**, 1697 (1965).

Kelley, K. K., *U.S. Bur. Mines Bull.* 383 (1935).

Kendall, J. T., in *Silicon Carbide—A High Temperature Semiconductor*, Ed. by J. R. O'Connor and J. Smiltens, Pergamon, New York, 1960, pp. 67ff.

Kerrigan, J. V., *J. Appl. Phys.* **34**, 3408 (1963).

Lely, J. A., *Ber. Deut. Keram. Ges.* **32**, 229 (1955).

Litton, F. B., and H. C. Anderson, *J. Electrochem. Soc.* **101**, 287 (1954).

McCarty, L. V., *J. Electrochem. Soc.* **106**, 1036 (1959).

Marcus, R. B., to be published, *J. Appl. Phys.*

Marinace, J. C., *IBM J. Res. Develop.* **4**, 248 (1960).

Mark, A., *J. Electrochem. Soc.* **108**, 880 (1961).

Mason, B. J., in *The Art and Science of Growing Crystals*, Ed. by J. J. Gilman, Wiley, New York, 1963, pp. 119ff.

Maunail, J. C., W. E. Baker, and D. M. J. Compton, in *Metallurgy of Elemental and Compound Semiconductors*, Vol. 12, Ed. by R. O. Grubel, Wiley-Interscience, New York, 1961, pp. 271ff.

McCarty, L. V., *J. Electrochem. Soc.* **106**, 1036 (1959).

Miller, K. J., and M. J. Grieco, *J. Electrochem. Soc.* **110**, 1252 (1963).

Moest, R. R., and B. R. Shupp, *J. Electrochem. Soc.* **109**, 1061 (1962).

Monchamp, R. R., W. J. McAleer, and P. I. Pollak, *J. Electrochem. Soc.* **111**, 879 (1964).

Nitsche, R., in *Crystal Growth*, Ed. by H. S. Peiser, Pergamon, New York, 1967, pp. 215ff.

O'Connor, J. R., in *The Art and Science of Growing Crystals*, Ed. by J. J. Gilman, Wiley, New York, 1963, pp. 93ff.

O'Connor, J. R., and J. Smiltens, Eds. *Silicon Carbide—A High Temperature Semiconductor*, Pergamon, New York, 1960.

Oxley, J. H., and J. M. Blocker, *J. Electrochem. Soc.* **108**, 460 (1961).

Piper, W. W., and S. J. Polich, *J. Appl. Phys.* **32**, 1278 (1961).

Powell, Carrol F., Joseph H. Oxley, and John M. Blocker, Ed. *Vapor Deposition*, Wiley, New York, 1966.

Price, P. B., *Phil. Mag.* **5**, 473 (1960); *J. Appl. Phys.* **32**, 1746 (1961).

Reed, T. B., and W. J. LaFleur, *Appl. Phys. Letters* **5**, 191 (1964).

Reynolds, D. C., in *The Art and Science of Growing Crystals*, Ed. by J. J. Gilman, Wiley, New York, 1963, pp. 62ff.

Riebling, E. F., and W. W. Webb, *Science* **126**, 309 (1957).

Rolsten, R. F., *Iodide Metals and Metal Iodides*, Wiley, New York, 1961.

Sagal, M. W., unpublished work, Bell Telephone Laboratories, 1966.

Samelson, H., and V. A. Brophy, *J. Electrochem. Soc.* **108**, 150 (1961).

Sangster, R. C., E. F. Maverick, and M. L. Croutch, *J. Electrochem. Soc.* **104**, 317 (1957).

Schäfer, H., *Chemical Transport Reactions*, translated by H. Frankfort, Academic Press, New York, 1964.

Schäfer, H., H. Jacob, and K. Etzel, *Z. Anorg. Allgem. Chem.* **286**, 42 (1956).

Schaffer, P. S., *J. Am. Ceram. Soc.* **48**, 508 (1965).

Schaffer, P. S., *Vapor Phase Growth of Ruby Monocrystals*, Semiannual Technical Report, Contract No. NONR-4574(00)-1 ARPA Order No. 306, Office of Naval Research, Physical Services Division, Karlington, February 1966, Lexington Laboratories, Cambridge, Mass.

Scholz, H., and R. Kluckow, in *Crystal Growth*, Ed. by H. S. Peiser, Pergamon, New York, 1967, pp. 475ff.

Schwartz, N., in *Transactions of the 10th National Vacuum Symposium, American Vacuum Society*, Macmillan, New York, 1963, p. 325.

Sears, G. W., *Acta Met.* **3**, 367 (1955).

Shapiro, Z. M., cited in B. Lustman and F. Kerze, *The Metallurgy of Zirconium*, McGraw-Hill, New York, 1955.

Sinclair, W. R., F. G. Peters, D. W. Stillinger, and S. E. Koonce, *J. Electrochem. Soc.* **112**, 1096 (1955).

Steinmaier, W., *Phillips Res. Rept.* **18**, 75 (1963).

Steinmaier, W., and J. Bloem, *J. Electrochem. Soc.* **111**, 206 (1964).

Theuerer, H. C., *J. Electrochem. Soc.* **108**, 649 (1961).

Theuerer, H. C., and J. J. Hauser, *J. Appl. Phys.* **35**, 554 (1964).

Theuerer, H. C., and J. J. Hauser, *Trans. AIME* **233**, 588 (1965).

Thorpe, J. F., and M. A. Whiteley, Eds., *Thorpe's Dictionary of Applied Chemistry*, Vol. II, Wiley, New York, 4th ed., 1961, p. 357.

Thurmond, C. D., and C. J. Frosch, *J. Electrochem. Soc.* **111**, 184 (1964).

Tung, S. K., and R. E. Caffrey, *Trans. Met. Soc. AIME* **233**, 572 (1965).

van Arkel, A. E., *Metallwritschaft* **13**, 405 (1934).

van Arkel, A. E., and J. H. de Boer, *Z. Anorg. Allgem. Chem.* **148**, 345 (1925).

Wajda, E. S., and R. Glang, in *Metallurgy of Elemental and Compound Semiconductors*, Vol. 12, Ed. by R. O. Grubel, Wiley-Interscience, New York, 1961, pp. 229ff.

Webb, W. W., *J. Appl. Phys.* **36**, 214 (1965).

Webb, W. W., and W. D. Forgeng, *J. Appl. Phys.* **28**, 1449 (1957).

Willardson, Robert K., and Harvey L. Goering, Eds., *Compound Semiconductors*, Vol. I, Reinhold, New York, 1962.

Williams, J. C., W. R. Sinclair, and S. E. Koonce, *J. Am. Ceram. Soc.* **46**, 161 (1963).

7

GROWTH FROM LIQUID SOLUTION

7.1 Introduction

Polycomponent growth methods involving a liquid phase have so many common elements that it is convenient to consider them in a single chapter. Because these methods involve an additional component or components besides the crystallizing phase, they are subject to the disadvantages associated with polycomponent growth (see Chapter 2 and, especially, Sec. 2.10). However, they also present the conventional advantages associated with lower-temperature processes (see Sec. 2.1). In all cases because growth involves crystallization from a liquid polycomponent phase, the growth can be considered growth from solution. Indeed, the main differences between the various methods are of technique rather than principle and hinge on the nature of the solvent employed. Several methods closely related to solution growth are also discussed at the conclusion of this chapter.

It is convenient to consider the growth methods on the basis of the solvent used, because the equipment, range of applicability, problems, and approach are to a large extent determined by the choice of solvent. However, it should be borne in mind that perhaps a more fundamental delineation of the methods could be made on the basis of the method of producing supersaturation. The growth methods listed on such a basis would be

I. Isothermal Methods (Constant-Temperature Methods)
   A. Solvent evaporation or solvent concentration charge (mainly used in aqueous and molten-salt growth)
   B. Temperature differential (mainly used in hydrothermal, aqueous, and molten-salt growth; also includes temperature-gradient zone melting when the gradient is moved through the sample)
   C. Chemical or electrochemical reaction (mainly used in aqueous growth)

II. Nonisothermal Methods (Temperature-Variation Methods)
   A. Slow cooling (mainly used in aqueous, liquid metal solvent, and molten-salt growth)
   B. Temperature-gradient zone melting (when the gradient is imposed across the whole sample)

Any of the solvents discussed in this Chapter (or, for that matter, any other solvent) could be used for growth by any of the above means. In earlier sections we have pointed out that the boundary between growth by reaction (see Sec. 2.11) and growth by solution is not clear and that, for instance, growth by vapor-phase reaction when ΔG is close to zero might be quite akin to growth from solution in the presence of a complexing agent (mineralizer; see Sec. 7.3). Isothermal methods, of course, have the advantage that any property of the crystal that is temperature dependent will be under better control. If the desired phase is stable over a narrow temperature range, it will be more likely that it can be prepared in an isothermal process. If an impurity, dopant, or component has a temperature-dependent distribution constant, more homogeneous crystals will result in an isothermal process. The principal advantage of slow-cooling processes is that the diffusion path is usually shorter, so reasonable rates are achieved without elaborate control or apparatus investment. Solvent evaporation or the addition of a second solvent to the system, which reduces solubility, has the disadvantage that if the distribution constant for impurities, etc., is solvent-concentration dependent, inhomogeneity will be a problem. Solvent addition is as hard to control as chemical reaction and some of the tricks employed therein (such as gel growth) may be advantageous. Temperature differential methods are subject to diffusion-path and dissolving-step problems, while chemical and electrochemical reactions are hard to control. In the above listing we have indicated the solvents wherein the method is presently most often applied. Further discussion of the methods may be found in the appropriate section devoted to each solvent. Temperature cycling, although so far not much applied to liquid-solution growth, should have great use in limiting spontaneous nucleation (see Sec. 6.4.1).

7.2 Aqueous-Solution Growth

The growth of crystals from aqueous solution is probably the oldest method of crystal growth. Indeed, the formation of crystalline products in scores of industrial processes ranging from the preparation of table sugar to copper sulfate employs this technique. The chemical-engineering literature discusses this sort of procedure from the theoretical, practical, and equipment viewpoints at length (Van Hook, 1961b). However, in such processes the object is to obtain economically a high yield of a uniform crystalline product where the crystallite size required is usually quite small. Often it is desired to obtain some uniformity in particle size but seldom is perfection of the crystals of any importance and never are large crystals obtained. In such practices growth is by spontaneous nucleation or by seeding where the seeds are powders. Thus these procedures are not directly applicable to the growth of large single crystals of high perfection.

Buckley (1951a) reviews virtually all of the literature on solution growth of large crystals up to about 1950. Discussions of the technique by Holden and Singer (1960) and by Holden and Thompson (1964), although actually intended for secondary-school students, are among the most recent and useful. (See also Petrov, et al.)

In aqueous-solution growth, as in all solution growth, the object is to supersaturate the solution without causing spontaneous nucleation and to make the supersaturation and hence the rate as high as is commensurate with controlled growth on a seed of material of the requisite perfection. The methods hinge essentially on the manner in which supersaturation is brought about, as discussed in Sec. 7.1. Figure 7.1 shows the expected dependence of solubility on temperature for most materials near ambient conditions. Figure 7.1 is, of course, derived from the solute–H_2O phase diagram and is part of a cut at constant pressure. In many cases the van't Hoff equation is obeyed so that

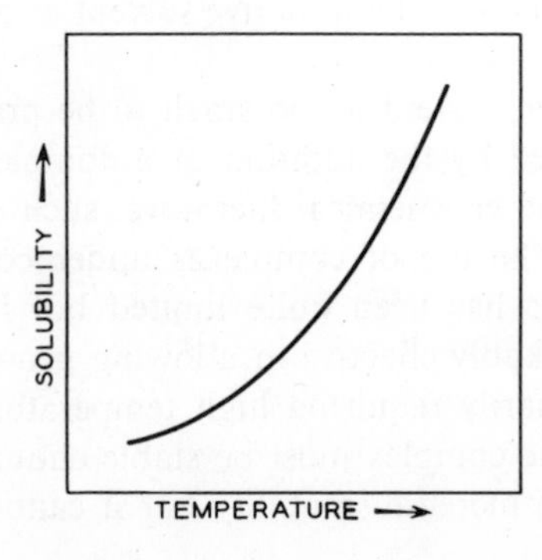

Fig. 7.1 Solubility vs. temperature for common materials.

$$\frac{d \ln s}{dT} = \frac{\Delta H}{RT^2} \tag{7.1}$$

where s is solubility (expressed as atom fraction in the most general case), T absolute temperature, ΔH the heat of solution, and R the gas constant. When the dissolving reaction is endothermic, the slope of the solubility vs. temperature curve is positive, as in Fig. 7.1. When the reaction is exothermic, the slope is negative (so-called *retrograde solubility*) but growth may still be achieved by the methods of Sec. 7.1, except that in method (II–A) the temperature is raised and in method (I–B) growth takes place at a temperature above the dissolving temperature.

In the case of growth from solution, the desired phase should be the stable solid, so that phase-equilibria experiments may be required. Undesired phases are often hydrates, hydroxides, or oxyhydroxides. In general, hydrolysis is not so prevalant at higher temperatures, so that, if hydroxides or oxyhydroxides are the observed phases, crystallization at higher temperatures is indicated. If hydrates are formed, higher temperatures will also often result in stabilizing the desired phase.

Phase-equilibria and solubility-determination methods in H_2O near ambient conditions are discussed in Vold and Vold (1949) and several standard compilations of solubilities are available (Linke, 1965; Stephen and Stephen, 1963). It should be clear that little modification of the techniques discussed here would be required for a solvent other than water, provided its liquid range, vapor pressure, and corrosive nature were similar to that of water.

If the solubility in pure water is too low, as will often be the case for covalently bonded nonionizable materials such as organics, then appropriate solvents, such as organic solvents, can be employed. If the solubility is too high (leading to high-viscosity solutions) then mixed solvents such as alcohol –water may be appropriate. If the slope of the solubility temperature curve is either too steep (minor temperature fluctuations cause spontaneous nucleation) or not steep enough (not enough material is deposited in growth by slow cooling or in growth by temperature differential, the differential required is impractically large) then an alternative solvent or a mixed-solvent system may be appropriate.

If the solubility in a given solvent is too small to be practical, then in some cases it can be improved by the addition of a complexing agent. In hydrothermal work and in the geochemical literature, such complexers are usually called mineralizers. The use of complexes under conditions other than hydrothermal conditions has been quite limited but in principle, at least, they could prove remarkably effective in allowing growth at ambient conditions of materials ordinarily requiring high temperatures, exotic solvents, and high pressures. The complex must be stable enough and exist in high-enough concentration to increase solubility but it cannot be so stable

Table 7.1

REPRESENTATIVE CRYSTALS GROWN FROM AQUEOUS SOLVENTS

Material	*Formula*	*Solvent*	*Method*	*Procedure*	*Reference*
Potassium alum	$KAl(SO_4)_2 \cdot 12\,H_2O$	H_2O	Rotary crystallizer	Saturated solution at 48°C, cooled ~0.2°/day for several weeks	Holden and Thompson, 1964
Chrome alum	$KCr(SO_4)_2 \cdot 12\,H_2O$	H_2O	Rotary crystallizer	~ to potassium alum	Holden and Thompson, 1964
Sodium chlorate	$NaClO_3$	H_2O	Rotary crystallizer	Saturated solution at 51°C, cooled ~0.2°/day for several weeks	Holden and Thompson, 1964
Nickel-sulfate hydrates	$NiSO_4 \cdot 6\,H_2O$ and $NiSO_4 \cdot 7\,H_2O$	H_2O	Rotary crystallizer	Cooling from ~50–35°C produces hexahydrate; below 35°C, heptahydrate forms	Holden and Thompson, 1964
Rochelle Salt (potassium-sodium tartrate)	$KNaC_4H_4O_6 \cdot 4\,H_2O$	H_2O	Rotary crystallizer	Similar to alums; dehydrates easily, must be stored in sealed jar	Holden and Thompson, 1964
Rock salt	NaCl	H_2O	Evaporation	Saturated brine containing 0.1% H_2SO_4 + 0.1% $Pb(NO_3)_2$ evaporated at 75°C about 12 hr, forms 6-mm NaCl cubes (270 ppm Pb in NaCl); without Pb^{2+}, cloudy hydrates result	Gibbs and Clayton, 1924

Table 7.1—*Cont.*

Material	*Formula*	*Solvent*	*Method*	*Procedure*	*Reference*
Rochelle salt (potassium-sodium tartrate)	$KNaC_4H_4O_6 \cdot 4\,H_2O$	H_2O	Mason jar or evaporation	Make solution containing 130 g/100 cm^3 H_2O, heat until all dissolves and grow by evaporation method or make solution containing 134 g/100 cm^3 H_2O, heat until all dissolves and grow by mason-jar method	Holden and Singer, 1960
Potassium alum	$KAl(SO_4)_2 \cdot 12\,H_2O$	H_2O	Mason jar or evaporation	See rochelle salt above; 20 g/100 cm^3 for evaporation, 24 g/100 cm^3 for mason-jar method	Holden and Singer, 1960
Chrome alum	$KCr(SO_4)_2 \cdot 12\,H_2O$	H_2O	Mason jar or evaporation	See rochelle salt above; 60 g/100 cm^3 for evaporation, 65 g/100 cm^3 for mason-jar method	Holden and Singer, 1960
Sodium borate	$NaBO_3$	H_2O	Mason jar or evaporation	See rochelle salt above; 50 g/100 cm^3 for evaporation, 52 g/100 cm^3 for mason-jar method	Holden and Singer, 1960
Sodium chlorate	$NaClO_3$	H_2O	Mason jar or evaporation	See rochelle salt above; 113.4 g/100 cm^3 for evaporation; 117.4 g/100 cm^3 for mason-jar method.	Holden and Singer, 1960

Table 7.1—*Cont.*

Material	*Formula*	*Solvent*	*Method*	*Procedure*	*Reference*
Sodium nitrate	$NaNO_3$	H_2O	Mason jar or evaporation	See rochelle salt above; 110 g/100 cm^3 for evaporation, 113 g/100 cm^3 for mason-jar method.	Holden and Singer, 1960
Ammonium dihydrogen phosphate (ADP)	$NH_4H_2PO_4$	H_2O	Rotary crystallizer	(001) plates grow poorly until (101) "capping" faces formed; if solution made pH ~ 5, rate on side faces increased	Walker and Kohman, 1948; Walker, 1947; Holden, 1949
Ethylene-diamine tartrate (EDT)	$[NH_2CH_2CH_2NH_2]$ $[OC(CHOH)_2CO]$	H_2O	Rotary crystallizer	Anhydrous form stable above 40.6 °C, hydrate below; anhydrous form exists metastably in absence of monohydrate nuclei below 40.6°C; introduction of nucleii into crystal-growth plant caused "blight" of hydrate until growth temperature was raised	Walker and Kohman, 1948; Walker, 1947; Holden, 1949

that it is the stable solid phase. Mineral acids, especially HF for refractory fluorides (which form complex fluoride species) and sulfides for sulfide solids (which form complex sulfides), are promising possibilities.

Relative growth rates on various crystallographic faces often depend on which atoms have been adsorbed on the particular face. Thus growth rates can be strongly impurity-ion and pH dependent. Similarly, morphology and perfection can exhibit a great sensitivity to the presence of impurities. This is especially true in growth methods, such as aqueous-solution growth, that take place at modest temperatures, where adsorption and chemisorption are comparatively easy. Perhaps the classic case of an impurity that permitted the growth of a good-quality crystal is Pb^{2+}, which allows good-quality NaCl growth (see Table 7.1). Yamamoto (1938) made a listing of over twenty impurities that are known to affect the growth of NaCl, KCl, KBr, KI, RbCl, NH_4Cl, $LiCl \cdot H_2O$, K_2SO_4, and KNO_3 crystals. When one has difficulty in getting good-quality growth from aqueous media (especially when growth is cloudy) impurity addition should be considered. Small highly charged cations with high polarizing power that are easily adsorbed or chemisorbed seem most effective.

7.2.1 HOLDEN'S ROTARY CRYSTALLIZER

For growth by slow cooling, perhaps the most useful apparatus is Holden's rotary crystallizer (Holden and Thompson, 1964). Figure 7.2 shows a schematic of this apparatus. Saturated or near-saturated solution is added to the tank at T_1. For ease in handling, it is usually convenient to transfer solution at a temperature slightly above its saturation temperature. Suitably oriented seeds are mounted on the seed holder. In many cases fragments of crystal are suitable for seeds in growth with this apparatus. The fragments are conveniently mounted by forcing them into a piece of plastic insulating tubing ("spaghetti") that has been forced over the crystal support end of the crystallizer (the "spider"; see Fig. 7.2).

The production of fragments for seeds can often prove troublesome. Sometimes crystals will form on the walls and stirrer of the rotary crystallizer during cooling and these may be reclaimed for subsequent seeded runs. Another technique for preparing seeds is solvent evaporation (Sec. 7.2.3). Evaporation of solution on a watch glass or in a Petri dish is a convenient means of preparing fragments useful for seeds. The mason-jar technique (see Sec. 7.2.2) may be used without any seeds. Nucleation will take place on the walls and bottom of the jar, and a single-crystal nucleus can be selected or cut from the crystals formed. It is sometimes useful to hang a string in the mason jar as a nucleation site, because growth there will be unrestricted.

A convenient material for the construction of support rods for the Holden crystallizer, which is inert to many solutions, is tungsten-rod stock.

It is often possible to check for saturation of an aqueous solution by inserting a small crystallite near the surface and, with an appropriately adjusted light, watching for schlieren lines (caused by index-of-refraction variations) associated with the concentration gradients that will exist near a crystallite not in equilibrium with its solution. If the current is descending, it is more dense than the surrounding fluid and the crystal is dissolving (the solution is undersaturated); if the current is rising, more dense fluid is approaching the crystal from the solution and it is growing (the solution is supersaturated). Depending on the results of this test, the temperature may be adjusted to bring the system closer to equilibrium.

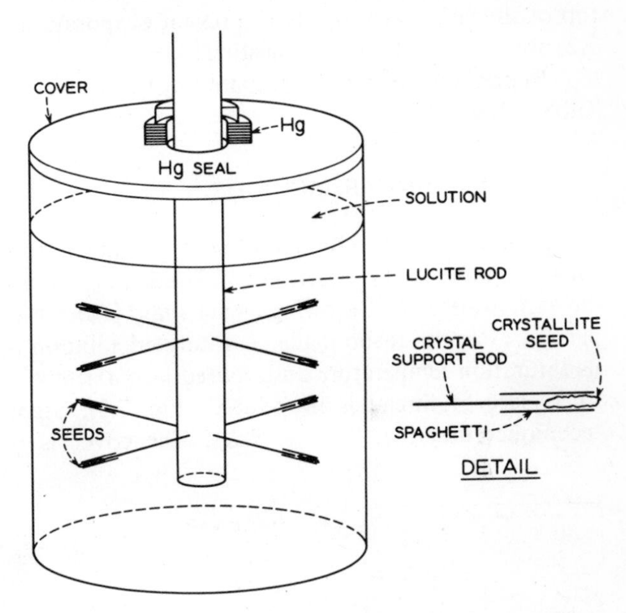

Fig. 7.2 Holden's rotary crystallizer.

Temperature is maintained at the desired level by a variety of heater–controller arrangements. Knife-style resistance heaters immersed in the solution (sometimes together with a hot plate) are commonly used. More rapid response is obtained when an infrared lamp focused on the solution is used as a cycled auxiliary heater. Temperature-sensing elements are usually mercury thermoregulators, although more sophisticated sensors have been used. Tank volume is dependent on the size crystal desired. To decrease diffusion effects it is usually necessary to slowly rotate the spider. So that vortices are not formed, it is generally common to reverse the rotation direction every few seconds. A complete description of growth with the Holden crystallizer is provided in Holden and Thompson (1964) and all the equipment for such growth can be purchased in kit form from the Dependable Printed Circuit Corp., Wayne, N.J.

Supersaturation is provided by slow cooling and the cooling rate is determined by the system. Usually 0.1–1.5°/day are appropriate. Motor-driven cam-type controllers have been used to establish the cooling rate, but it is usually sufficient to set down the control point of the mercury thermoregulator by a predetermined amount once each day. Clearly such growth is not isothermal, so any growth-temperature-dependent property of the crystal will not be uniform throughout. If the temperature is dropped in steps, the rate will not be uniform and rate-dependent properties of the crystal will not be uniform. The cover is generally sealed as tightly to the top of the vessel as possible to prevent evaporation and Hg seals (see Fig. 7.2) are often used around the stirrer.

Representative crystals prepared by the rotary crystallizer are included in Table 7.1.

7.2.2 MASON-JAR METHOD

A more simple technique of slow cooling is the "mason-jar method." This technique is best described by Holden and Singer (1960). It is perhaps the easiest method of growing crystals and is cheap enough to permit many parallel attempts to be made. A saturated solution is heated slightly above its saturation temperature and poured into a screw-cap jar and a seed tied to a piece of thread is introduced (Fig. 7.3). Seeds are obtained by the techniques described in Sec. 7.2.1. The cover is placed on the jar and

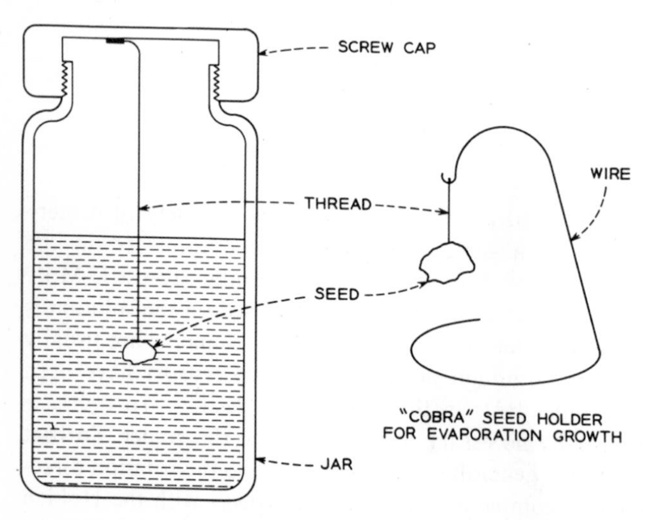

Fig. 7.3 Mason-jar growth method.

it is allowed to cool to room temperature. Provided the solubility is high enough and the cooling rate slow enough, well-formed crystals will often grow on the seed. In this technique the cooling rate is, of course, faster than in the rotary-crystallizer method and no stirring is possible; but the simplicity still makes it attractive, particularly in preliminary work and where modest-size crystals are satisfactory.

The mason-jar method may easily be adapted to growth by evaporation. The procedure is as before except that the seed is suspended on a wire "cobra" seed holder (Fig. 7.3). This seed holder does not project above the solution, so it will not provide sites for nucleation at the interface, which during evaporation is, of course, highly supersaturated. Instead of closing the jar with a tight cap it is covered with a cloth (often fastened at the lip with a rubber band) that excludes dust but allows evaporation. The first growth is by slow cooling but, once the solution reaches room temperature, subsequent growth is by evaporation. If it is desired that evaporation proceed at a higher rate or that growth take place at a higher temperature, the jar may be heated (a hot plate, water bath, or water thermostat is convenient). A constant-temperature room is convenient for such growth, but the temperature stability found in the cellars of many buildings is often sufficient to considerably improve growth.

If growth by evaporation at elevated temperature is desired but the evaporation rate must be limited, an Erlenmeyer flask closed with a one-hole stopper containing a glass tube drawn to a capillary of the appropriate size may be used. In all evaporative growth excessive nucleation at the interface is a problem. Sometimes petroleum jelly smeared around the inside of the container just above the interface will prevent a ring of spurious nucleation starting there and growing toward the top. Such growth is caused by solution moving by capillarity up through crystallites and then being evaporated at the top of the ring of nucleation. If solubility is not appropriate or evaporation rate is either too high or too low, changing to a nonaqueous solvent or using a mixed solvent may be appropriate.

Table 7.1 lists a variety of crystals grown by slow cooling and evaporation. Holden and Singer (1960) give many recipes for growth. Many additional geometries have been used for growth (Buckley, 1951a) by slow cooling and evaporation and the number of possible modifications (most of them trivial) is so near to limitless as to make summarization here inappropriate.

7.2.3 TEMPERATURE DIFFERENTIAL METHOD

Growth by the temperature differential method was first reported by Kruger and Finke (Buckley, 1951b). A modification of their apparatus is shown in Fig. 7.4. Excess solute (nutrient) is placed in the dissolving vessel and solution is circulated between the dissolving and growth vessel by convection and (usually) by a mechanical stirrer. A convenient way to begin is to bring the growth and dissolving vessels to the same temperature (heating can be by knife heaters, thermostated bath, hot plate, etc.) and test the solution in the growth vessel until it is saturated. Then raise the temperature in the growth vessel slightly ($T_1 > T_2$) and insert the seed. Seeds can be prepared by the methods of Sec. 7.2.1. The solution will be undersaturated and surface damage on the seed, attached nucleii, floating nucleii, etc., will be destroyed. Then gradually lower T_1 ($T_2 > T_1$) until $\Delta T = T_2 - T_1$ is the desired value. The tubes connecting the vessels serve as heat exchangers, (heat input in the lower tube and heat loss from the upper tube) so that heating tape around the lower tube and a cool water jacket around the upper may be useful. Rotation of the seeds as in the rotary crystallizer may be helpful, as may direct pumping of solution. A convenient pump that squeezes

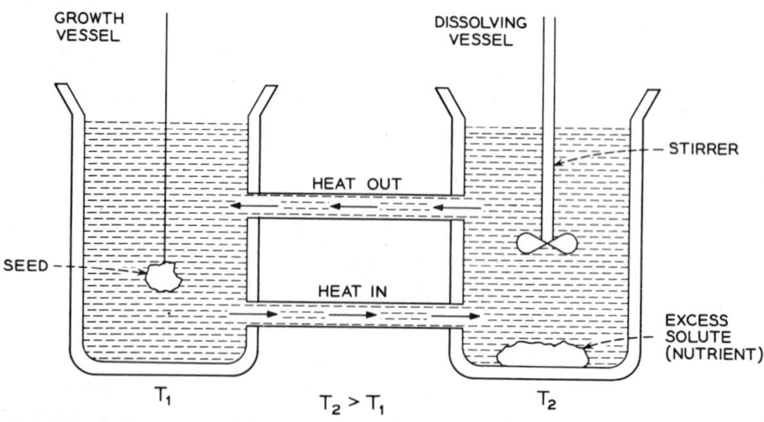

Fig. 7.4 Temperature-differential method.

plastic tubing rhythmically to effect liquid motion and thus presents no contact of the moving parts with the solution is the "peristaltic pump," which is available in several sizes from J. H. Beize, Inc., So. Plainfield, N.J. Filters in the upper tube can be used to retain small undissolved crystallites or the additon of a third temperature equlibration chamber midway in the upper tube may be necessary. If it is desired to eliminate convection as a driving force for circulation, the system may be arranged so that circulation is from dissolving vessel to growth vessel in the bottom tube and from growth vessel to dissolving vessel in the upper tube.

Walker and Kohman (1948) have described a production size facility for the growth of ammonium dihydrogen phosphate (ADP) and ethylenediamine tartrate (EDT) for piezoelectric applications by the temperature-differential method. Figure 7.5 shows one version of their apparatus. Tank (1) is the crystallization tank and the seeds are rotated as in the rotary crystallizer. Tank (2) is the saturator and tank (3) is a "superheater"; that is, in tank (3) the solution is warmed to a temperature a few degrees above its saturation temperature to ensure that nuclei are destroyed. The solution is pumped from tank (3) to tank (2), where it cools rapidly [the thermal mass of solution in tank (3) is very large] and isothermal growth takes place because of the supersaturation corresponding to $\Delta T = T_2 - T_1$. Flow between tanks (1) and (2) and tanks (2) and (3) is by gravity. Further details of growth are given in Table 7.1.

An interesting problem that occurred in the growth of EDT in production-size equipment was the formation of "barnacles" (Kohman, 1950). Growth was originally at 38°C and, with the process well underway at this temperature, suddenly undesired "barnacles" of foreign material began to appear on the growing crystals. Shortly thereafter, "barnacles" began to appear in growth runs at the laboratory location some miles distant. After some investigation it was found that the "barnacles," which were shown to

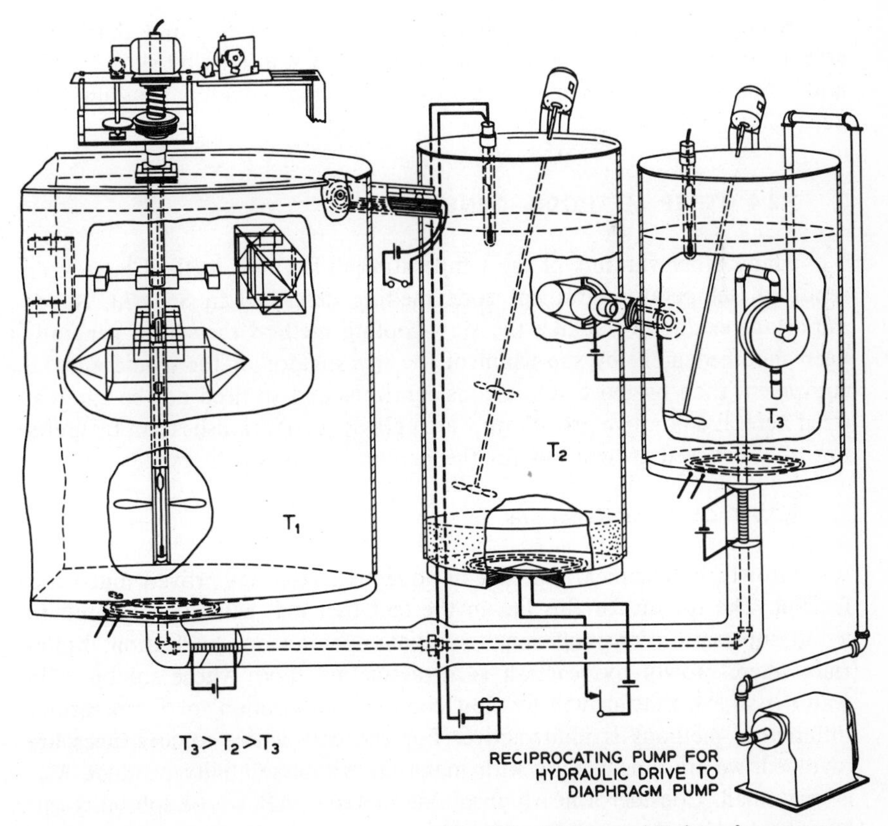

Fig. 7.5 Walker and Kohman's apparatus.

be EDT·1 H_2O were stable below 40.6°C and the desired phase (EDT) above that temperature. EDT was metastable for some degrees below the transition and could be grown there provided no nuclei of EDT·1 H_2O were present. Once nuclei of EDT·1 H_2O were present, they were carried inadvertently by individuals who traveled between the production facility and the laboratory, where they nucleated EDT·1 H_2O. The problem was solved by carrying on all growth above 40.6°C and being careful to keep grown EDT out of contact with a moist environment. This experience suggested the plot for a well-known science-fiction novel (Vonnegut, 1963), which describes the plight of society when a new polymorphic form of ice, whose melting point is close to 25°C, is discovered.

To a large extent the ready availability of synthetic quartz has decreased the commercial interest in EDT and ADP but the techniques used are readily modifiable to the commercial production of other water-soluble materials. Such facilities have produced over 40,000 lb of EDT in a single year. Slow-cooling methods are also appropriate and have been used for EDT and ADP (Buckley, 1951b).

Kolb (1969) has shown that Se is complexed by S^{2-} and that small crystals of trigonal Se can be grown from Na_2S solutions by slow cooling and evaporation. This is an example of complexing providing solubility (see Sec. 7.2).

7.2.4 OTHER METHODS—GENERAL

Many other variants of the temperature-differential method have been reported (temperature-gradient zone melting described in Sec. 7.6 is one important example) but like the slow-cooling method they have generally been modified more by the whim of the investigator or the availability of equipment than by more subtle considerations and so need not be summarized here. Evaporative growth on watch glasses or Petri dishes can be useful not only for seed growth but for the preparation of small crystals.

7.2.4-1 Chemical Reactions

Growth by chemical reaction in aqueous media has proved much less fruitful. The reason for this lies in the fact that it is extremely difficult to avoid supersaturations sufficient to initiate spontaneous nucleation. Materials where growth by reaction is attractive are those whose solubility in water is so low that growth by slow-cooling, evaporation, or temperature-differential methods is unattractive. For the case where ionic species are involved, we are thus dealing with materials whose solubility product, K_{SP}, is very small. Consider a nearly insoluble material, AB, whose solubility can be considered to be caused mainly by ionic species. Thus

$$AB_s \rightleftharpoons A^+ + B^- \tag{7.2}$$

where the solubility product

$$K_{SP} = (A^+)(B^-) \tag{7.3}$$

(parentheses indicate concentrations of the appropriate species). One might consider that growth by reaction could be achieved if aqueous solutions of the soluble salts, AC and DB, were mixed. However, most mixing schemes will result in local $(A^+)(B^-)$ products that exceed K_{SP} by an amount to cause spontaneous nucleation. It is for this reason that small crystals or crystalline powders are almost inevitably the product of such growth attempts. Clearly any method where a solution of AC is directly added to a solution of DB, or vice versa, is doomed to failure. However, techniques where solutions of AC and DB are both added to pure water containing a seed may be workable, because exceeding K_{SP} locally may be more easily prevented. The growth rates provided by such methods will probably be slow because the rate at which AB can be brought to a growing seed will be limited by the supersaturation that can be tolerated without spontaneous nucleation and thus will be small because $K_{SP} \ll 0$. However, techniques such as those illustrated in Fig. 7.6 might be worth exploring.

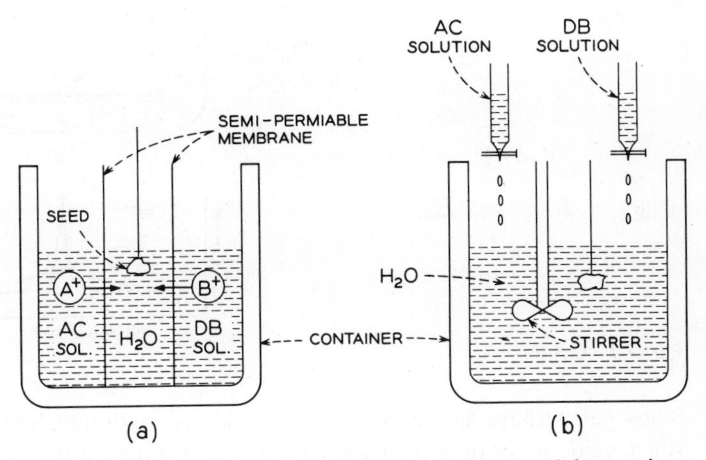

Fig. 7.6 Schemes for growth by reaction.

7.2.4-2 Gel Media

Perhaps the best controlled growth of materials by reaction in aqueous solution makes use of gel media. Crystallization in gel media was first studied seriously by Liesegang (1914), who attempted explanations for the periodically spaced precipitates that can be found in nature and formed in the laboratory and that bear his name (Liesegang rings). Other early work was by Bradford (1926), Holmes (1926), Ostwald (1897), and Rayleigh (1919).

Interest in gel media for crystal growth has been greatly stimulated by the recent work of Henisch et al. (1965a). Gel media provide a milieu that is advantageous for growth by reaction because K_{SP} need only be exceeded in a local region where growth is desired. In general two geometries for gel growth have been used. These are illustrated for the growth of calcium tartrate ($CaC_4H_4O_6$) in Fig. 7.7. The reaction of highly soluble tartaric acid and calcium chloride produces insoluble $CaC_4H_4O_6$:

$$H_2C_4O_6 + CaCl_2 \longrightarrow CaC_4O_{6(s)} + 2\,HCl \qquad (7.4)$$

Silica gel can be formed by acidifying a soluble silicate

$$H^+ + SiO_3^{2-} \longrightarrow SiO_{2(gel)} + H_2O \qquad (7.5)$$

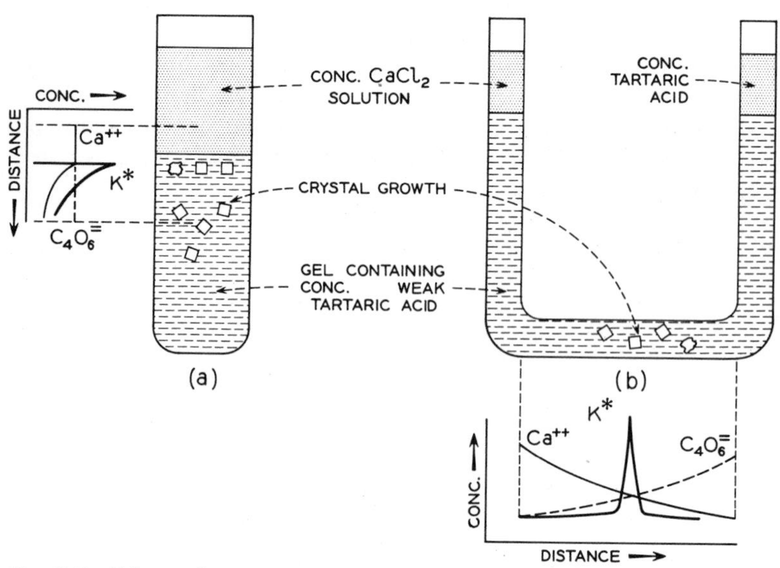

Fig. 7.7 Gel growth.

Silica gel perhaps might better be considered hydrated SiO_2 or hydrated silicic acid; it forms a network structure entrapping water. This gelatinous structure provides an ideal medium for the diffusion of ions and can be used to keep reactive ions separated until reaction is desired. In a typical experiment in the apparatus of Fig. 7.7(a) a solution of sodium silicate [Henisch et al. (1965a, b) suggest specific gravity 1.06 g/cm^3 or less] is mixed with, for instance, 0.8–2 N $H_2C_4O_6$ and allowed to gel at constant temperature. Then a solution of 1.0 N $CaCl_2$ is placed on top of the gel. It supplies Ca^{2+}, which reacts with $C_4O_6^{2-}$ at the gel–liquid interface and within the gel to form the slightly soluble calcium tartrate. The solution also prevents the gel from drying out while growth takes place. However, by the scheme of Fig. 7.7(a) K^*† is found to be greatest at the gel–liquid interface, and the crystals will form in the presence of a large excess of Ca^{2+}.

The scheme of Fig. 7.7(b) is superior because Ca^{2+} and $C_4O_6^{2-}$ diffuse toward each other from opposite sides of a "U" tube and K^* will be greatest at some point well within the gel medium. In addition, K^* will peak more sharply than in the scheme of Fig. 7.7(a), so a much smaller local region will be the site of nucleation (or, if a seed is placed in this region, it may be the sole site of growth). For the scheme of Fig. 7.7(b), the gel is formed initially from very weak acid solution. The speed of crystal formation depends on

1. Gel density (which affects ionic diffusion constants)‡
2. Concentration of reagents in gel
3. Concentration of reagents diffused into gel
4. Temperature, length of diffusion path, etc.

Crystals of millimeter to centimeter size form in days to weeks. Table 7.2 lists some materials that have been grown from gels. Calcium tartrate has perhaps been most studied. Some observations concerning the mechanism have been advanced. The points of general agreement (Dennis and Henisch, 1967; Henisch et al., 1965a; Wakim et al., 1965; Henisch et al., 1965b) seem to be

1. The supersaturation within the gel is probably greater than would ordinarily exist in free solution without runaway growth.§

2. Growth can take place even when crystal size results in gel rupture but properties are then not so good.

3. Any medium that helps to control diffusion is helpful, but large-surface-area media and, in particular, gels give best results.

4. Gels may inhibit embryo formation, affect the diffusion layer at a growing interface, or affect turbulence and convection in liquids but definitive experiments establishing their role in these areas remain to be done.

In short, the possibilities of gel media for the growth of crystals by reaction both from the practical viewpoint of producing useful crystals and from the viewpoint of the new opportunities they present for mechanism studies have been so far virtually untapped. Similarly, the effects of electric fields in accelerating and controlling diffusion in gels and across semipermeable membranes, such as shown in Fig. 7.6(a), have not been exploited.

---

†K^* is the product of actual concentrations raised to the same power as in K_{SP}. When $K^* > K_{SP}$ precipitation occurs.

‡Best crystals seem to result from low gel densities.

§This is more often asserted than proved. Convincing proof really has not been made.

Table 7.2

SOME CRYSTALS GROWN FROM GEL MEDIA

Material	*Remarks*	*Reference*
Calcium tartrate	Silica gel media—tartaric acid + calcium chloride	Henisch et al., 1965a
Calcium tungstate	Silica gel media—sodium tungstate + calcium chloride	Henisch et al., 1965a
Lead iodide	Silica gel media—potassium iodide + lead acetate	Henisch et al., 1965a
Silver acetate	Silica gel media—acetic acid + silver nitrate	Ellis and Boulin
Cuprous Chloride	Silica gel media—dilution of CuCl complexed in HCl	O'Connor et al., 1966

7.2.4-3 Electrochemical Reactions

Growth by electrochemical reaction has been less explored than gel growth but probably has an even greater potential. The medium need not be aqueous and nonaqueous media (see Secs. 7.1 and 7.4) must be used for the growth of compounds that react with water or where the reduction potential is such that water will be decomposed during the reaction. In aqueous media the most common materials deposited in polycrystalline form are metals. Copper is typical where the cathode reaction is

$$Cu^{2+}_{aq} + 2e^- \longrightarrow Cu_{(s)} \tag{7.6}$$

Electrochemical reactions of this type might be considered a special case of growth by chemical reaction, where the electron is a reactant. Among the advantages they have over most other aqueous-growth reactions are

1. The "concentration" of the electron is easily controlled by adjusting the current density on the electrode where deposition is taking place.
2. High electron "concentration" occurs on the electrode seed, which is just the region where growth is desired.
3. By controlling the deposition voltage, separations and purification can take place during growth.
4. A continuous process is possible with "dissolving" at the anode and growth at the cathode.

Perhaps the main drawback beyond the requirements of nonreactivity of the grown material with water and the need that the voltage required not decompose water is that for continuous growth the simplest arrangements will probably require that the deposited material be electrically conductive.

From the above it would seem that metals would be ideal candidates for electrochemical growth. Property studies on metals grown so far from their melting points would be especially informative and growth would be easy on single-crystal seeds (electrodes). Surprisingly, although this method has been used occasionally to study the mechanism of electrochemical deposition, it has been almost entirely neglected as a crystal-growth technique. Perloff and Wold (1967) have grown crystals of MoO_2 by the electrolytic reduction of $Na_2MoO_3 + MoO_3$ at 675°C. Examples of mechanism papers include those of Harrison et al. (1966), Damojanovic (1965) and Price et al. (1958).

7.3 Hydrothermal Growth

Closely related to growth from aqueous solution at ambient or near-ambient conditions is growth from hydrothermal solution. If solution growth at ambient conditions is unsuccessful because of low solubility or solvent reaction, it is natural to attempt solution growth from a solvent with greater solvent power or from a solvent that does not react with the material being grown. As we have seen above, one solution of this problem is the use of solvents other than water at ambient conditions. Another solution discussed above is to find a complexing agent (mineralizer) that forms additional species (not present in the solvent originally) that thus increase the overall solubility. Provided the complexes are not stable solids, the addition of a mineralizer can be very helpful. A third solution is to increase the solubility by conducting growth at temperatures above ambient conditions. In hydrothermal growth both temperature and pressure are considerably above ambient and mineralizers are generally employed, but the solvent is usually water. In molten-salt growth (see Sec. 7.4) the solvent is changed, the temperature is raised, and complexing agents are sometimes employed. Hydrothermal crystal growth has been reviewed (Ballman and Laudise, 1963; Laudise, 1962) and hydrothermal equipment has been discussed (Laudise and Nielsen, 1961) in the literature.

Geologists showed the earliest interest in hydrothermal crystal growth. They felt that a variety of minerals had formed in nature in the presence of a high-temperature high-pressure aqueous phase and made attempts to reproduce these conditions in the laboratory in order to confirm theories concerning the genesis of minerals. Early experiments included studies of phase equilibria between the components involved in mineral assemblages. No attempts to produce crystal growth are made in such experiments, because microcrystals are sufficient for the deduction of phase diagrams. De Senarmont (1851) and G. Spezia (1905) made the first attempts to produce macrocrystals in the system SiO_2–H_2O where they grew single crystals of α-quartz. Since that time hydrothermal systems involving α-quartz have been studied extensively because of the economic importance of piezoelectric quartz devices, and a great part of our knowledge of hydrothermal synthesis has been derived from these studies.

The system employed for the growth of most materials is similar to that used for quartz and is as follows: Relatively finely divided particles of α-quartz *nutrient* are placed in the bottom of the vessel and suitably oriented single-crystal seed plates of α-quartz are suspended in the upper *growth* region. The vessel is filled to some predetermined fraction of its free volume, typically 0.80 (often referred to as 80% fill, or a degree of fill of 0.80) with a basic solution such as 0.50 *M* NaOH. With its long axis vertical, the vessel is placed in a furnace that has been designed to heat the lower *dissolving* section isothermally hotter than the upper growth region, which is also maintained isothermal. A perforated metal disc called a "baffle" is often placed within the vessel separating the dissolving and growth regions to aid in localizing the temperature differential.

As the temperature is raised, the liquid level rises, the pressure increases, and, finally, at some temperature below the critical temperature of water (374°C), the vessel fills completely with the liquid phase.† Consequently, for dilute solutions the solvent density is about equal to the degree of fill. The pressure is fixed by the temperature, temperature distribution, and initial degree of fill of the vessel. For a growth-zone temperature of 400°C and a dissolving temperature of 350°C, the pressure is ~2000 bars, and reasonable rates of crystallization can be achieved. Under these conditions the solution will saturate at 400°C and move by convection to the growth zone, where the solubility is lower and the solution is supersaturated. Seed plates whose principal faces are (0001) will grow at about 1.0 mm/day in the $\langle 0001 \rangle$ direction.

The criteria that must be fulfilled for the growth of a crystal under hydrothermal conditions in an experiment with the above apparatus configuration are:

1. A combination of solvent, pressure, and temperature must be discovered in which the crystal is thermodynamically stable and has sufficient solubility to permit a reasonable supersaturation so that appreciable rates of crystallization can be obtained without excessive wall or homogeneous nucleation. For substances so far studied, solubilities of 2–5% are generally required.
2. Sufficiently large values of the ratio of the surface area dissolving to that of the seeds so that dissolving is not rate limiting (dissolving area ~5 × growing area is usually sufficient).
3. Sufficiently large $(\partial p/\partial T)_{\bar{p}}$, temperature coefficient of solution density, at constant average solution density, that, with an appropriate temperature differential, convective circulation will be sufficiently rapid not to be rate limiting.
4. Temperature coefficient of solubility, $(\partial s/\partial T)_{\bar{p}}$, such that an appropriate temperature differential will produce a satisfactory supersaturation.‡
5. Vessel suitable to contain the pressure–temperature conditions of the experiment without excessive corrosion.

We can thus see that the principal difficulties in hydrothermal crystal growth are

†Even above 374°C the system is probably still liquid because the critical end point in the system SiO_2–Na_2O–H_2O has not been reached. Even in systems above the critical point where the fluid is by definition gaseous, its density, solvent power and many other properties make it more like a liquid.

‡Occasionally metastable solutes may be used as nutrient. Growth will then occur even with no temperature differential (metastable phases have a higher solubility than the stable phase) but will cease when the nutrient has converted in situ to the stable phase. Some older (unsuccessful) processes for quartz growth used vitreous silica as nutrient.

1. Design of leakproof inert vessels
2. Making a priori predictions about solvent power and growth rates
3. Difficulties associated with the long times required to grow reasonable-size crystals without any observation possible during growth.†

7.3.1 EQUIPMENT

Because equipment requirements are rather unusual, it is worthwhile to discuss autoclave design in some detail at this point (Laudise and Nielsen, 1961). Figure 7.8 shows a schematic of a system in which α-quartz has been crystallized. The autoclave is a steel or a special high-temperature alloy vessel whose strength is great enough to sustain the pressure–temperature conditions expected in the experiment. For moderate conditions vessels may

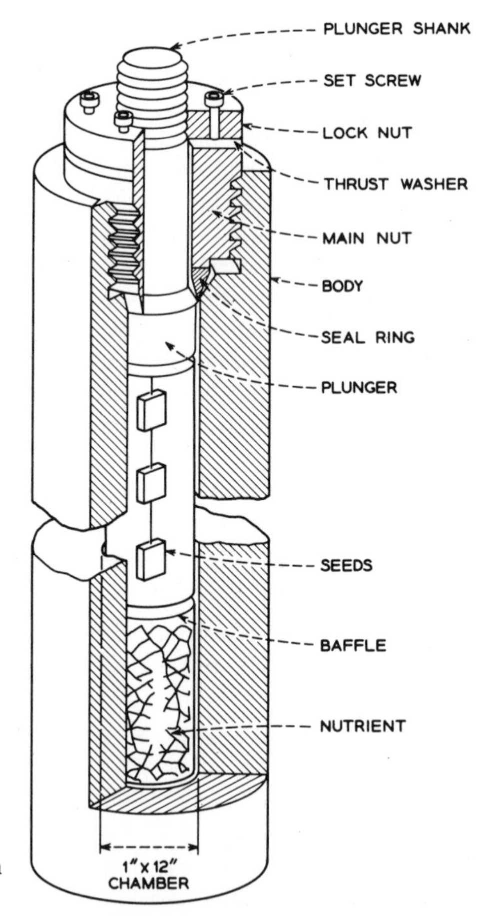

Fig. 7.8 Modified Bridgman hydrothermal autoclave.

†Gamma graphs have been used to observe growth, but resolution is poor and the method is cumbersome.

be purchased from a variety of sources (for example, Autoclave Engineers Co., Erie, Pa., and Tem Pres Research Co., State College, Pa.). The vessel is generally sealed by a closure employing the principle of unsupported area discussed by Bridgman (1949). Figure 7.8 shows the so-called modified Bridgman closure (Autoclave Engineers, Erie, Pa.), which has been found to be generally satisfactory for pressures above 500 atm. For pressures below 600 atm, gasket closures of the kind shown in Fig. 7.9, which are modifications of a design suggested by Morey and Niggli (1913) are satisfactory and may be readily fitted with noble-metal liners.

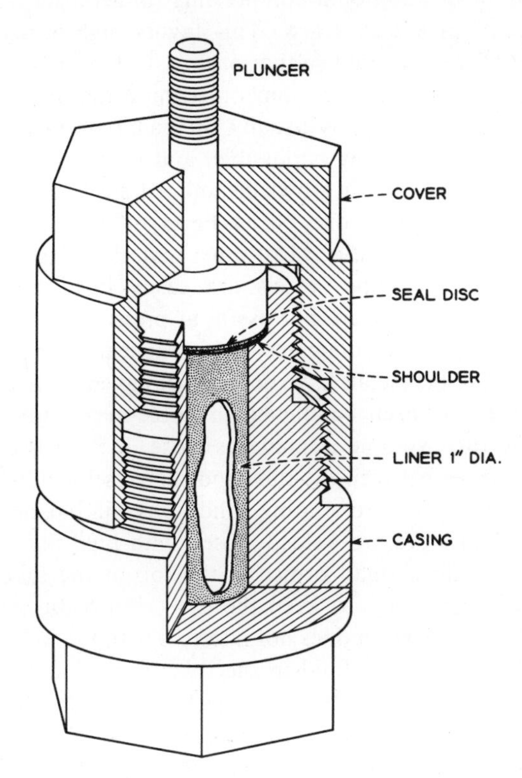

Fig. 7.9 Morey hydrothermal autoclave.

In the modified Bridgman closure the pressure in the vessel is transmitted through the plunger to the seal surfaces, which initially are nearly line contacts. Consequently the pressure in the seal surface greatly exceeds the pressure in the vessel, because most of the area of the piston is unsupported.

As the pressure in the vessel increases, the pressure in the seal surfaces increases more rapidly, and hence the seal is said to be "self-energized." The initial line contact, of course, deforms because of the elastic deformation of the seal ring to a surface contact, but, even in the limiting case where both the sealing surfaces are in complete contact, the seal is still self-energized, because it is unsupported by the amount of the area of the shank of the plunger. Thus the pressure contained is limited by the tensile properties of the autoclave.

In the vessel shown in Fig. 7.9, the seal is not self-energized. Because the internal pressure does not serve to increase the pressure in the seal, there is no excess pressure in the gasket. In contrast, the gasket pressure is generated by the torque applied in closing the vessel and consequently will be exceeded, with resultant leaks, at rather low pressures.

Purity of the grown crystals in hydrothermal growth presents special problems. On one hand, the growth temperature is low and for a polycomponent system the concentration of additional components (mineralizer + unwanted contaminants) except for water is low. This favors high-purity growth. However, the activity of water is high so that proton, OH, and H_2O (King et al., 1960; Laudise and Kolb, 1969) are sometimes impurities in the grown crystal. Prepurification of materials is usually difficult (contrasted with melt-grown crystals where zone refining is possible) and reaction with the vessel can be a problem. Quartz can be grown in unlined steel vessels from NaOH solution because the mineral acmite, $Na_2O \cdot Fe_2O_3 \cdot 4SiO_2$, forms and protectively coats the steel autoclave walls. Most other materials require a noble-metal-lined autoclave. Morey-type vessels are easily lined but, because the seal is not self-energized, they are not useful at pressures much above 600 atms except when the vessel diameter is impractically small. Bridgman-type closures made of noble metals have not been designed, so that for high pressures most growth has been conducted in noble-metal tubes placed in unlined vessels. The tube volume is made as close to the vessel volume as practicable and the space between the tube and the vessel is filled to a degree of fill such that the pressure at operating conditions will balance the pressure within the tube. Tubes are arc-welded closed with final filling by a hypodermic syringe inserted in a small orifice in the top of the tube that is welded last. It is usually necessary to keep the tube chilled during the final welds so that solution volatilization does not interfere with the weld. In general the equation of the state of the fluid in the tube is not known, but most mineralizers and solutes depress the pressure, so that the degree of fill is less than that to give equivalent pressure for pure water [Kennedy's pressure–volume–temperature data (1950) for pure water are of great use in hydrothermal work]. The pressure of the solution in the tube is thus less than that of the water between the tube and the vessel at all temperatures below the operating temperature. Temperature gradients in the system further complicate the situation, but in most instances average temperatures allow satisfactory estimates to be made for pressure-balance experiments. The small volume between the tube and the vessel is advantageous (Monchamp et al., 1968) because the tube need deform only slightly to affect the pressure in that region by a large amount. Thus deformations insufficient to rupture the tube are sufficient to bring the pressures in the tube and between the tube and the vessel into balance.

7.3.2 PHASE EQUILIBRIA AND SOLUBILITY

The first steps in an orderly approach to hydrothermal synthesis of a particular material are phase-equilibria and solubility studies. Although it is true that one can begin using the configurations for growth described above by choosing a likely solvent and conditions and hoping for the best, experience has taught that in most cases an orderly approach ultimately results in faster progress.

Phase equilibria and solubilities are most conveniently studied in commercially available Tuttle-type cold-seal vessels (Roy and Tuttle, 1956; Tem Press Research Company, State College, Pa.). One of these is shown in Fig. 7.10 together with the noble-metal capsule that contains the material and

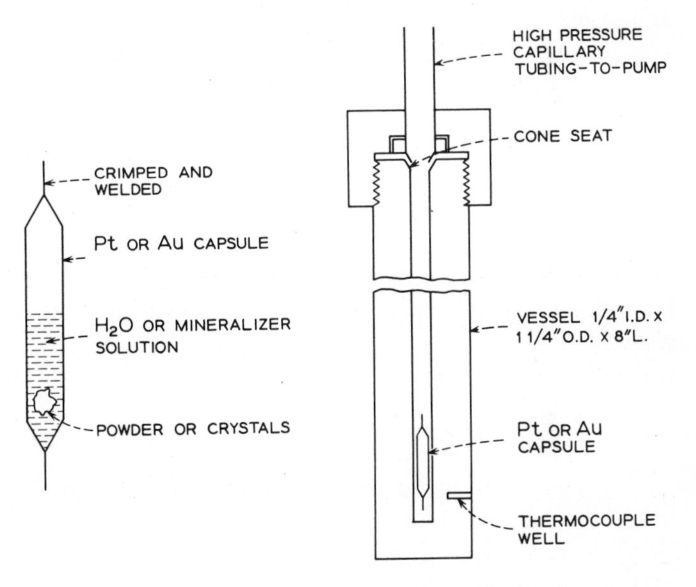

Fig. 7.10 Tuttle autoclave.

solution to be studied. In phase-equilibria studies, the components to be studied are weighed into a capsule that has been made from noble-metal tubing (Pt, Au, or Ag) by crimping and/or welding one end shut. Welding is easily accomplished with a microarc welder (Tem Pres Research Company, State College, Pa.) using a pointed graphite electrode. The desired volume of the solution to be studied (water or water plus an appropriate mineralizer) is added to the capsule with a microburet attached to a hypodermic needle or a hypodermic syringe. The open end of the capsule is then crimped and welded closed. Crimping with a three-jawed chuck, such as is used in a lathe, is advantageous because this keeps the crimped length less than the tube diameter and allows the use of tubes whose diameters are closer to the

ID of the vessel. The vessel is closed and pumped to the desired pressure. The pressure-transmitting fluid is ordinarily water, and the pump is usually an air-driven intensifier (Pressure Products Industries, Hatboro, Pa.). The air pressure to the intensifier (from a tank or laboratory supply line) is controlled with a needle valve and this establishes and holds constant the water pressure. The vessel is then brought to the desired temperature (a tube furnace is arranged so that it can be raised over the vessel) and held there for a time longer then the equilibrium time. The furnace is lowered and the vessel is air-blasted or water-quenched. The capsule contents are identified by X-ray diffraction, petrographic microscopy, or other means. The result is a knowledge of the stable phase at one point in p–T-composition space. The process is repeated to determine a phase diagram. If a weighed single crystal or crystals are placed in the capsule under conditions where the phase is stable and where the fluid phase is liquid alone or gas alone, the weight loss will be a measure of solubility. However, special care must be exercised to avoid overshooting the desired temperature during warm-up, loosing material before weighing, or weighing material that has precipitated during the quench. In many systems the solubility obeys the van't Hoff equation and is linearly dependent on solution density (Barns et al., 1963; Laudise, 1962). The ratio of solubility to mineralizer concentration often gives clues to the species present (Marshall and Laudise, 1966; Laudise and Kolb, 1969). Occasionally $(\partial s/\partial T)p$, the temperature coefficient of solubility, is negative (so-called *retrograde solubility*) requiring growth in a reversed temperature gradient.

7.3.3 KINETICS—QUARTZ

Growth-rate kinetics have been studied (Laudise, 1959) and crystals have been prepared using vessels of the sort shown in Fig. 7.8. Commercial production of quartz (Laudise and Sullivan, 1959) uses vessels as large as 10ft by 10 in. ID of the sort shown in Fig. 7.11. In rate studies the increase in seed thickness/time of the experiment is generally considered to be the rate. For quartz studies the important variables that affect rate have been found to be

1. The nature and concentration of mineralizer.
2. The degree of fill of the vessel at room temperature, d;

$$d = \frac{V_s}{V_f} \tag{7.7}$$

where V_s is the volume of solution at room temperature and V_f is the free volume of the autoclave.

$$V_f = V_0 - \sum V_i \tag{7.8}$$

where V_0 is the internal autoclave volume and $\sum V_i$ is the total of the volumes of other solids in the autoclave (seeds, frame, baffle, nutrient, etc.). The percent fill, f, is

$$f = 100d \tag{7.9}$$

The value of d is approximately equal to the density or reciprocal of the specific volume of the solution in the one-fluid-phase region.

3. Crystallization temperature.
4. The temperature difference (ΔT) between the dissolving and crystallization zones.
5. The percent of open area in the baffle.
6. The seed orientation.
7. The ratio dissolving surface area/growing surface area.

From quartz studies variables 1–7 are found to be interrelated in a manner to give a simple rate equation

$$\mathscr{R}_{hkl} = \alpha k_{hkl} \Delta s \tag{7.10}$$

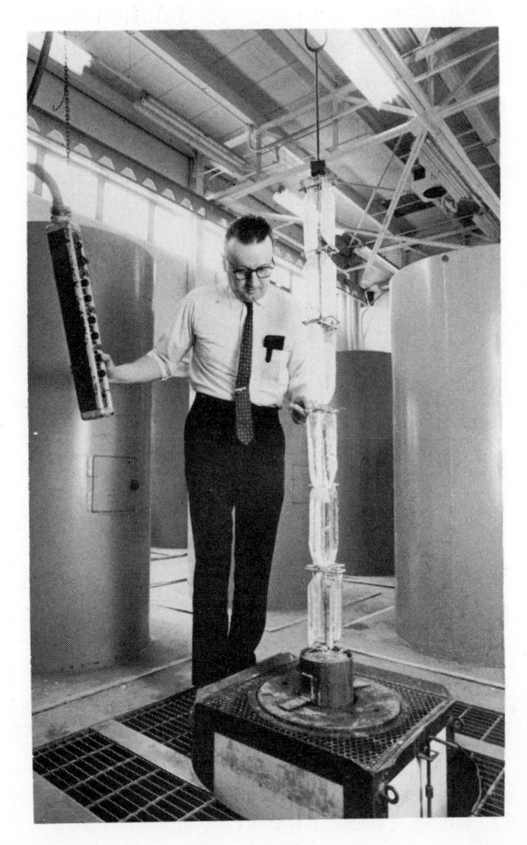

Fig. 7.11 Commercial sized hydrothermal autoclave and grown quartz crystals. (Courtesy of D. L. Rudd, Western Electric Co., Merrimack Valley Works, North Andover, Mass.)

where $\mathscr{R}_{hkl}$ is the rate in a particular crystallographic direction, k_{hkl} is the velocity constant in that direction, Δs is the supersaturation and α is a dimensional conversion constant. The fact that a simple rate equation is obeyed, however, does not establish that the mechanism of growth is simple. By reference to Eq. (7.10) the dependence on variables 1–7 is rather easily explained.

1. The nature and concentration of the mineralizer establishes Δs. In NaOH between about 0.5 and 5 M the mineralizer does not affect k_{hkl}. In mineralizers that are chemically different from $(OH)^-$, it might be expected that k_{hkl} would be mineralizer dependent.
2. The degree of fill affects the solubility and slope of the solubility curve, and thus its effect is mainly on Δs. In NaOH between $d = 0.65$ and 0.87 the effect is only on Δs.
3. The effect of the crystallization temperature is to alter k_{hkl}. In $(OH)^-$ and (CO_3^-), k_{hkl} obeys an Arrhenius dependence, as shown in Fig. 7.12. The energies of activation calculated from the Arrhenius plots are of the order of 20 kcal/mole (some differences are observed for different growth directions; Laudise, 1959). These are rather large ΔE's for a pure diffusion process.
4. ΔT affects only Δs. Thus as Fig. 7.13 shows, $\mathscr{R}_{hkl}$ is linear with ΔT over regions where Δs is linear with ΔT.
5. The percent of open area in the baffle between 2–50% changes the rate as shown in Fig. 7.14. However, when small-heat-capacity thermocouples were used to measure the internal temperatures to determine the effect of

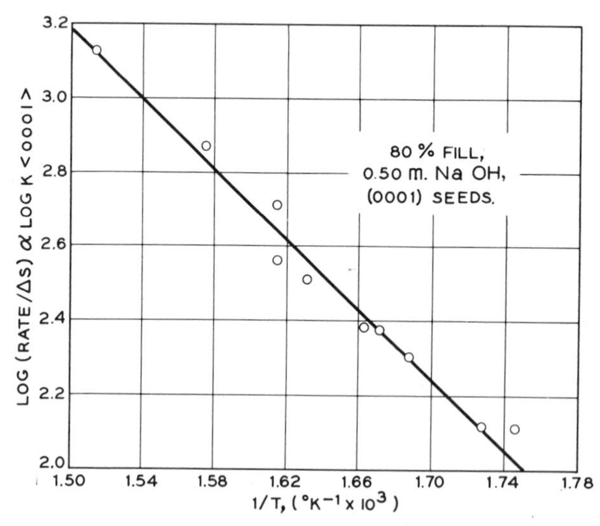

Fig. 7.12 Arrhenius plot—quartz growth (after Laudise, 1959). (Reprinted from journal of the American Chemical Society, **81**, 565 (1959). Copyright 1959 by the American Chemical Society. Reprinted by permission of the copyright owner.)

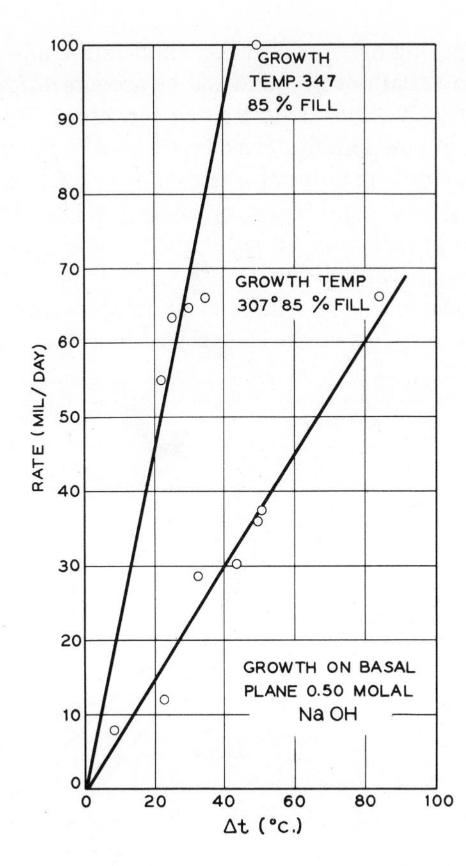

Fig. 7.13 Quartz growth rate vs. ΔT (after Laudise, 1959). (Reprinted from Journal of the American Chemical Society, **81**, 565 (1959). Copyright 1959 by the American Chemical Society. Reprinted by permission of the copyright owner.)

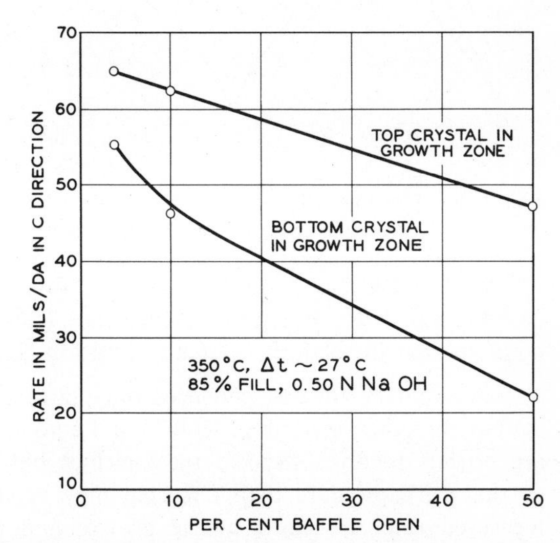

Fig. 7.14 Percent open area in baffle vs. quartz growth rate (Laudise, 1963). (Courtesy of John Wiley & Sons, Inc.)

variations in percent baffle opening on crystallization temperature and ΔT, it was found that the observed variations in rate could be accounted for by changes in T and ΔT (Laudise, 1959). This indicates that convective circulation is rapid enough that it is not rate limiting. The conventional baffle opening used is 5–10%, and its main role is to localize temperature gradients in the baffle region so that the growth and dissolving zones are isothermal. This causes all crystals in the growth zone to grow at the same rate and prevents crystallization in the nutrient zone.

6. Seed orientation as shown in Fig. 7.15 has a great influence in affecting rate through its influence on k_{hkl}, as might be expected in a noncentrosymmetric crystal such as quartz.

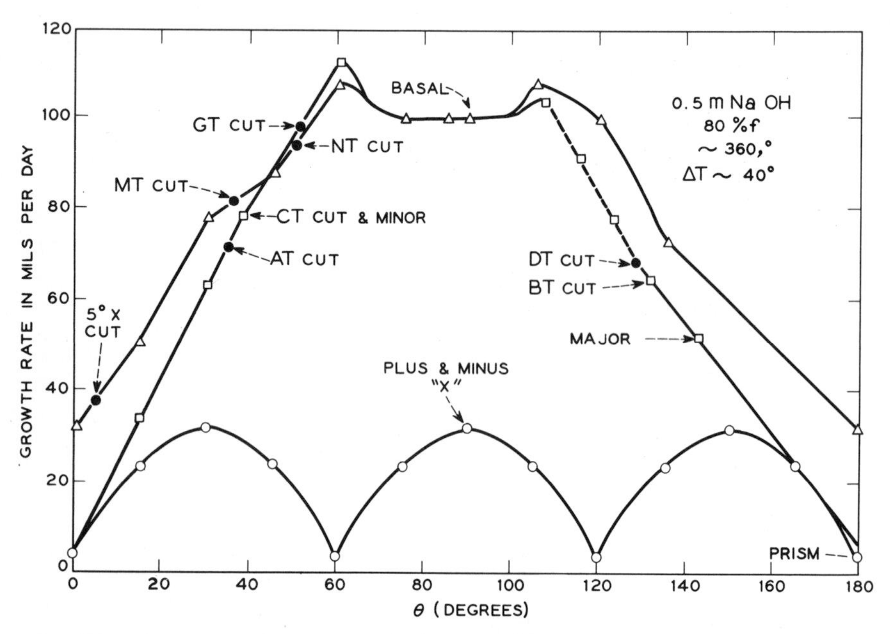

Fig. 7.15 Quartz growth rate vs. seed orientation (after Ballman and Laudise, 1963). (Courtesy of John Wiley & Sons, Inc.)

7. Provided the ratio (surface area dissolving/surface area growing) > 5 the rate does not depend on the ratio. Thus when the dissolving area is large enough dissolving is not rate limiting (Laudise, 1959).

In summary, the best evidence from quartz studies and from more limited studies on other materials (Kolb and Laudise, 1966) is that the rate-limiting process involves steps at the interface but the reaction is consecutive so that perhaps more than one step may be involved in establishing the measured rate. In short, quartz growth and probably all hydrothermal growth is mechanistically similar to aqueous-solution growth at ambient conditions.

7.3.4 PERFECTION

The perfection of hydrothermally grown crystals has not been extensively studied. However, several results in the quartz system that have general implications should be mentioned. In the first place, hydrothermal growth is an ideal method for preparing low-temperature polymorphs. If an undesired polymorph is stable at high temperature, it is sometimes possible to prepare it by some conventional growth techniques (melt growth, vapor growth, flux growth, etc.) and then obtain the desired phase by solid–solid transformation. However, as we have seen in Chapter 4, solid–solid transformations often lead to polytypes, stacking faults, twins, low-angle grain boundaries, and polycrystalline growth. Hydrothermal growth circumvents these problems by directly crystallizing the desired phase. Thus for α-quartz it is an ideal preparation technique. However, growth must take place below about 573°C, the α–β-quartz transition. Growth at higher temperatures would produce other modifications, such as cristobalite and tridymite.

Similarly, cubic ZnS with a low stacking-fault density and absence of polytypism is easily prepared far from the cubic–hexagonal transition (1080°C) by hydrothermal crystallization (Laudise and Ballmen, 1960).

However, hydrates and oxyhydroxides can be stable phases in hydrothermal growth. For instance, sapphire (Al_2O_3) must be crystallized above about 400°C to avoid diaspore (AlOOH). Indeed, proton, OH, or H_2O contaminants from water and $(OH)^-$ mineralizers can be troublesome in hydrothermal materials. In hydrothermal quartz (King et al., 1960) and $Y_3Fe_5O_{12}$ (yttrium–iron garnet, YIG; Kolb et al., 1967b) a strong absorption at about 3 μ corresponding to the OH stretching frequency has been observed. A reduction of the proton concentration (H^+ resides near oxygen in the lattice, giving rise to the OH absorption) can often be accomplished by understanding the charge-compensation mechanism associated with its entry. For instance, in quartz, interstitial protons may

1. Charge-compensate for substitutional Al^{3+}'s at Si^{4+} sites (Al is a contaminant in natural quartz nutrient).
2. Charge-compensate for Na^+'s at Si^{4+} sites (Na^+ comes from the NaOH mineralizer).
3. Charge-compensate for Fe^{2+} or Fe^{3+} at Si^{4+} sites (Fe comes from the steel autoclave).

A reduction in concentration in the system of any of the elements for which H^+ charge-compensates will thus reduce H^+ in the crystal. In the case of quartz it has proven more convenient in commercial procedures to add an ion that directly influences the distribution constant for H^+. Such an ion is Li^+. In the presence of Li^+ and especially $LiNO_2$ the H^+ concentration in quartz is reduced by more than a factor of five (Ballman et al., 1966). The Li^+ concentration in the crystal is not appreciably altered nor is (within the rather poor precision of trace analysis) the concentration of other possible charge-compensating ions. Apparently $LiNO_2$ is adsorbed on the surface of the growing crystal in a manner to block H^+ or OH^- entrance without being included itself. Because the H^+ concentration is proportional to the acoustic loss in the material and because low-acoustic-loss material is important for piezoelectric applications, $LiNO_2$ growth is of considerable practical importance.

In the case of YIG, H^+ appears to be charge compensated by Ca^{2+} (a contaminant in KOH or NaOH mineralizers and in flux-grown seeds) or Na^+ (from the NaOH mineralizer; Kolb et al., 1967b). Growth from a KOH mineralizer and the use of high-purity seeds greatly reduces the H^+ concentration. H^+ broadens the ferromagnetic resonance line width in this material and results in an absorption in the region of interest for laser transmission, so its reduction is of considerable interest.

The homogeneity of hydrothermally grown crystals is influenced by fluctuations in growth rate brought on by fluctuations in growth conditions within the autoclave. In the case of quartz, periodic banding in, for instance, Fe-doped crystals can be observed directly. The bands are parallel to the growth face and are spaced at a distance corresponding approximately to one day's growth. Such fluctuations are probably associated with poor temperature control during daily line voltage and ambient temperature fluctuations. It is interesting to point out that small-heat-capacity thermocouples show a rapid (sec–min) appreciable (0.5–2°) temperature fluctuation in the tops of hydrothermal vessels associated with irregular convection. Whether these fluctuations produce smaller-scale inhomogeneities has not been investigated. Repressing turbulent convection in hydrothermal growth would prove difficult because a large temperature differential from bottom to top of the vessel is required to produce the supersaturation necessary for growth. This differential combined with the large $(\partial\rho/\partial T)_p$ (temperature coefficient of solution density at constant pressure) of most hydrothermal solutions results in rapid convection. This convection is generally irregular and probably often turbulent. Its rapidity allows rapid growth in comparison with most polycomponent systems but does not promote highly uniform growth.

Figure 7.16 shows inhomogeneities in hydrothermal quartz. These bands are caused by Al^{3+} and Na^+ concentration variations. Al + Na produce a color center when exposed to an X-ray beam (analogous to smoky quartz) and the specimen of Fig. 7.16 was X-irradiated.

Fig. 7.16 Inhomogeneities in hydrothermal quartz revealed by x-irradiation. (Courtesy of D. L. Wood.)

Etch pits suggestive of dislocations or of tubelike impurity inclusions have been observed in both natural and synthetic quartz (Nielsen and Foster, 1960).

7.3.5 GROWTH OF OTHER MATERIALS

In addition to quartz, a variety of other materials have been grown hydrothermally. Table 7.3 gives a summary of representative materials.

7.3.5-1 Sapphire

The second material after quartz to be grown in any size hydrothermally was α-corundum (Al_2O_3, sapphire; Laudise and Ballman, 1958). The Al_2O_3–H_2O diagram (Fig. 7.17) was known but was partially redetermined in the presence of the mineralizer NaOH. This work showed that a process above 400°C would be required for corundum growth because the reaction

$$Al_2O_{3\,(s,\ corundum)} + H_2O \rightleftharpoons 2\,AlOOH_{(s,\ diaspore)} \tag{7.11}$$

proceeds to the right below about 400°C. As might be expected, increasing water pressure raises the dehydration temperature of Al_2O_3. This phase boundary is not appreciably altered by the presence of alkaline mineralizers. Later solubility studies (Barns et al., 1963) showed that growth in pure water was impractical because of low solubilities and growth in $(OH)^-$ would be quite slow because the slope of the solubility curve in $(OH)^-$ does not allow a large Δs except at ΔT's so high they would make the dissolving temperature too high to be practicable. The solubility results also showed that previous synthesis in $(CO_3)^{2-}$ mineralizers (Fig. 7.18) had been successful because the slope of the solubility curve was large enough to give a reasonable Δs at moderate ΔT's. Inert conditions are required, so a lined vessel or a noble metal can (Ag, usually) must be used. Ruby may be prepared by including a soluble chromium salt in the system. Dichromate ion was used in early work. Very large-scale growth of sapphire has been undertaken by R. R. Monchamp, R. C. Puttbach, and J. W. Nielsen (Monchamp and Putbach, 1964) with considerable success. Figure 7.19 shows crystals grown

Table 7.3

REPRESENTATIVE HYDROTHERMALLY GROWN CRYSTALS

Material	*Solvent*	*Typical Conditions*		*Fill (%)*	*Rate*	*References*
		Cryst. Temp. (°C)	ΔT (°C)			
SiO_2	1 *M* NaOH	380	50	82	80 mils/day	Laudise and Sullivan, 1959
Al_2O_3	1 *M* K_2CO_3	490	50	80	10 mils/day	Laudise and Ballman, 1958
ZnO	5 *M* KOH	350	10	85	10 mils/day	Kolb and Laudise, 1966
$Y_3Fe_5O_{12}$	20 *M* KOH	350	10	88	5 mils/day	Kolb et al., 1967b
$YbFeO_3$	20 *M* KOH	350	10	88	5 mils/day	Kolb et al., 1968
ZnS	5 *M* NaOH	350	10	85	2 mils/day	Laudise and Ballman, 1960; Laudise et al., 1965

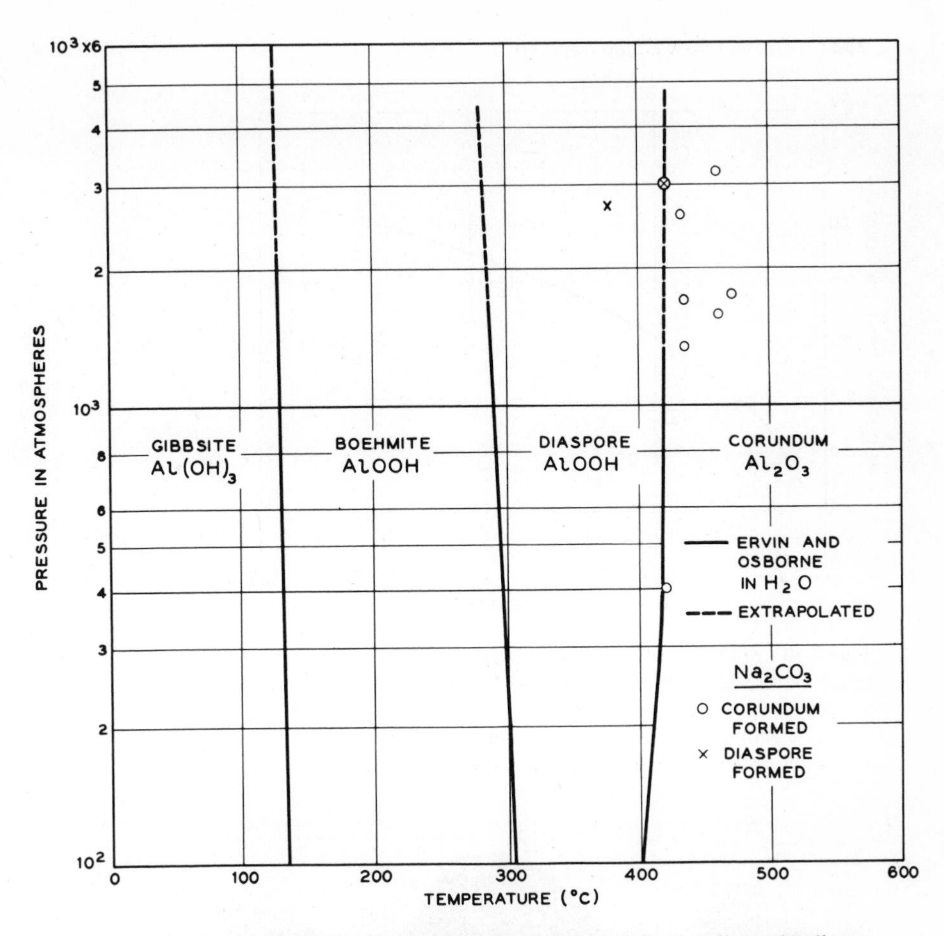

Fig. 7.17 Al_2O_3–H_2O phase diagram (after Laudise and Ballman, 1958). (Reprinted from the Journal of the American Chemical Society, **80**, 2656 (1958). Copyright 1958 by the American Chemical Society. Reprinted by permission of the copyright owner.)

in their laboratory. Bubbles and veils in the grown material seem to present some problems. The details of growth conditions are given in Table 7.3.

7.3.5-2 Yttrium–Iron Garnet

$Y_3Fe_5O_{12}$ (YIG or yttrium–iron garnet) is ordinarily grown from the flux (see Sec. 7.4.3-1) but hydrothermal growth presents some advantages. Such growth is easier to control on a seed and takes place at a lower temperature (where Fe^{2+} formation is not favored) than flux growth. The phase diagram Fe_2O_3–Y_2O_3–H_2O has been investigated with various mineralizers and some conditions where YIG is congruently saturating have been found (Laudise and Kolb, 1962). Growth is considerably simplified if the ratio of

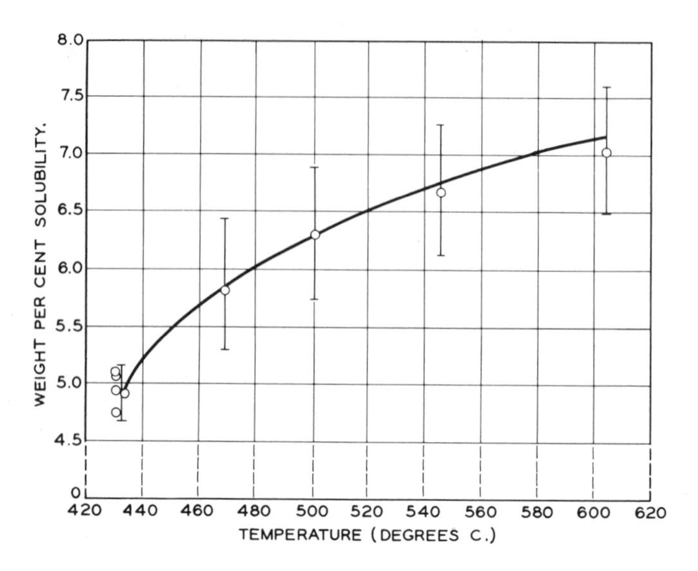

Fig. 7.18 Solubility of sapphire in 3.4 m Na_2CO_3 at 1450 bars (after Barns, Laudise, and Shields, 1963). (Reprinted from the Journal of Physical Chemistry, **67**, 837 (1963). Copyright 1963 by the American Chemical Society. Reprinted by permission of the copyright owner.)

Fig. 7.19 Hydrothermally grown ruby. (Courtesy of R. C. Puttbach and D. La Pore, Airtron Inc., A Division of Litton Industries, Morris Plains, N. J.)

Fe_2O_3/Y_2O_3 in the solution under conditions where YIG crystallizes is the same as in the solid, i.e., $= \frac{5}{3}$. Large high-quality garnets whose magnetic and infrared transmission properties are the equal of the best flux-grown garnet can be grown with comparative ease (see Sec. 7.3.4 for discussion of YIG perfection). Conditions of growth are given in Table 7.3.

7.3.5-3 Zinc Oxide

The common method of ZnO growth is from the vapor phase but this technique usually produces small *c* axis needles or *a* axis plates and controlled growth of larger crystals is difficult. Flux growth produces thin (0001)–$(000\bar{1})$ plates. The largest crystals of ZnO have been produced by hydrothermal growth (Kolb and Laudise, 1966; Monchamp et al., 1967). The ZnO–H_2O phase diagram has been determined and shown to be unaltered in the presence of (OH^-) (Laudise and Ballman, 1960). The solubility of ZnO in a variety of mineralizers has been measured (Laudise and Kolb, 1963) and ZnO has been grown from NaOH, KOH, and NH_4Cl solutions. The highest-perfection crystals result from KOH solutions (Laudise et al., 1964). Dendritic growth on (0001)–$(000\bar{1})$ plates is repressed in the presence of Li^+. This cation is apparently adsorbed onto the (0001) and $(000\bar{1})$ surfaces, reducing the specific surface free energy so that cusps in the Wulff plot occur in $\langle 0001 \rangle$ and $\langle 000\bar{1} \rangle$ rather than the lobes that are present in the absence of Li^+ (Kolb et al., 1967a). ZnO ordinarily grows with excess interstitial Zn and is consequently an n-type semiconductor. Li^+ acts as an acceptor so that by adjusting the Li^+ concentration in the growth solution the Li concentration in the grown crystal can be varied between $\sim$1–20 ppm. Under these conditions, the resistivity varies between 1.5 and 10^9 Ω cm. The piezoelectric properties of hydrothermally grown ZnO are identical to vapor-growth material (Kolb et al., 1967). Table 7.3 lists the growth conditions.

7.4 Molten-Salt Growth

Perhaps the most generally applicable polycomponent method is molten-salt crystallization, because if one searches long enough it should almost always be possible to find a molten-salt solvent for a given crystal. Flux growth has been reviewed in the literature (Laudise, 1963; White, 1965). Molten-salt, flux or fluxed-melt growth thus makes use of the considerable solvent power for refractory crystals that inorganic oxides and salts exhibit above their melting points. The common molten-salt solvents include KF, PbO, PbF_2, B_2O_3, and mixtures of these. The common technique is to dissolve sufficient amounts of the components to form the crystal at a temperature slightly above the saturation temperature and then to slowly cool the crucible (usually Pt). Growth is thus on spontaneously formed nuclei. When the appropriate cooling cycle is complete, it is sometimes possible to remove the crucible from the furnace, pour off excess flux, and mechanically recover the grown crystals. However, it is usually necessary that the frozen flux be dissolved (leached) away from the grown crystals by a solvent that dissolves the flux and not the crystals. Strong mineral acids are often appropriate solvents. It is clear that seeded growth would greatly increase the power and usefulness of flux growth, but, so far, with minor exceptions (Laudise et al., 1962) all growth has been in the absence of deliberately added seeds.

Historically, molten-salt growth was first practiced by French and German chemists and mineralogists about the turn of the century (see, for instance, Hautefeuille and Perrey, 1888). This initial work received impetus from the then newly available electric furnace, but further work was desultory until the desire for refractory crystals in the 1950's caused several researchers, notably J. P. Remeika (1954) to turn again to the technique. Remeika grew $BaTiO_3$ from KF, and since that time the technique has been applied to several hundred materials of solid-state interest. The principal advantages and disadvantages are those associated with polycomponent growth. Impurity control may prove especially difficult if flux components are soluble in the grown crystal. Also, molten salts are often more viscous than most solvents, so diffusion difficulties can be large.

7.4.1 EQUIPMENT

Perhaps the main novel requirement in flux growth is the ready availability of platinumware. Platinum crucibles of 150 cm^3 are suitable for solubility, phase equilibria, and small-scale crystal-growth experiments. But crucibles up to several liters have been used for the growth of, for instance, large garnet crystals. The care of platinumware is well discussed in Wilson (1959). It is especially important to avoid reducing conditions when cations such as Bi^{3+} and Pb^{2+} are present. When such cations are reduced to their metals, the metal alloys with Pt and the resulting alloy has a low melting point, which leads to crucible leakage. Most flux work has been below $\sim$1300°C, so platinum presents no melting-point difficulties. For higher temperatures, Rh or Ir, of course, could be used.

Ordinary resistance-type muffle furnaces are usually used for heating. With SiC heaters such furnaces are suitable up to about $\sim$1400°C. "On-off" controllers or proportional controllers provide $\pm 2°$ control, which is generally satisfactory. Saturable core transformers and more sophisticated control equipment have been used where better control is desired. Furnaces and controllers are discussed in the literature (Paschkis and Persson, 1960) and are available from a variety of suppliers (Hevi Duty Electric Co., Milwaukee, Wis.; Burrell Corp, Pittsburgh, Pa.; Harper Electric Co., Buffalo, N.Y.;

Leeds and Northrup Co., Philadelphia, Pa.). It is clear that cooling rates lower than the precision in temperature control are not profitable. For instance, when temperature varies $\pm 2°$ over an hour period, a cooling rate of 0.5°/hr would be inappropriate. Where particular thermal profiles are desired, ceramic cores wound with kanthal (trade mark, the Kanthal Corp., Stanford, Conn.) or a similar high-resistivity alloy, Pt, Pt–Rh, or Pt–Ir alloys are useful. Suitable taps on the windings allow the thermal gradient to be varied. If the flux is volatile, it is necessary to crimp or weld on the platinum cover. To accelerate equilibrium and even out radial thermal inhomogeneities, it is often useful to rotate the crucible during growth. Equipment for flux growth is illustrated in Figs. 7.20 and 7.21.

7.4.2 PHASE EQUILIBRIUM AND SOLUBILITY

The choice of a solvent for molten-salt growth is perhaps the greatest difficulty the grower faces. The requirements of a good solvent are

1. That the solute is the stable solid phase at growth conditions.
2. Solute solubility of 10–50%.
3. An appreciable temperature coefficient of solubility (~1 wt%/10°) so that slow cooling is practical (isothermal growth in a thermal gradient will relax this requirement, as will growth by flux evaporation).
4. Low volatility (covered or welded crucibles will relax this requirement).
5. Unreactivity with platinum (other crucible materials could relax this requirement, but few other materials have been found practical).
6. Low solvent solubility in the grown crystals (a common ion between crystal and solvent is helpful in repressing solvent contamination).
7. Good-quality growth at reasonable rates in the solvent (usually, growing-interface shape, growth rates, etc., are solvent dependent). Low viscosity is especially important here.

Chemical considerations are often useful in solvent choice so that, for instance

1. Acidic oxides would be good solvents for basic crystals and vice versa provided salts are not the stable solid phase under growth conditions.
2. Complex formers will be useful solvents provided complexes are not solids.
3. Chain-breaking cations are useful in lowering melt viscosity.

Solvent choice is usually guided by intuition and analogy with known systems. A good compilation of phase-equilibria data is especially helpful in this respect (Levin et al., 1964). A discussion of the sort of phase diagrams that will be especially useful in crystal growth is available in the literature (Laudise, 1963).

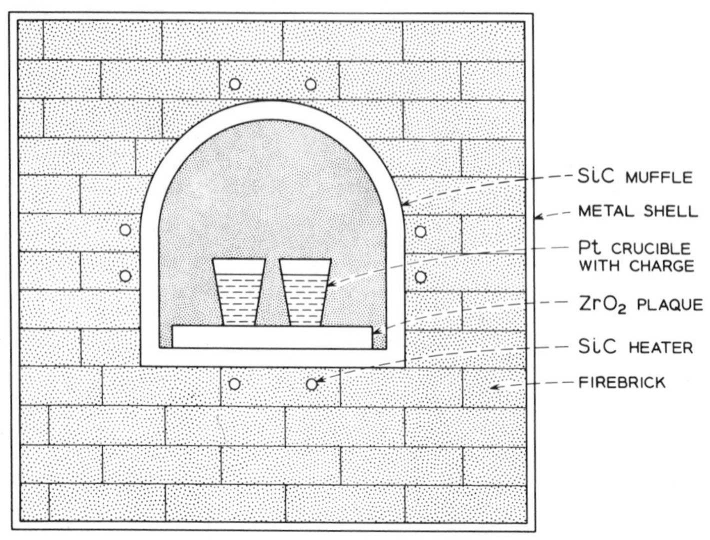

Fig. 7.20 Schematic of small scale flux furnace (after Laudise, 1963). (Courtesy of John Wiley & Sons, Inc.)

Fig. 7.21 Furnace used for the large scale growth of crystals from the flux. (Courtesy of L. G. Van Uitert and W. H. Grodkiewicz.)

The determination of phase diagrams is adequately covered in the literature (Levin et al., 1964), so little discussion is required here. Preliminary information about melting points as a function of composition can often be obtained in a microscope-equipped strip furnace (typical supplier, Tem Pres Research, State College, Pa.) and more sophisticated measurements may be made by differential thermal analysis. Many phase diagrams have been determined by the examination of quenched melts. Solubilities can sometimes be measured by simple weight-loss experiments. Generalizations concerning solubility and systematic solubility studies are rare. Ballman and Laudise (1965) have reported the effect of flux acidity on zircon and phenacite solubility.

7.4.3 GROWTH BY SLOW COOLING—TYPICAL CRYSTALS GROWN

7.4.3-1 Garnet

Perhaps more work has been expended on yttrium–iron garnet than any other flux-grown crystal. $Y_3Fe_5O_{12}$ (YIG) was discovered to be ferromagnetic by Bertaut and Forrat (1956) and by Geller and Gilleo (1957) and the demand for single crystals of this material and its isomorphs stimulated crystal-growth efforts in laboratories throughout the world. As in all flux growth, the first requirement (ideally) is a knowledge of the pertinent phase diagram. The Fe_2O_3–Y_2O_3 phase diagram was first investigated by Nielsen and Dearborn (1958). In a strict sense the system can only be described by the use of three components such as Fe_2O_3–Fe_3O_4–$YFeO_3$ because the reductive reaction

$$3\,Fe_2O_3 \longrightarrow 2\,Fe_3O_4 + \tfrac{1}{2}\,O_2 \tag{7.12}$$

may take place. Van Hook (1961a) investigated the system at several oxygen pressures, and the ternary model of Fig. 7.22 is qualitatively correct. The line of intersection of the 0.21-atm oxygen isobar and the liquidus surface determines the composition of melts in equilibrium with various crystalline phases in air. It should be pointed out that, although the traces of each of the line segments $A'A$, AB, and BB' in the Fe_3O_4–Fe_2O_3–$YFeO_3$ system are a straight line, the line $AA'BB'$ is not straight. The diagram lying along this isobar as determined by Van Hook is shown in Fig. 7.23. It has the appearance of a binary diagram with a "eutectic" at A and a "peritetic" at B. The "eutectic" and "peritetic" points are actually the intersections of the air isobar with the invariant boundary curves of the ternary system. Van Hook has determined the isobar for 1 atm of O_2 pressure and the "isobar" for the O_2 pressure in equilibrium with CO_2. This phase-equilibria work shows that garnet melts incongruently at 1555°C and is only stable in the melt-composition range A–B of Fig. 7.23. The compositions given for A and B in Fig. 7.23 are starting compositions; the true melt compositions are undoubt-

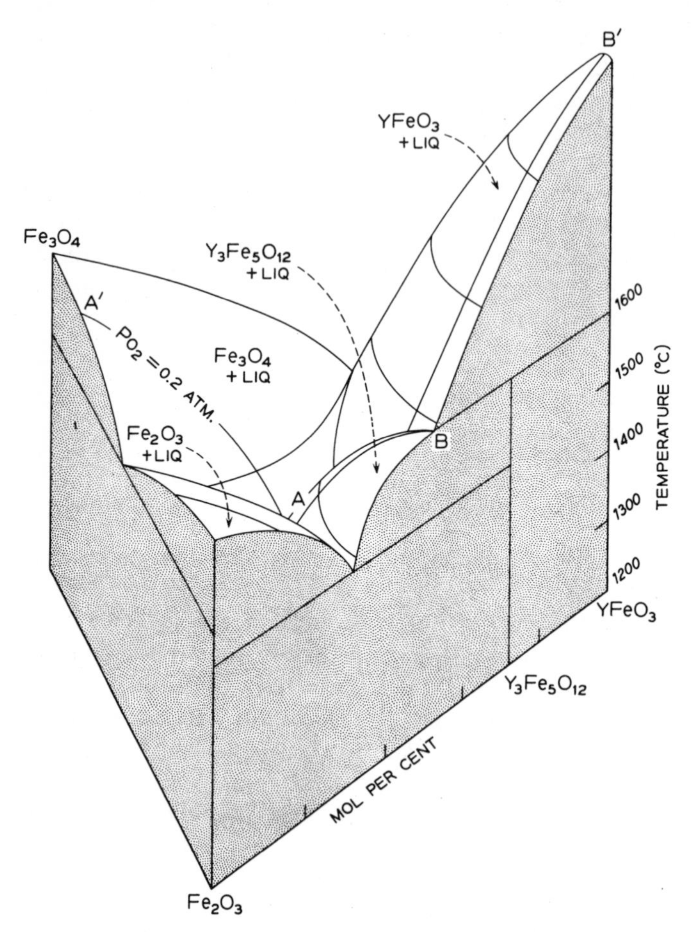

Fig. 7.22 YIG phase diagram (after Van Hook, 1961). (Courtesy of the American Ceramic Society.)

edly oxygen deficient, as is the grown garnet. Consequently, the Fe^{2+} composition in grown garnet is probably a function of the temperature and O_2 pressure. Because of incongruent melting and reduction problems, it was recognized early that flux growth at lower temperatures was an appropriate garnet-growth method. In their first experiments, Nielsen and Dearborn (1958) used molten PbO as a solvent. In PbO, YIG is incongruently saturating between 1350°C and 950°C but when $Fe_2O_3/Y_2O_3 > 5/3$ (the oxide ratio in the solid) YIG can be crystallized. The best YIG crystals were grown with $Fe_2O_3/Y_2O_3 \simeq 12.6$. PbO is very volatile and when Fe_2O_3 is in large excess, the first phases to precipitate are Fe_2O_3 and $PbFe_{12}O_{19}$ (magnetoplumbite), which inhibit YIG growth at the interface. YIG thus nucleates

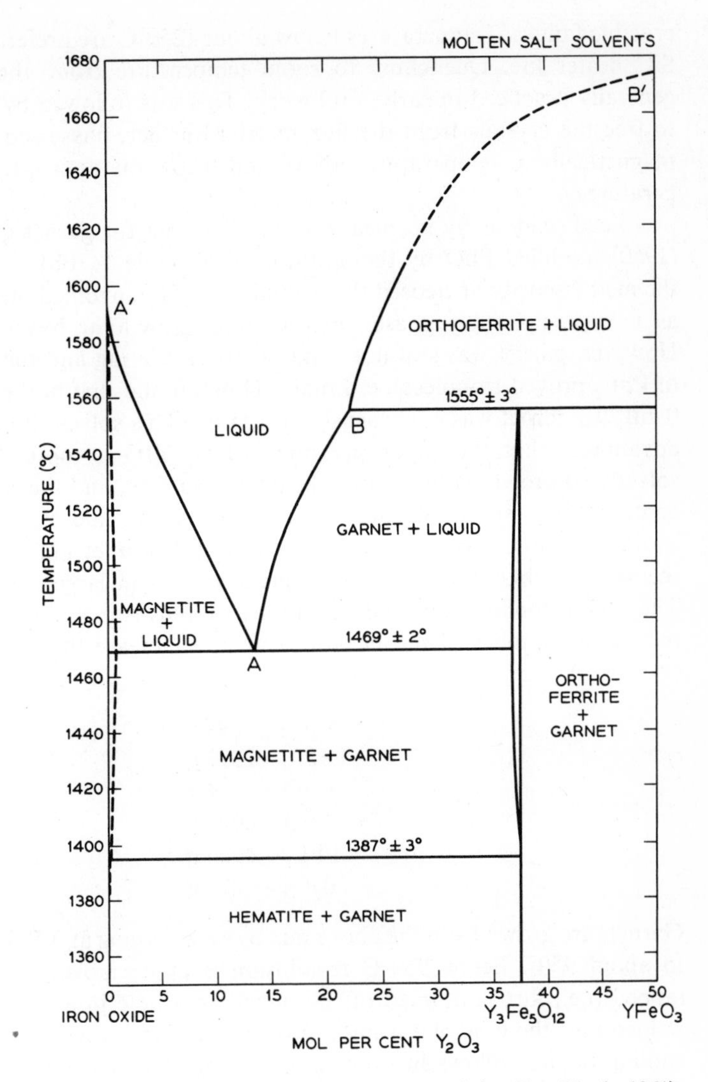

Fig. 7.23 Detail of YIG phase diagram (after Van Hook, 1961). (Courtesy of the American Ceramic Society.)

and grows below the interface under smaller supersaturation conditions, with the result that better-quality and larger crystals are produced. Crimped-on Pt lids repress volatilization (welded lids would probably be better but their use is not general). The temperature should not be much above 1350°C because the reaction

$$PbO \longrightarrow Pb + \tfrac{1}{2}O_2 \qquad (7.13)$$

is appreciable above that temperature and Pb alloys with Pt, with resultant crucible failure. Temperatures below about 1250°C are preferable to increase SiC heater life. Quenching to room temperature from about 900°C was generally practiced in early YIG work. This was followed by a HNO_3 leach to free the crystals from the flux matrix. Further, phase separation is done magnetically at temperatures above and below the appropriate Curie temperatures.

Lead oxide is by no means an ideal solvent for garnet growth. Nielsen (1960) modified PbO by the addition of 57 mole % PbF_2. This decreased the melt viscosity, increased the solubility of YIG, avoided magnetoplumbite as a coprecipitating phase, and allowed growth at lower temperatures. However, garnet was still not congruently saturating and the high volatility of PbF_2 proved troublesome. Linares (1962) found that in the solvent, $BaO{-}0{\cdot}6B_2O_3$, garnet was congruently saturating. This solvent has the additional advantages that the vapor pressure is low, YIG is more dense than the solvent, so growth is more often below the surface, and the solvent does not contain Pb^{2+}, so the hazards of noxious vapors and Pb^{2+} reduction are avoided. Van Uitert and coworkers (Grodkiewicz et al., 1967) have used the solvent $PbO{-}PbF_2{-}B_2O_3$ with great success in large-scale experiments. They have found that small additions of plus-two cations such as Ca^{2+} reduce the number of nuclei that grow and increase the size of the grown crystals. A typical mixture for large-scale garnet growth is

Y_2O_3—1694 g
Fe_2O_3—2397 g
CaO— 4 g
PbO—6021 g
PbF_2—4926 g
B_2O_3— 279 g

Garnets are grown from the above mix by slow cooling at 0.5°/hr from 1300°C to about 950°. Below 950°C re-solution is appreciable, so the crucible is tapped (i.e., flux is drained off by punching a hole in a thin Pt diaphragm welded into the crucible) using a furnace like that shown in Fig. 7.21. Slow cooling in situ follows in order not to strain the grown garnets thermally. Rotation of the crucible in the furnace to accelerate dissolving during the equilibration period and even out thermal asymmetries during growth is helpful. A small negative differential ($\sim$2.5°), arranged so that the bottom of the crucible is coolest, helps to localize growth and increase size. Typical crucibles are 8×10 in. in diam with a 0.060-in. wall thickness. Oxygen flowing through the furnace cools the drain tube, sweeps Pb vapors out of the furnace, provides the negative temperature gradient, and represses Pb^{2+} reduction. Typical crystals grown by this technique are shown in Fig. 7.24.

Similar results have been obtained with $Y_3Al_5O_{12}$: Nd (yttrium–aluminum garnet, YAG, doped with Nd) when the composition is

Y_2O_3— 720 g
Al_2O_3—1220 g
Nd_2O_3— 253 g
PbO—3556 g
PbF_2—4346 g
B_2O_3— 279 g

Perfection studies in flux-grown crystals have not been extensive. Flux inclusions, dendritic growth, hopper growth, and layered growth are common because constitutional supercooling occurs easily in the relatively viscous polycomponent melts, where diffusion is difficult. Large thermal gradients, close to the growing interface, that would repress constitutional supercooling are ordinarily not present in conventional geometries. Flux inclusions (often parallel to the growing faces) are common in YIG and YAG. The best grown material often occurs late in the cooling cycle because at that time the large growing surface helps to reduce supersaturation. A nonuniform cooling cycle with very slow cooling at the beginning to reduce the supersaturation when the growing surface area is small would improve perfection. A lower cooling rate in the beginning would also reduce the number of nuclei and thus increase the average size of grown crystals. Temperature cycling (see Sec. 6.4.1), which results in alternate growth and dissolving, may be useful in reducing the number of nuclei present, because during the heating part

Fig. 7.24 Flux grown yttrium–iron garnet crystals and crucible from which growth took place. (Courtesy of L. G. Van Uitert and W. H. Grodkiewicz.)

of a cycle smaller nuclei will completely dissolve. The result would be a higher yield of large crystals.

In YIG oxidation states of the Fe other than plus three have a strong degrading effect on the magnetic properties and on the transmission at 1.14 μ (Wood and Remeika, 1966). Charge-compensating impurities can affect the oxidation state of the iron in the grown crystal. Plus-two impurities increase the distribution constant for Fe^{4+}

$$Ca^{2+}_{(melt)} + Fe^{4+}_{(melt)} \longrightarrow Ca^{2+}_{(Fe\ site)} + Fe^{4+}_{(Fe\ site)} \tag{7.14}$$

while plus-four impurities increase the distribution constant for Fe^{2+}

$$Si^{4+}_{(melt)} + Fe^{2+}_{(melt)} \longrightarrow Si^{4+}_{(Fe\ site)} + Fe^{2+}_{(Fe\ site)} \tag{7.15}$$

Wood and Remeika (1966) have studied these reactions making use of the 1.14 μ absorption, which has been shown to be a measure of non-plus-three iron. Ca^{2+} and Si^{4+} are common contaminants in starting materials and in muffle furnace environments. Ca^{2+}-doped garnet has been shown to be *p*-type, while Si-doped garnet is *n*-type, as might be expected. The lowest-conductivity, highest-optical-transmission, lowest-ferromagnetic-resonance garnet occurs when 0.006 atom of Si^{4+} is present per $Y_3Fe_5O_{12}$. This Si^{4+} apparently compensates Fe^{4+} produced by uncontrolled traces of Ca^{2+} or by other processes.

7.4.3-2 Barium Titanate

Barium metatitanate in its cubic modification has been known since 1942 (Wanier and Solomon, 1942) as an interesting dielectric material. The desire for large single crystals of the cubic modification of this material for ferroelectric studies was a stimulus for crystal-growth activities. The pertinent phase transformations according to Mertz (1949) are

$$\text{Liquid (congruent melting)} \xrightarrow{1618^\circ C}$$
$$\text{Hexagonal (nonferroelectric)} \xrightarrow{1460^\circ C}$$
$$\text{Cubic perovskite (nonferroelectric)} \xrightarrow{120^\circ C}$$
$$\text{Tetragonal (ferroelectric)} \xrightarrow{5^\circ C}$$
$$\text{Orthorrhombic (ferroelectric)} \xrightarrow{-80^\circ C}$$
$$\text{Rhombohedral (ferroelectric)}$$

Passing through the hexagonal–cubic transition produces strains, twins, and polycrystals, so that a growth method that directly produces the cubic modification is desired. Rase and Roy (1955) have determined the pertinent region of the BaO–TiO_2 phase diagram (Fig. 7.25) and show that the cubic–hexagonal transition temperature is depressed by TiO_2 additions, so that cubic $BaTiO_3$ is the equilibrium phase from about 55 to 69% TiO_2.

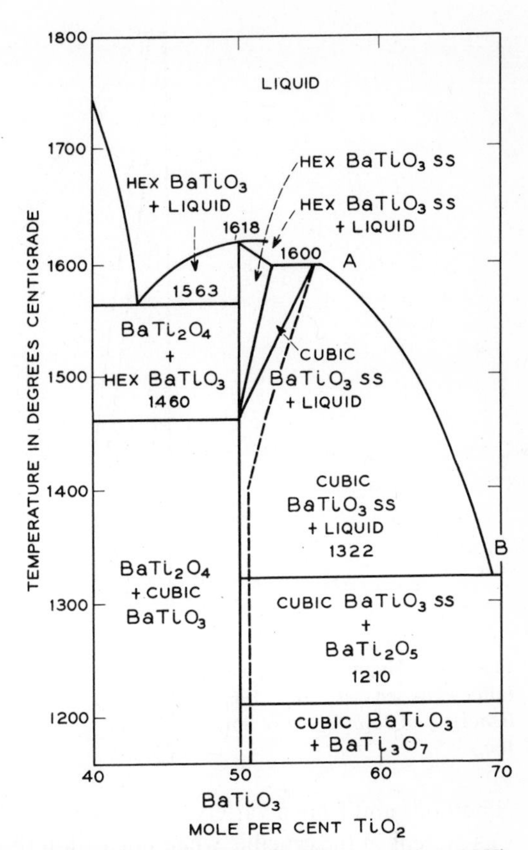

Fig. 7.25 BaO–TiO_2 phase diagram (after Rase and Roy, 1955). (Courtesy of the American Ceramic Society.)

Linz (see Sec. 5.4) has pulled cubic $BaTiO_3$ from melts containing about 5% excess TiO_2 over the $BaTiO_3$ stoichiometry. However, most of the early work on $BaTiO_3$ was performed on solution-grown material prepared according to Remeika's technique (1954), where the flux was KF. Other solvents including $BaCl_2$ and BaF_2 have been used, but KF is preferable because (1) it does not attack Pt excessively; (2) $BaTiO_3$ is more dense than KF, so nucleation at the air–melt interface is lessened; and (3) it is easily leached with water. In the standard Remeika recipe, a KF melt containing 30 wt% $BaTiO_3$–70 wt% KF with a tightly crimped cover, is soaked at from 1150 to 1200°C for about 8 hr. The furnace is then cooled comparatively rapidly (20–50°/hr), to between 900 and 1000°C. The flux is decanted off and the crystals are furnace-annealed by cooling at 10–50°/hr to room temperature. Crystals are mechanically separated from the flux and then hot-water-leached. The typical butterfly twin habit is shown in Fig. 7.26.

Nielsen et al., (1962) found that the twin could form under a variety of

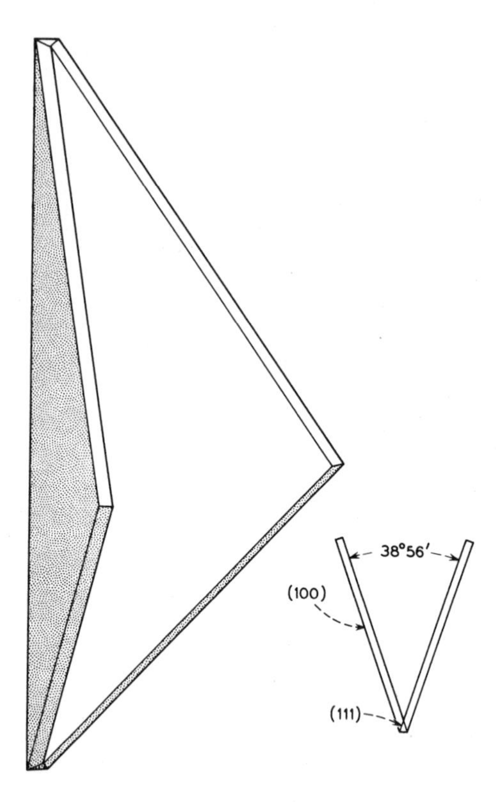

Fig. 7.26 Habit of $BaTiO_3$ (after Nielsen et al., 1962). (Courtesy of John Wiley & Sons, Inc.)

conditions and from a variety of solvents, provided an excess of solid $BaTiO_3$ was present in the crucible when nucleation began. It was also necessary that an appreciable number of particles in this excess be smaller than 10 μ. The recipe of Remeika provides considerable excess solid $BaTiO_3$ at the beginning of the cooling cycle as a result of extensive evaporation during the soak period (see Fig. 7.27). Unless this excess is present, small cubic crystals are the inevitable product. The dashed regions of Fig. 7.27 are speculative† and there probably is some solid solution of other phases in the crystallized $BaTiO_3$, particularly at higher temperatures. DeVries (1961) and DeVries and Sears (1961) have discussed the nucleation and growth of the twin in some detail and point out that the butterfly twin is probably formed when (111) twinning of the (100) habit takes place. Such a twin will have reentrant angles that grow out of existence through the formation of the butterfly habit. It is probable that the original nucleus is a cube that has been cleaved to reveal a (111) plane and unless small enough $BaTiO_3$ particles are present

†Recent evidence (J. W. Nielsen, private communication) indicates Fig. 7.27 is valid only in inert atmospheres. Even in air, slow reactions with crucibles occur.

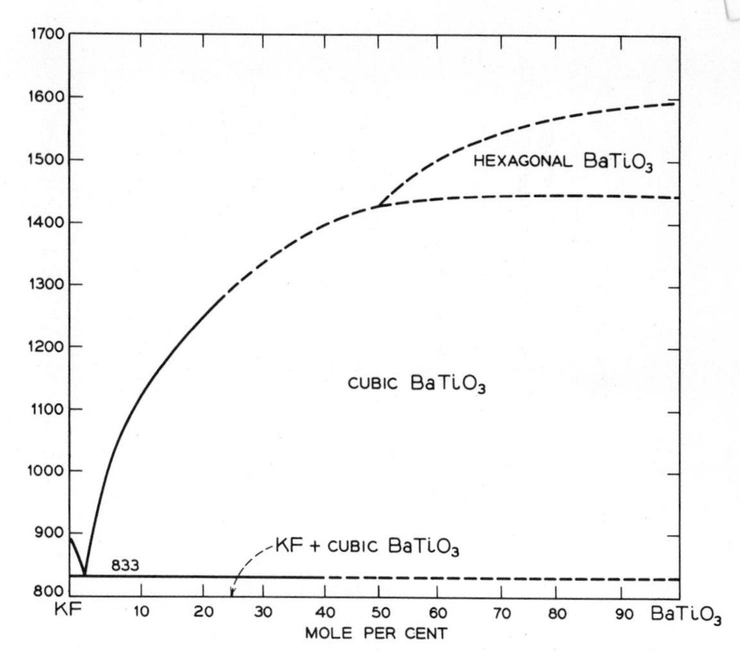

Fig. 7.27 Solubility of $BaTiO_3$ in KF (after Karan and Skinner, 1953; Karan, 1954).

to make the probability of exposure of this (111) plane high, the steps necessary for twin nucleation cannot begin.

Recipes for the growth of a variety of other flux-grown crystals by slow cooling are given in Table 7.4.

7.4.4 GROWTH BY EVAPORATION

Flux evaporation has the advantages of isothermal growth. If a material is stable over a narrow temperature range in a given flux so that slow-cooling is impractical, it can sometimes be grown by evaporation provided the vapor pressure of the flux is high enough. If the material is incongruently volatile after appreciable volatilization, another phase may become stable and will begin to precipitate. If a dopant or impurity has a sharply temperature-dependent distribution constant, isothermal growth may be advantageous, but if the distribution constant is greatly dependent on melt composition, stoichiometry, or dopant or impurity concentration, then evaporative growth may produce less homogeneous crystals.

It is quite probable that some of the growth, for instance, in KF and PbO fluxes is caused by flux evaporation. R. Roy has used flux evaporation of $(Na \cdot K)_2O \cdot 3B_2O_3$, halide, and PbO fluxes to grow TiO_{2-x}, Cr_2O_3, (Al,

Table 7.4

TYPICAL CRYSTALS GROWN FROM MOLTEN-SALT SOLVENTS

Material	*Formula*	*Solvent*	*Method*	*Conditions*	*Remarks*	*References*
Yttrium–iron garnet (YIG)	$Y_3Fe_5O_{12}$	PbO	Slow-cooling	Equilibrate at 1370°C; cool 1–5°/hr.	Crystals up to several cm; magnetoplumbite crystallizes first, can substitute other rare earths for Y; see Secs. 7.3.5–2 and 7.4.5 for other methods.	Nielsen and Dearborn, 1958
Barium titanate	$BaTiO_3$	KF	Evaporation and cooling	Equilibrate at 1200°C; cool 20–40°/hr to ~850°C, pour off flux; leach H_2O.	Some supersaturation partly caused by KF evaporation; butterfly twin habit.	Remeika, 1954
Barium titanate	$BaTiO_3$	TiO_2	Pulled from melt†	$TiO_2/BaO > 1$ in melt.	Several cm growth on seeds.	Von Hippel, 1963
Yttrium–aluminum garnet (YAG)	$Y_3Al_5O_{12}$	$PbO–PbF_2$	Slow-cooling	Equilibrate at 1150°C; cool at 4–5°/hr to 750°C.	Crystals several cm; see Sec 5.4.2–2 for Czochralski growth method.	Lefever et al., 1961
Sapphire or gallia	Al_2O_3 Ga_2O_3	PbF_2	Slow-cooling	Equilibrate at 1200°C; cool to 900°C at 3°/hr.	Plates up to 1 cm; method works for end members; gives β Al_2O_3 structure for solid solutions.	Hart and White, unpublished work reported in White, 1965

†Since the melt composition is different from the solid, this is best considered as seeded growth from the flux (see Sec. 7.4.5).

Table 7.4—*Cont.*

Material	*Formula*	*Solvent*	*Method*	*Conditions*	*Remarks*	*References*
Sodium niobate	$NaNbO_3$	$NaF–Nb_2CO_3$	Slow-cooling	Equilibrate at 1300°C; cool to 200°C at 20°/hr.	Plates 1 × 1 × .3 mm.	Cross and Nicholson, 1955
Beryllia	BeO	$Li_2O–MoO_3$ or PbF_2	Evaporation and slow-cooling; temperature gradient	Varying; see references.	Crystals up to 2 cm.	Austerman, 1963; Kelly, 1963
Magnesium oxide	MgO	PbF_2	Evaporation	Temperature above 1150°C.	Crystals up to 2 cm.	Nielsen and Dearborn, 1958
Sapphire	Al_2O_3	PbF_2	Slow-cooling	Equilibrate at 1400°C; cool to 900°C at 1°/hr.	Plates up to 3 cm.	Reported in White, 1965
Nickel oxide	NiO	PbF_2	Slow-cooling	Equilibrate at 1200°C; cool to 900°C at 3°/hr.	Crystals up to 0.5 mm.	Hart and White, unpublished work reported in White, 1965
"Barium ferr-oxdure"	$BaO \cdot 6Fe_2O_3$	Na_2CO_3	Slow-cooling	Equilibrate at 1375°C; cool to 1100°C at 0.75°/hr.	Crystals up to 1.2 cm.	Gambino and Leonhard, 1961
Beryl	$BeAl_2Si_6O_{16}$	$Li_2O–MoO_3$, B_2O_3, or $PbO–PbF_2$	Slow-cooling	Equilibrate at 975°C; cool to 790°C at 6°/hr.	Seeded growth works.	Lefever et al., 1962

Table 7.4—*Cont.*

Material	*Formula*	*Solvent*	*Method*	*Conditions*	*Remarks*	*References*
Magnesium aluminate	$MgAl_2O_4$	PbF_2	Evaporation	Temperature above 1450°C.	Crystals up to 2 cm.	Nielsen and Dearborn, 1958
Yttrium vanadate	YVO_4	V_2O_5	Slow-cooling	Equilibrate at 1200°C; cool at 3°/hr to 900°C.	Crystals up to 2 mm.	Hart and White, unpublished work reported in White, 1965
Lead zirconate	$PbZrO_3$	PbF_2 or $PbCl_2$	Evaporation	Temperature 1200°	Crystals up to 0.3 mm	Jona et al., 1955
Magnesium ferrite	$MgFe_2O_4$	PbP_2O_7	Slow-cooling	Equilibrate at 1310°C; cool at 4.3°/hr to 900°C.	Crystals up to 3 mm.	Wickham, 1962
Various oxides	HfO_2, TiO_2 ThO_2, CeO_2, $YCrO_3$, and Al_2O_3	PbF_2 or BiF_3+ B_2O_3	Evaporation	Evaporate at 1300°C.	Crystals 1–10 mm in size.	Grodkiewicz and Nitti, 1966
Rare earth ortho-ferrites	$RFeO_3$ R = rare earth	PbO	Slow-cooling	Equilibrate at 1300°C; cool 30°/hr to ~850°C.	"Small" crystals	Remeika, 1956

$Cr)_2O_3$, VO_2, and In_2O_3 by isothermal flux evaporation at temperatures between 1500 and 1100°C (Peiser, 1967).

W. H. Grodkiewicz and D. J. Nitti (1966) have grown HfO_2, TiO_2, ThO_2, CeO_2, $YbCrO_3$, and Al_2O_3 by flux evaporation at 1300°C. The conditions for growth are given in Table 7.4. L. M. Viting (1965) has grown $MgFe_2O_4$ from PbF_2 by flux evaporation.

7.4.5 SEEDED GROWTH

Seeded growth in comparison with spontaneous nucleation can give control over orientation, growth rate, perfection, and doping. In addition, in one of its modifications it can give all the advantages of isothermal growth. In spite of the obvious attractiveness of seeded growth, very few flux-grown crystals have been prepared by the technique. This is probably mainly because of the fact that growth rates are necessarily slow because diffusion in any polycomponent system and particularly in fluxes is not rapid. In uncontrolled random nucleation growth, the rates are probably no faster but, because they are not so easily measured and are generally not known, the experimenter generally is not discouraged so much with their slowness. In addition, seeded growth requires more elaborate equipment than spontaneous nucleation growth, so for this reason it has been less widely exploited.

However, three classes of seeded growth are possible, and the first two have been reduced to practice:

1. Growth by slow-cooling in the presence of a seed.
2. Growth on a seed in the cool part of a system, while excess solute (nutrient) is in contact with the solvent in a hotter part of the system. This method is analogous to hydrothermal growth and is sometimes called thermal gradient or Krüger–Finke growth (Buckley, 1951b).
3. Growth on a seed where supersaturation is caused by solvent evaporation.

The principal advantage of methods 1 and 3 is that no dissolving step is required and the diffusion path can in some cases be shorter. The best crystals have so far been grown by method 1. Method 3 has not, within the author's knowledge, been deliberately attempted. Methods 2 and 3 have the advantage that the growth takes place isothermally. This can allow growth in systems where the desired phase is stable only over narrow temperature ranges, and where the distribution constant for a dopant, impurity, or component (in the case of nonstoichiometric compounds) is temperature dependent. In such cases, methods 2 and 3 will produce more homogeneous crystals. However, when the solvent volatilizes incongruently or when the distribution constant is solvent-concentration dependent, evaporation will not produce crystals of as high homogeneity as a thermal-gradient process.

Thermal-gradient processes are useful where large crystals are desired, because the crystal size is not limited by the total amount of solute in the solvent.

Growth in the presence of a seed has been used by several researchers, including Miller (1958), to grow $KNbO_3$, Reynolds and Guggenheim (1958), to grow a ferrite; Linares (1964), to grow YIG; and Kestigian (1967), to grow YIG. Perhaps the most complete study of the technique reported so far is that of Laudise et al. (1962), who grew YIG from $BaO \cdot 6B_2O_3$. Figure 7.28 shows the solubility of YIG in the flux, and Fig. 7.29 shows the furnace arrangement. Both slow-cooling and thermal-gradient growth were investigated. Rotation was 30 sec clockwise followed by 30 sec counterclockwise. Figure 7.30 shows the effect of rotation rate on YIG growth in $\langle 110 \rangle$ at two different cooling rates. In slow-cooling experiments in the absence of YIG nutrient, when the cooling rate, $d\theta/dt$, was 10°/hr the flat region of *AB*

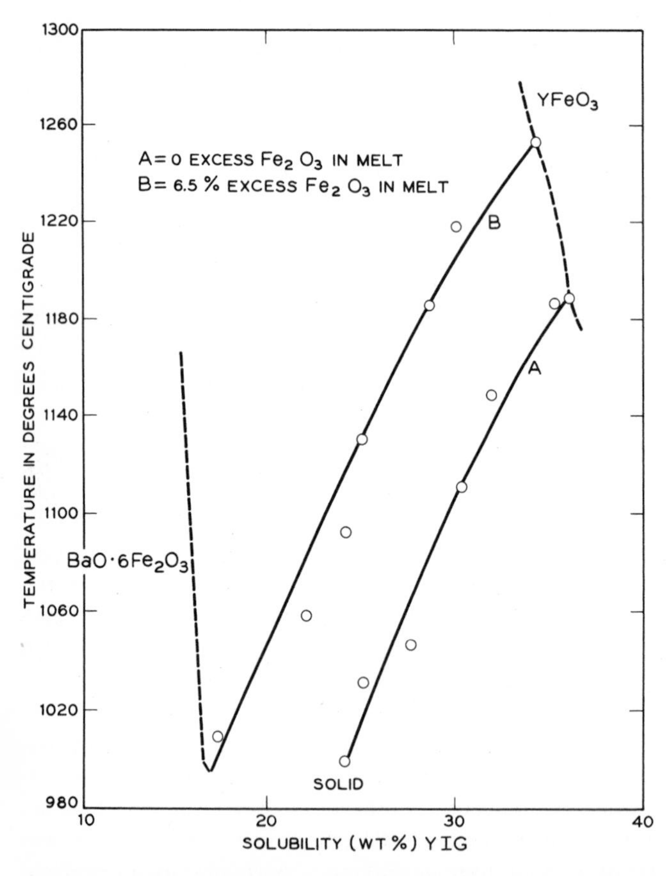

Fig. 7.28 Solubility of YIG in $BaO \cdot 6B_2O_3$ (after Laudise, 1963). (Courtesy of John Wiley & Sons, Inc.)

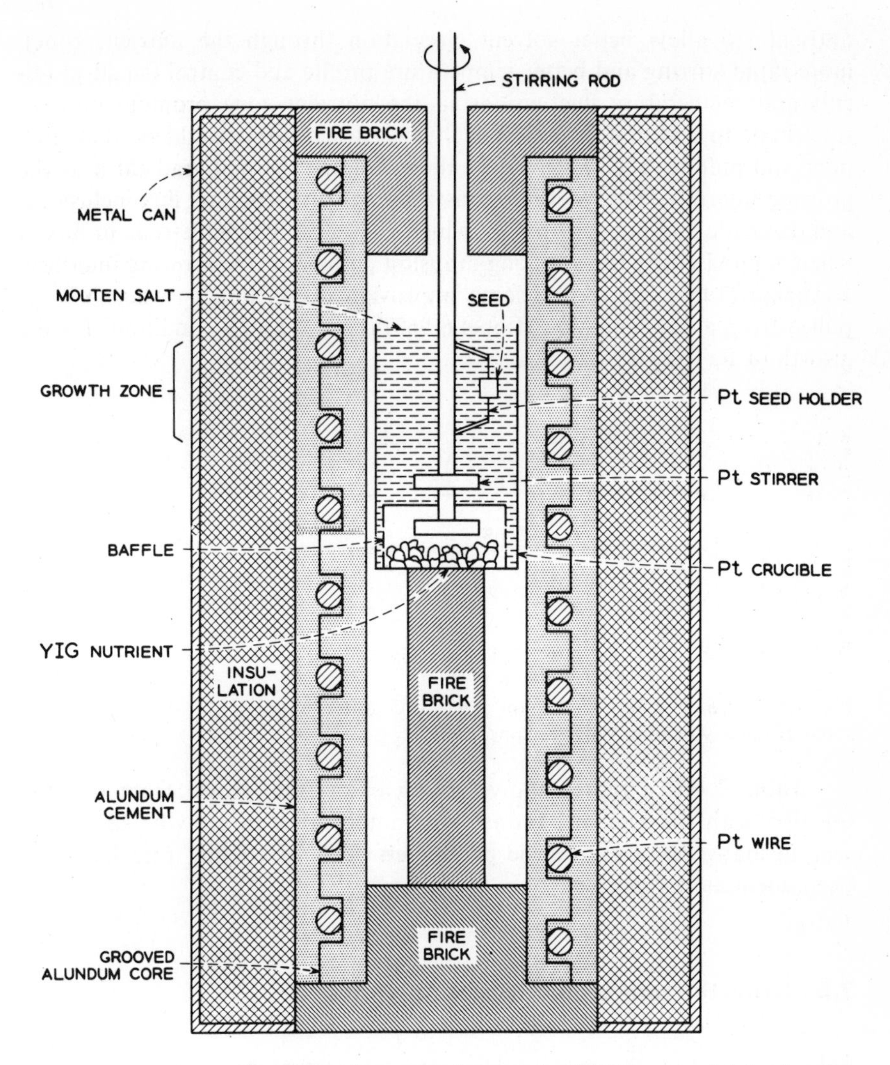

Fig. 7.29 Apparatus for growth of YIG on a seed (after Laudise, 1963). (Courtesy of John Wiley & Sons, Inc.)

began at 200 rpm. The value of A, Fig. 7.30, has been called the critical stirring speed. It increases with $d\theta/dt$ and the solution mass. The greater $d\theta/dt$, the greater ds, the supersaturation, and the more important diffusion becomes. Similarly, the greater the melt mass, the longer the diffusion path becomes, hence the dependence of A on $d\theta/dt$ and melt mass.

In thermal gradient growth, even at rates of stirring as high as 200 rpm, rate tended to fall off during the run. Apparently dissolving of YIG became rate limiting as YIG nutrient sintered together and its surface area decreased. Improvements could be expected with a larger nutrient volume, larger-size nutrient (to allow better solvent circulation through the nutrient zone), more rapid stirring and better temperature profile and control (small gradients and temperature fluctuations in the nutrient zone promote nutrient sintering). In addition, a more effective seed geometry would be that used in crystal pulling. Such a geometry allows a large thermal gradient near the growing interface. This prevents constitutional supercooling, flux inclusions, and dendritic and hopper growth, which is especially troublesome in fluxes, where high viscosity leads to long diffusion paths near the growing interface. Kestigian (1967) has shown some improvements in quality when YIG is pulled from a flux, and Von Hippel (1963) reports Linz's conditions for the growth of $BaTiO_3$ from excess TiO_2.

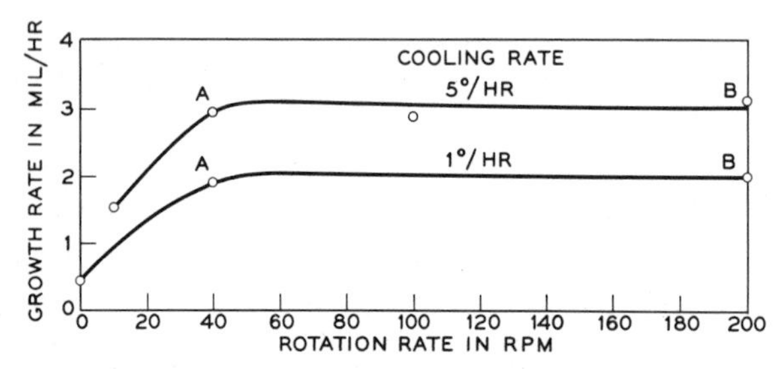

Fig. 7.30 Growth rate vs. rotation rate—YIG (after Laudise, 1963). (Courtesy of John Wiley & Sons, Inc.)

Table 7.4 is a representative listing of crystals grown from the flux together with the known parameters of importance in growth. As can be seen in many cases, even basic parameters such as cooling rate have not been adequately reported.

7.5 Growth from Liquid Metal Solvents

Solution growth with a metal solvent presents advantages and disadvantages similar to other solution-growth processes and a number of metals and semiconductors have been grown from metallic solvents. Molten transition metals as solvents for graphite in diamond crystallization have already been discussed (see Sec. 4.4.5-1). In diamond crystallization the supersaturation is provided because the dissolving phase (graphite) is metastable and thus has a higher solubility than the stable phase (diamond), which crystallizes.

The most widely used materials presently grown from metal solution are probably the III–V semiconductors, especially GaP, where the solvent is often Ga. The growth takes place at low enough temperatures that the volatility of As and P, which cause difficulties in melt growth, is low. The solvent is especially convenient because it does not introduce impurities into the system and it melts at low temperature.

In the case of GaP, one of the first and simplest solution-growth methods was described by Wolff et al. (1954, 1958; see also Miller, 1962) and Ga-grown GaP has been of high importance in electroluminescence studies because it produces pair spectra with many sharp lines (Thomas et al., 1964).

Molten gallium containing 5–10 at.% phosphorus is sealed in an evacuated vitreous-silica ampoule and cooled slowly from temperatures between 1000 and 1275°C at rates of from 3 to 5°/hr down to temperatures 200–300° below the initial growth temperature. In some cases, AlN, BN, and Al_2O_3 crucibles sealed inside the quartz ampoules are used to contain the melt. The system is cooled more rapidly once a temperature of from 900–700°C is reached. The resulting crystals are recovered by a combination of mechanical separation and leaching away of the Ga in concentrated HCl. Grown crystals are predominantly {111} platelets up to ~1 cm in maximum dimension (when growth is from 50-g melts; Thomas et al., 1964).

In recent years the highest electroluminescent efficiencies in GaP have been obtained with Ga-grown crystals where semiconductor diodes are made by *liquid-phase epitaxy* or *tipping*. This method has been used to grow GaAs and GaP (Nelson, 1963; Lorenz and Pilkuhn, 1966; Trumbore et al., 1967; Logan et al., 1967). In the liquid-phase epitaxial growth of GaP, a nearly saturated solution of GaP (Ga solvent) is brought into contact with a GaP seed (substrate) and the temperature is programmed to produce supersaturation and the growth of an appropriately doped, relatively thin layer on the seed. Figure 7.31 illustrates a typical apparatus.

In addition to GaAs and GaP, tipping has been used to grow epitaxial layers of Ge (Nelson, 1963) and yttrium–iron garnet (YIG; Linares, 1968). The usual solvents are

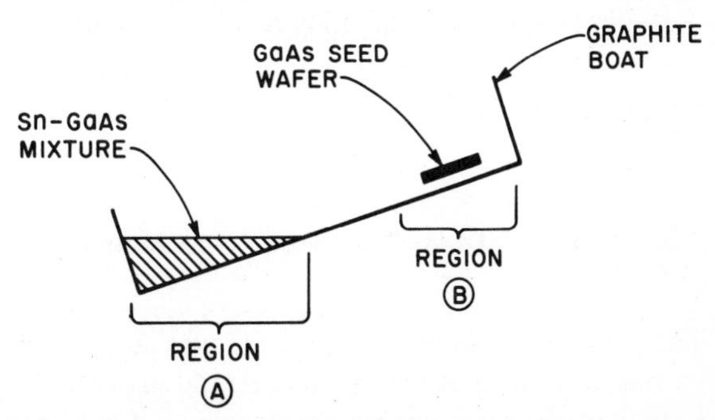

Fig. 7.31 Apparatus for liquid phase epitaxial growth of GaP—"tipping furnace."

GaP–Ga, (+Te ⟶ *n*-type material; +Zn ⟶ *p*)
GaAs–Sn, (+Sn ⟶ *n*-type; +Zn ⟶ *p*)
Ge–In, (+Ga ⟶ *p*-type)
YIG($Y_3Fe_5O_{12}$)–PbO, PbF_2, and other fluxes used for garnet. The substrate is yttrium–aluminum garnet.

Common boat materials are graphite for Ge, BN for GaP, and Pt for YIG.

For illustrative purposes, a GaAs run will be described. At the start of a run, region *A* of Fig. 7.31 is lower than region *B* and a Sn–GaAs mixture is placed at *A*, while a GaAs seed wafer is fastened at *B*. The graphite boat is fixed at the center of a constant-temperature zone in an appropriately designed furnace. The boat is heated to about 640°C and held for a period to dissolve GaAs in the Sn. Then the boat is tipped so that *B* is lower than *A*. It is probably important that, when contact is made with the substrate, the solution be slightly undersaturated. Seed dissolution removes surface damage and under some conditions helps to ensure that the growing interface starts off planar.† The furnace is cooled from ~640 to 500°C in about 30 min, the solution is decanted off, and the substrate with its doped epitaxial layer is recovered.

Appropriate heating cycles for other materials are as follows (Nelson, 1963):

GaAs from Ga—contact seed at 900°C, cool to 400°C in 30-min, decant
Ge from In—contact seed at 500°C, cool to 400°C in 30 min
Ge from Sn-Pb—contact seed at 400°C, cool to 300°C in 10 min
GaP from Ga—contact seed at ~1100°C, cool to 700°C in 40 min (Lorenz and Pilkuhn, 1966)

It might be expected that liquid-phase epitaxy will find increasing use throughout solvent growth. A logical extension of liquid metal solvent growth is to temperature-gradient zone melting, which is discussed in Sec. 7.6. Smakula (1962) lists a number of other materials grown from liquid metal solvents.

7.6 Temperature-Gradient Zone Melting

Temperature-gradient zone melting was suggested by Pfann and first applied by him to both purification and crystal growth (Pfann, 1955; 1966). It essentially consists of moving a narrow zone of solvent through a material by the

†The solvent should ideally be a polish etch for the substrate.

use of a temperature gradient. Figure 7.32 shows a schematic of temperature-gradient zone melting together with an appropriate phase diagram. Consider the phase diagram AB of Fig. 7.32, where A, the solute, is the crystal to be grown, and B is the solvent. If we sandwich a thin layer of solid B between blocks of pure A and place the whole in a temperature gradient where the temperature of the layer is above the lowest melting point in the system, and the hottest temperature is below the melting point of A, the layer will dissolve some A and expand vertically in length. Now if $T_1 > T_2 > T_3 > T_4$, the zone will move toward T_1. When it is at the position T_2–T_3, the cool interface will dissolve enough A to correspond to the equilibrium liquidus concentration at C_3. Dissolving of A at the hot interface will continue until C_2 is reached ($C_2 > C_3$). The concentration gradient will cause A to diffuse toward the cool interface, where the solution will be supersaturated (supercooled) and A, whose concentration of B is kC_3, will freeze (where k is the distribution constant). Continuation of this solution–diffusion–freezing process causes the molten layer to move through the block.

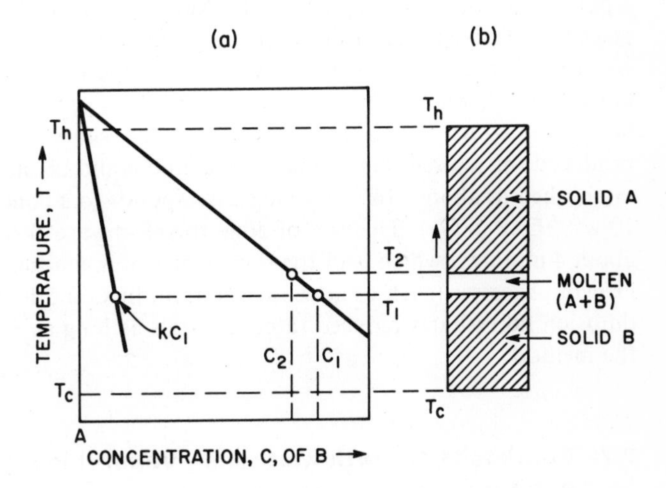

Fig. 7.32 Thermal gradient zone melting (after Pfann, 1966).

Pfann (1966) analyzed the factors that influence zone length, shape, and the composition changes to be expected down a frozen ingot. For purposes of crystal growth it is often important to grow crystals of low impurity content. Thus systems where the solid solubility of the solvent is low may be advantageous. Homogeneity of impurities is often desirable. If the gradient is uniformly imposed from one end of the ingot, as in Fig. 7.32, the zone advances by compositional changes that move it along the liquidus from T_4 to T_1. Hence the compositional range in the solid will be from kC_4 to kC_1. A smaller gradient will give a more homogeneous crystal but a slower rate of zone movement. A zone can be held at essentially constant temperature during its entire travel by heating it with a ring-form heater in the conventional zone-refining configuration and moving the heater [and the zone containing (A + B)] along the ingot. This will provide a high gradient near the melting and freezing interfaces for rapid growth and may produce crystals of rather high homogeneity. This is a variant of zone leveling. Among the single crystals grown by temperature-gradient zone melting are GaAs using gallium as solvent (Mlavsky and Weinstein, 1963), αSiC using Cr as solvent (Griffiths and Mlavsky, 1964), germanium using Pb as solvent (Pfann, 1966), and GaP using GaAs as solvent (Broder and Wolff, 1963; Plaskett et al., 1967).

Perhaps typical of the conditions and results obtained is the work of Plaskett et al., (1967) on GaP. They moved a (Ga + GaP) zone through a GaP ingot by means of a travelling heater. Their procedure was a refinement of that used by Broder and Wolff (1963) and has sometimes been called the *travelling-solvent method.* A small seed of GaP was placed in the bottom of a pointed-tip BN cylindrical crucible. Above it a $\frac{1}{4}$-in.-thick layer of Ga was placed, and above that a dense polycrystalline ingot of GaP. The crucible was placed in an evacuated, sealed quartz ampoule and inserted in a resistance furnace, which helped to control the slope of the profile. The furnace had a single-turn rf coil that provided a local sharp temperature gradient, produced the molten (Ga + GaP) zone, and could be moved. The zone was established at about 1160°C (which corresponds to a concentration of about 10 wt% GaP in Ga). The rate of zone travel upward through the ingot was about 4 mm/day, which is of the same order of magnitude as other solution-growth processes (such as aqueous solution, flux, and hydrothermal), where diffusion necessitates reduced rates. Crystals as long as 5 cm were grown by the method.

7.7 Composite Growth Methods—Vapor–Liquid–Solid Growth

In polycomponent growth the equilibrium may involve more than two phases. Vapor–liquid–solid (V–L–S) growth, which was described by Wagner and Ellis (1965), is the only known example but mechanisms such as liquid–liquid–solid (where two liquids are immiscible) and others are conceivable. Vapor–liquid–solid growth is illustrated by the schematic of Fig. 7.33, which describes the method used in Si growth. Deposition from the vapor is by iodide disproportionation (see Sec. 6.4.2-2):

$$2\,SiI_{2(g)} \rightleftharpoons SiI_{4(g)} + Si_{(soln)} \qquad (7.16)$$

or by the hydrogen reduction of $SiCl_4$ (see Sec. 6.4.2-2):

$$SiCl_{4(g)} + 2\,H_{2(g)} \rightleftharpoons 4\,HCl_{(g)} + Si_{(soln)} \tag{7.17}$$

The liquid phase is a solution of Si in Au. In the process of growth, the vapor-phase reaction does not directly form solid Si, instead it forms Si in liquid solution. The solid Si is formed by growth from the liquid solution:

$$Si_{(soln)} \rightleftharpoons Si_{(s)} \tag{7.18}$$

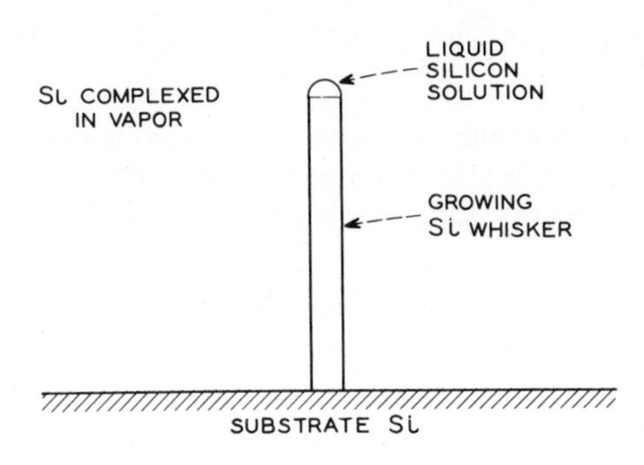

Fig. 7.33 V–L–S growth.

Conditions for vapor deposition (flow rate, reactant rates, temperature, etc.) are generally similar to those used in vapor deposition with the same reactions (see Sec. 6.4.2-2). For instance, for $SiCl_4$ reduction the H_2 flow rate was typically 1000 cm^3/min and the $SiCl_4/H_2$ mole ratio was 0.02. Liquid drop temperature was between 875 and 1200°C, which corresponds to a Si solubility in gold of between 55 and 70 at.% Si (Wagner and Doherty, 1967). The energy of activation for V–L–S growth was 7.5 kcal/mole as compared to about 31 kcal/mole for vapor growth on a Si substrate. The rate was about five times greater than vapor–solid growth (about 10 μ/min) under similar conditions. Apparently a liquid surface is ideally atomically rough in comparison to a solid single-crystal surface, so that the vapor–liquid reaction proceeds much easier than the analogous vapor–solid reaction. Deposition from the liquid (the liquid–solid reaction) is aided by the fact that a large thermal gradient at the growing interface is relatively easy to obtain and represses constitutional supercooling. The initial growth mechanism is somewhat obscure but growth is easy to initiate. Typically a small particle of gold is placed on a clean {111} surface of a single-crystal Si wafer and heated to 950°C, where it forms a small liquid Au–Si alloy droplet. Then the desired vapors are introduced and growth of the whisker begins. The alloy droplet rides atop the whisker and under certain conditions morphologies more complicated than a single whisker can be formed. A small concentration of Au remains behind in the grown whisker, and eventually, if the whisker is long enough, all the Au will be consumed.

Other materials besides Si, including Ge, GaAs, GaP, and Ga(AsP), have been grown by the technique (Barns and Ellis, 1965: Wolfe et al., 1965) but so far crystals large in three dimensions have not been grown.

Wagner (1969) has written an extensive review of mechanism and materials grown. An appropriate component to form the liquid drop should

1. Be liquid at a convenient temperature and convenient concentration of solute
2. Wet the substrate
3. Not be included at high concentration in the grown crystal
4. Not react with the vapor-phase components except to take up the solute

In Ge, Au has been used as a solvent and Ga has been used for GaAs growth. It appears that V–L–S growth may be the operative mechanism in several other systems, including α-Al_2O_3, SiC, and BeO, where the morphology of the whisker includes a rounded tip. The grown crystals are dislocation-free (in contrast to the single spiral dislocation found in normal whiskers). The mechanism thus accounts for the absence of Eshelby twist (Eshelby, 1953) in some whiskers. Indeed, as more V–L–S investigations proceed, the evidence that most whisker growth is by V–L–S and the spiral-dislocation mechanism is a rarity grows stronger (Wagner, 1969).

The potential applications of V–L–S growth are so far untapped. Microcircuits could be prepared by a suitable initial deposition of, for instance, Au on Si followed by V–L–S growth. Altering vapor-phase composition has been used to grow semiconductor junctions.

REFERENCES

Austerman, S. B., International Conference on BeO, Newport, Sydney, Australia, 1963, Paper No. 54.

Ballman, A. A., and R. A. Laudise, in *The Art and Science of Growing Crystals*, Ed. by J. J. Gilman, Wiley, New York, 1963, p. 231.

Ballman, A. A., and R. A. Laudise, *J. Am. Ceram. Soc.* **48**, 130 (1965).

Ballman, A. A., R. A. Laudise, and D. W. Rudd, *Appl. Phys. Letters* **8**, 53 (1966).

Barns, R. L., and W. C. Ellis, *J. Appl. Phys.* **36**, 2296 (1965).

Barns, R. L., R. A. Laudise, and R. M. Shields, *J. Phys. Chem.* **67**, 835 (1963).

Bertaut, F., and F. Forrat, *Compt. Rend. Acad. Sci.* (*Paris*) **242**, 382 (1956).

Bradford, S. C., *Colloid Chemistry*, Ed. by J. Alexander, Reinhold, New York, 1926, p. 790.

Bridgman, P. W., *The Physics of High Pressure*, Bell, London, 1949.

Broder, J. D., and G. A. Wolff, *J. Electrochem. Soc.* **110**, 1150 (1963).

Buckley, H. E., *Crystal Growth*, Wiley, New York, 1951a, pp. 43–67.

Buckley, H. E., *Crystal Growth*, Wiley, New York, 1951b, p. 49.

Cross, L. E., and B. J. Nicholson, *Phil. Mag.* **46**, 453 (1955).

Damojanovic, A., *Plating* **52**, 1017 (1965).

Dennis, J., and H. K. Henisch, *J. Electrochem. Soc.* **114**, 263 (1967).

de Senarmont, H., *Ann. Chim. Phys.* **32**, 129 (1851).

De Vries, R. C., *J. Am. Ceram. Soc.* **42**, 457 (1961).

De Vries, R. C., and G. W. Sears, *J. Chem. Phys.* **34**, 618 (1961).

Ellis, W. C., and D. M. Boulin, private communication.

Eshelby, J. D., *J. Appl. Phys.* **24**, 176 (1953).

Gambino, R. J., and F. J. Leonhard, *J. Am. Ceram. Soc.* **44**, 221 (1961).

Geller, S., and M. A. Gilleo, *Acta Cryst.* **10**, 239 (1957).

Gibbs, W. E., and W. Clayton, *Nature* **113**, 492 (1924).

Griffiths, L. B., and A. I. Mlavsky, *J. Electrochem. Soc.* **111**, 805 (1964).

Grodkiewicz, W. H., and D. J. Nitti, *J. Am. Ceram. Soc.* **49**, 576 (1966).

Grodkiewicz, W. H., E. F. Dearborn, and L. G. Van Uitert, *Crystal Growth*, Ed. by H. S. Peiser, Pergamon, New York, 1967, p. 441.

Harrison, J. A., S. K. Rangarajon, and H. R. Thirsk, *J. Electrochem. Soc.* **113**, 1120 (1966).

Hautefeuille, P., and A. Perrey, *Compt. Rend.* **106**, 1800 (1888).

Henisch, H. K., J. Dennis, and J. I. Hanoka, *J. Phys. Chem. Solids* **26**, 493 (1965a).

Henisch, H. K., J. I. Hanoka, and J. Dennis, *J. Electrochem. Soc.* **112**, 627 (1965b).

Holden, A. N., *Discussions Faraday Soc.* **5**, 312 (1949).

Holden, Alan, and Phylis Singer, *Crystals and Crystal Growing*, Anchor Books-Doubleday, New York, 1960.

Holden, Alan, and R. H. Thompson, *Growing Crystals with a Rotary Crystallizer*, Bell Telephone Laboratories, New York, 1964.

Holmes, H. N., *Colloid Chemistry*, Ed. by J. Alexander, Chemical Catalog Co., New York, 1926, p. 796.

Jona, F., G. Shirane, and R. Pepinsky, *Phys. Rev.* **97**, 1584 (1955).

Karan, C., *J. Chem. Phys.* **22**, 957 (1954).

Karan, C., and B. J. Skinner, *J. Chem. Phys.* **21**, 2225 (1953).

Kelly, J. W., International Conference on BeO, Newport, Sydney, Australia, 1963, Paper No. 56.

Kennedy, G. C., *Am. J. Sci.* **248**, 540 (1950).

Kestigian, M., *J. Am. Ceram. Soc.* **50**, 165 (1967).

King, J. C., D. L. Wood, and D. M. Dodd, *Phys. Rev. Letters* **4**, 500 (1960).

Kohman, G. T., *Bell Lab. Record* **28**, 13 (1950).

Kolb, E. D., *The Physics of Selenium and Tellurium*, Pergammon Press, New York, 1969, pp. 155ff.

Kolb, E. D., and R. A. Laudise, *J. Am. Ceram. Soc.* **49**, 302 (1966), and references therein.

Kolb, E. D., A. S. Coriell, R. A. Laudise, and A. R. Hutson, *Mater. Res. Bull.* **2**, 1099 (1967a).

Kolb, E. D., D. L. Wood, E. G. Spencer, and R. A. Laudise, *J. Appl. Phys.* **38**, 1027 (1967b), and references therein.

Kolb, E. D., D. L. Wood, and R. A. Laudise, *J. Appl. Phys.*, **39**, 1362 (1968).

Laudise, R. A., *J. Am. Chem. Soc.* **81**, 562 (1959).

Laudise, R. A., in *Progress in Inorganic Chemistry*, Vol. III, Ed. by F. A. Cotton, Wiley-Interscience, New York, 1962, p. 1.

Laudise, R. A., in *The Art and Science of Growing Crystals*, Ed. by J. J. Gilman, Wiley, New York, 1963, pp. 252ff, 261.

Laudise, R. A., in "Crystal Growth," Ed. by H. S. Peiser, Pergamon, New York, 1967, p. 3., and references therein.

Laudise, R. A., and A. A. Ballman, *J. Am. Chem. Soc.* **80**, 2655 (1958).

Laudise, R. A., and A. A. Ballman, *J. Phys. Chem.* **64**, 688 (1960).

Laudise, R. A., and E. D. Kolb, *J. Am. Ceram. Soc.* **45**, 51 (1962).

Laudise, R. A., and E. D. Kolb, *Am. Mineralogist* **48**, 642 (1963).

Laudise, R. A., and E. D. Kolb, *Endeavour* **28** (105), 114 (1969).

Laudise, R. A., E. D. Kolb, and A. J. Caporaso, *J. Am. Ceram. Soc.* **47**, 9 (1964).

Laudise, R. A., and J. W. Nielsen in *Solid State Physics*, Vol. XII, Ed. by F. Seitz and D. Turnbull, Academic Press, New York, 1961.

Laudise, R. A., and R. A. Sullivan, *Chem. Eng. Progr.* **55**, 55 (1959), and references therein.

Laudise, R. A., R. C. Linares, and E. F. Dearborn, *J. Appl. Phys. Suppl.* **33**, 1362 (1962).

Laudise, R. A., E. D. Kolb, and J. P. DeNeufville, *Am. Mineralogist* **50**, 382 (1965).

Lefever, R. A., J. W. Torpy, and A. B. Chase, *J. Appl. Phys.* **32**, 962 (1961).

Lefever, R. A., A. B. Chase, and L. E. Sodon, *Am. Mineralogist* **47**, 1450 (1962).

Levin, E. M., C. R. Robbins, and H. F. McMurdie, *Phase Diagrams for Ceramists*, American Ceramic Society, Columbus, Ohio, 1964.

Liesegang, R. E., *Z. Physik. Chem. (Leipzig)* **88**, 1 (1914).

Linares, R. C., *J. Am. Ceram. Soc.* **45**, 307 (1962).

Linares, R. C., *J. Appl. Phys.* **35**, 433 (1964).

Linares, R. C., *J. Cryst. Growth* **3**, **4**, 443 (1968).

Linke, W. F., *Solubilities of Inorganic and Metal Organic Compounds* (Revision of Seidell's *Solubilities*), Vol. II, Van Nostrand, Princeton, N. J., 4th ed., 1965.

Logan, R. A., H. G. White, and F. A. Trumbore, *Appl. Phys. Letters* **10**, 206 (1967).

Lorenz, M. R., and M. Pilkuhn, *J. Appl. Phys.* **37**, 4094 (1966).

Marshall, D. J., and R. A. Laudise, in *Crystal Growth*, Ed. by H. S. Peiser, Pergamon, New York, 1966, p. 557, and references therein.

Mertz, W. J., *Phys. Rev.* **76**, 1221 (1949).

Miller, C. E., *J. Appl. Phys.* **29**, 233 (1958).

Miller, J. F., in *Compound Semiconductors*, Vol. I, Ed. by R. K. Willardson and H. L. Goering, Reinhold, New York, 1962, pp. 200ff.

Mlavsky, A. I., and M. Weinstein, *J. Appl. Phys.* **34**, 2885 (1963).

Monchamp, R. R., and R. C. Puttbach, *Hydrothermal Growth of Large Ruby Crystals*, ASD Project No. 8–132, AF Contract 33 (657) 10508, September 1964.

Monchamp, R. R., R. C. Puttbach, and J. W. Nielsen, *Hydrothermal Growth of Zinc Oxide Crystals*, Final Report for Contracts AF33(657)–8795 and AF33(615)–2228 Technical Report AFML–TR–67–144, Air Force Materials Laboratory, Wright–Patterson Air Base, Ohio, June 1967, pp. 119–21 and 143.

Monchamp, R. R., R. C. Puttbach, and J. W. Nielsen, *J. Cryst. Growth* **2**, 178 (1968).

Morey, G. W., and P. Niggli, *J. Am. Chem. Soc.* **35**, 1086 (1913).

Nelson, H., *RCA Rev.* **24**, 603 (1963).

Nielsen, J. W., *J. Appl. Phys.* **31**, 51S (1960).

Nielsen, J. W., and E. F. Dearborn, *J. Phys. Chem. Solids* **5**, 202 (1958).

Nielsen, J. W., and F. G. Foster, *Am. Mineralogist* **45**, 299 (1960).

Nielsen, J. W., R. C. Linares, and S. E. Koonce, *J. Am. Ceram. Soc.* **45**, 12 (1962); taken from Laudise, 1963, courtesy of John Wiley & Sons, Inc.

O'Connor, J. J., M. A. DiPietro, A. F. Armington, and B. Rubin, *Nature* **212**, 68 (1966).

Ostwald, W., *Z. Phys. Chem.* **27**, 365 (1897).

Paschkis, V., and J. Persson, *Industrial Electric Furnaces and Appliances*, Vol. I, Wiley-Interscience, New York, 2nd ed., 1960, pp. 295ff.

Peiser, H. S., Ed. *Crystal Growth*, Pergamon, New York, 1967, p. 505.

Perloff, D. S., and A. Wold, in *Crystal Growth*, Ed. by H. S. Peiser, Pergamon, New York, 1967, p. 361.

Petrov, T., E. Treivus and A. Kasatkin, *Growing Crystals from Solution*, Consultants Bureau, New York, 1969.

Pfann, W. G., *Trans. AIME* **203**, 961 (1955).

Pfann, W. G., *Zone Melting*, 2nd Ed., Wiley, New York, 1966, pp. 254ff.

Plaskett, T. S., S. E. Blum, and L. M. Foster, *J. Electrochem. Soc.* **114**, 1303 (1967).

Price, P. B., D. A. Vermilyea, and M. B. Webb, *Acta Met.* **6**, 524 (1958).

Rase, D. E., and R. Roy, *J. Am. Ceram. Soc.* **38**, 102 (1955).

Rayleigh, Lord, *Phil. Mag.* **38**, 738 (1919).

Remeika, J. P., *J. Am. Chem. Soc.* **76**, 940 (1954).

Remeika, J. P., *J. Am. Chem. Soc.* **78**, 4259 (1956).

Reynolds, G. F., and H. Guggenheim, *J. Phys. Chem.* **65**, 1655 (1958).

Roy, R., and O. F. Tuttle, *Phy. Chem. Earth* **1**, 138 (1956).

Smakula, A., *Einkristalle*, Springer, Berlin, 1962, pp. 154–155.

Spezia, G., *Acad. Sci. Torino Atti* **40**, 254 (1905).

Stephen, H., and T. Stephen, *Solubilities of Inorganic and Organic Compounds*, 5 Vols., Macmillan, New York, 1963.

Thomas, D. G., M. Gershenzon, and F. A. Trumbore, *Phys. Rev.* **133**, A269 (1964).

Trumbore, F. A., M. Kowalchik, and H. G. White, *J. Appl. Phys.* **38**, 1987 (1967).

Van Hook, H. J., *J. Am. Ceram. Soc.* **44**, 208 (1961); *Single Crystal Growth of Garnet Type Oxides*, Parts 1 and 2, AF (19) (604) 5511, prepared for Electronics Research Directorate, U.S. Air Force, Bedford, Mass., January 1961a.

Van Hook, Andrew, *Crystallization*, Reinhold, New York, 1961b, pp. 209ff.

Viting, L. M., *Vestn. Mosk. Univ. Ser. II Khim.*, **20**(4), 54 (1965).

Vold, R. D., and M. J. Vold, in *Techniques of Organic Chemistry*, Vol. I, Part I, Ed. by Arnold Weissberger, Wiley-Interscience, New York, 2nd ed., 1949, Chap. VII.

Von Hippel, A., *Tech. Rept. 178*, Laboratory for Insulation Research, M.I.T., March 1963, p. 44.

Vonnegut, K., Jr., *Cat's Cradle*, Dell, New York, 1963.

Wagner, R. S., in *Whisker Technology*, Ed. by A. P. Levite, Wiley, New York, 1969.

Wagner, R. S., and C. J. Doherty, *J. Electrochem. Soc.* **113**, 1300 (1967).

Wagner, R. S., and W. C. Ellis, *Trans. AIME* **233**, 1053 (1965).

Wakim, F. G., H. K. Henisch, and H. Atwater, *J. Phys. Chem.* **42**, 2619 (1965).

Walker, A. C., *Bell Labs. Record* **25**, 357 (1947).

Walker, A. C., and G. T. Kohman, Trans. *Am. Inst. Elec. Eng.* **67**, 565 (1948).

Wanier, E., and A. N. Solomon (1942). Reported by R. A. Laudise in *The Art and Science of Growing Crystals*, Ed. by J. J. Gilman, Wiley, New York, 1963, p. 259.

White, E. A. D., in *Techniques of Inorganic Chemistry*, Ed. by H. B. Jonassen and A. Weissberger, Wiley-Interscience, New York, 1965, p. 31.

Wickham, D. G., *J. Appl. Phys.* **33**, 3597 (1962).

Wilson, C. L., *Comprehensive Analytical Chemistry*, Ed. by C. L. Wilson and D. W. Wilson, American Elsevier, New York, 1959, pp. 27–32.

Wolfe, C. M., C. J. Nuese, and N. Holonyak, *J. Appl. Phys.* **36**, 3790 (1965).

Wolff, G., P. H. Keck, and J. D. Broder, *Phys. Rev.* **94**, 753 (1954).

Wolff, G., R. A. Herbert, and J. D. Broder, *Semiconductors and Phosphors*, Wiley-Interscience, New York, 1958, pp. 463ff.

Wood, D. L., and J. P. Remeika, *J. Appl. Phys.* **37**, 1232 (1966).

Yamamoto, T., *Bull. Inst. Phys. Chem. Res.* (*Tokyo*) **17**, 1278 (1938).

Appendix

BIBLIOGRAPHICAL NOTES

Chapter 1

1. Characterization is extensively discussed in "Characterization of Materials", prepared by the Committee on Characterization of Materials, Materials Advisory Board (Division of Engineering) National Research Council, National Academy of Sciences, Washington, 1967.
2. The scanning electron microscope (see D. R. Thornton, *Scanning Electron Microscopy*, Chapman and Hall, London, 1968) has become an extremely powerful tool for studying crystalline surfaces.

Chapter 2

1. The use of phase diagrams in crystal growth is reviewed by J. W. Nielsen and R. R. Monchamp in *The Use of Phase Diagrams in Ceramic, Glass and Metal Technology*, Vol. III, A. M. Alper, Ed., Academic Press, New York, 1970.
2. Stoichiometry, existence regions and their role in crystal growth have been discussed by W. Albers and C. Haas [*Phillips Technical Rev.* **30**, 82, 107, 142 (1969)].

Chapter 3

1. Kinetics of crystal growth is reviewed in R. F. Strickland-Constable *Kinetics and Mechanism of Crystallization*, Academic Press, New York, 1968.
2. Morphological stability is further discussed in a series of papers by J. W. Cahn; R. F. Sekerka; S. R. Coriell and R. L. Parker; L. A. Tarshis and W. A. Tiller; G. R. Kotler and W. A. Tiller; R. G. Seidenstickers; and R. B. Williamson and B. Chalmers in *Crystal Growth*, Ed. by H. S. Peiser, Pergamon, New York, 1967.
3. Modern concepts of interface stability and crystal shape (morphological stability) rest to a great extent upon the work of Mullins and Sekerka [W. W. Mullins and R. F. Sekerka, *J. Appl. Phys.* **34**, 323 (1963); **35**, 444 (1964)].
4. Theory of melt growth is discussed in K. A. Jackson "Current Concepts in Crystal Growth from the Melt", *Progress in Solid State Chemistry*, Vol. 4, Ed. H. Reiss, Pergamon, New York, 1967.

Chapter 4

1. Gem quality diamond of several carat size has been grown by the General Electric group [*New York Times*, May 29, 1970, page 1].

Chapter 5

1. J. H. E. Jeffes [*J. Cryst. Gr.* **3**, **4**, 13 (1968)] gives a good review of the physical chemistry of vapor transport processes.
2. R. Nitsche [*Fort. Minerol.* **44**, 231 (1967)] gives an especially complete list of conditions, reactions and crystals grown by chemical vapor transport.
3. Papers by J. E. Mee and coworkers [J. E. Mee, L. Archer, R. H. Meade and T. N. Hamilton, *J. Appl. Phys.* **10**, 289 (1967); J. E. Mee, *IEEE Trans. on Magnetics*, **3-MAG**, 190 (1967)] describe the chemical vapor deposition of epitaxial garnet films.
4. The so-called "close spaced growth" geometry [F. H. Nicoll, *J. Electrochem. Soc.* **110**, 1165 (1963); P. A. Hoss, L. A. Murray and J. J. Rivera, *J. Electrochem. Soc.* **115**, (1968)] has proven especially useful in the epitaxial growth of semiconductors.

Chapter 6

1. A. Witt and H. Gatos [*J. Electrochem. Soc.* **4**, 511 (1969) and references contained therein] have developed techniques for high-resolution etching which are especially powerful in revealing striae in Czochralski crystals. They have shown that careful control of thermal symmetry, especially by means of *no* seed rotation with crucible rotation, virtually eliminates banding.
2. A. G. Fisher [*J. Electrochem. Soc.* **117**, 41C (1970)] gives a good review of melt-pressure growth of luminescent semiconductors.
3. The hollow cathode has been extended to the growth of carbides with melting points above 3500°C [R. N. Storey and R. A. Laudise, *J. Cryst. Gr.* **6**, 261 (1970)].
4. Single-crystal sapphire ribbon, rod, and tubing are routinely grown by crystal pulling through an annulus by the Tyco Co. [H. E. LaBelle, Jr. and A. I. Mlavsky, *Nature* **216**, 574 (1967)].

Chapter 7

1. Water soluble iodates are becoming increasingly interesting as nonlinear optical materials [S. K. Kurtz, J. G. Bergman, Jr., and T. T. Perry, *Bull. Amer. Phys. Soc.* **13**, 388 (1968)].
2. The role of "cusps" in containing a local aqueous region about the growing crystal in gel growth is discussed by J. I. Hanoka [*J. Appl. Phys.* **40**, 2694 (1969)].

AUTHOR INDEX

SUBJECT INDEX

† This index was made with the SUPER-BOOKIE post-processor to the BELDEX computer indexing program. I would like to acknowledge the assistance of R. L. Barns and I. C. Ross.